OCEANOGRAPHY AND OCEAN ENGINEERING

DEEP-SEA

MARINE BIOLOGY, GEOLOGY, AND HUMAN IMPACT

Oceanography and Ocean Engineering

Additional books in this series can be found on Nova's website
under the Series tab.

Additional e-books in this series can be found on Nova's website
under the e-book tab.

Marine Biology

Additional books in this series can be found on Nova's website
under the Series tab.

Additional e-books in this series can be found on Nova's website
under the e-book tab.

OCEANOGRAPHY AND OCEAN ENGINEERING

DEEP-SEA

MARINE BIOLOGY, GEOLOGY, AND HUMAN IMPACT

DENNIS R. BAILEY
AND
SUSAN E. HOWARD
EDITORS

nova publishers
New York

Library of Congress Cataloging-in-Publication Data

Deep-sea : marine biology, geology, and human impact / editors, Dennis R. Bailey and Susan E. Howard.
p. cm.
Includes index.
ISBN 978-1-62257-173-4 (hardcover)
1.Deep-sea biology. 2. Deep-sea ecology. 3. Foraminifera, Fossil. I. Bailey, Dennis R., 1972- II. Howard, Susan E.
QH91.8.D44D44 2012
578.77'99--dc23

2012020010

Published by Nova Science Publishers, Inc. † New York

CONTENTS

Preface		**vii**
Chapter 1	Morphology, Taxonomy, Stratigraphical Distribution and Evolutionary Classification of the Schackoinid Planktic Foraminifera (Late Albian-Maastrichtian, Cretaceous) *M. Dan Georgescu*	**1**
Chapter 2	A Review of the Knowledge Related to the Nomenclature, Etymology, Morphology, Distribution, and Biological Characteristics of the Skilfish, Erilepis Zonifer (Anoplopomatidae), in the North Pacific Ocean *A. M. Orlov, A. M. Tokranov and B. A. Megrey*	**63**
Chapter 3	Evolutionary Classification of the Late Cretaceous (Coniacian-Maastrichtian) Planktic Foraminifera Archaeoglobigerina Pessagno, 1967 and Gandolfia New Genus/Directional Lineage *M. Dan Georgescu*	**101**
Chapter 4	Protein Degradation of Deep-Sea Sediment: Collagenolytic Enzymes Produced from a Novel Marine Bacterium *Atsushi Kurata, Tohru Kobayashi and Koki Horikoshi*	**131**
Chapter 5	Primary Production and Carbon Cycling in the Deep Sea: Archaeal Carbon Fixation and Methane Oxidation *Nianzhi Jiao and Chuanlun Zhang*	**149**
Chapter 6	The Deep-Mediterranean Sea: Habitats, Biodiversity and Ecosystem Functioning *Roberto Danovaro*	**171**
Index		**193**

PREFACE

The deep-sea ecosystems include waters and sediments at the lowest layer in the ocean, existing below the thermocline and above the seabed, at a depth of 1000 fathoms (1800 m) or more. They represent the world's largest biome, covering more than 65% of the world surface and including more than 95% of the global biosphere. This book examines the marine biology, geology and human impact of the deep-sea environment. Topics discussed include the stratigraphical distribution and evolutionary classification of the schackoinid planktic foraminifera; the biological characteristics of the skilfish, erilepis zonifer in the North Pacific Ocean; protein degradation at deep-sea sediment; primary production and carbon cycling in the deep-sea; and biodiversity and ecosystem function in the deep-Mediterranean Sea.

Chapter 1 – The taxonomic revision of the late Albian-Maastrichtian planktic foraminifera traditionally assigned to the genus *Schackoina* reveals the existence of five directional lineages. Five genera are defined to accommodate these lineages; four of them are new (*Groshenyia*, *Asymetria*, *Pseudohastigerinoides* and *Neoschackoina*) and one is emended (*Schackoina*). *Groshenyia* is of late Albian age, and includes two species: *G. groshenyae* - new species and *G. pentagonalis*. *Asymetria* includes the new species *A. asymetrica* of late Albian-earliest Cenomanian age and *A.* ? sp., which is known only from lower Albian sediments. The only schackoinid species with distal bulbous projection that passes the Albian/Cenomanian boundary is *A. asymetrica*. *Schackoina* is of late Albian-Maastrichtian age, and includes *S. trituberculata*, *S. cenomana*, *S. bicornis* and *S. multispinata*. Two directional lineages evolved from *Schackoina*: *Pseudohastigerinoides* in the early Cenomanian and *Neoschackoina* in the late Santonian. *Pseudohastigerinoides* is of Cenomanian-early Turonian age and includes *P. gandolfii* and *P. moliniensis*. *Neoschackoina* of the late Santonian-early Maastrichtian consists of *N. tappanae* and *N. sellaeforma*. Test coiling, chamber shape, radially elongate termination and ornamentation development are the morphological features of paramount importance in the new schackoinid classification.

Chapter 2 – In this chapter the authors review the nomenclature, morphology, distribution, and biological characteristics including aquarium observations of the skilfish, *Erilepis zonifer*, in the North Pacific Ocean. Results are based on the analysis of all published sources and new unpublished data from 600 captures taken in the North Pacific between 1956 and 2001. New data sources were used to derive life history parameters, namely relationships between capture depth vs. length, length-age, weight vs. length, and an index of maturity. Empirical methods using life history parameters were used to derive estimates of natural mortality and growth rates. This new information was compared to previously published information.

Chapter 3 – The genus *Archaeoglobigerina* Pessagno, 1967 is reviewed and redefined in evolutionary classification as branched lineage, which evolved in late Coniacian and became extinct in the late Maastrichtian. A new directional lineage, *Gandolfia*, is defined for the double keeled planktics of late Campanian-Maastrichtian age with plano-convex test and ornamentation that concentrated in chamber central portion on the spiral side. This attempt to develop an evolutionary classification framework for a group of globotruncanids planktic foraminifera is equally based on the test gross architecture features as well as on high resolution observations on the wall ultrastructure, ornamentation, and pore size and distribution patterns.

Chapter 4 – Collagen, a well-characterized polypeptide, constitutes the extracellular matrix in higher organisms, such as whales. Deep-sea biological communities were found around whale bones near the summit of the Torishima Seamount (4,026 m deep). From the whale bone community, the authors obtained a novel collagenolytic enzyme-producing bacterium belonging to genus *Alkalimonas*. The bacterium produced two proteases, which degrades collagen, gelatin, and casein. The bacterial proteases could play an important role in the degradation of whale carcasses.

Chapter 5 – Collagen, a well-characterized polypeptide, constitutes the extracellular matrix in higher organisms, such as whales. Deep-sea biological communities were found around whale bones near the summit of the Torishima Seamount (4,026 m deep). From the whale bone community, the authors obtained a novel collagenolytic enzyme-producing bacterium belonging to genus *Alkalimonas*. The bacterium produced two proteases, which degrades collagen, gelatin, and casein. The bacterial proteases could play an important role in the degradation of whale carcasses.

Chapter 6 – The deep-Mediterranean basin is a peculiar system characterized by high species richness. The unexpectedly high biodiversity of the deep-sea benthos is apparently related to the presence of strong environmental gradients and the richness of deep-sea habitats. Despite the increasing sampling effort and the advancements made in the last three decades, the knowledge of the deep Mediterranean Sea and its biodiversity is still poor, as highlighted by recent estimates reporting that approximately 75% of Mediterranean deep-sea biodiversity is still undiscovered. Here is provided an overview of the main deep-sea habitat types, their distribution, ecological characteristics and biodiversity. These habitats include: open slope systems, submarine canyons, deep basins, seamounts, deep-water corals, cold seeps, hydrothermal vents and permanent anoxic systems. Most of these habitats are extremely vulnerable and under serious threats related to the increasing direct and indirect human impact of deep-sea ecosystems.

In: Deep-Sea
Editors: D. R. Bailey and S. E. Howard
ISBN: 978-1-62257-173-4

Chapter 1

MORPHOLOGY, TAXONOMY, STRATIGRAPHICAL DISTRIBUTION AND EVOLUTIONARY CLASSIFICATION OF THE SCHACKOINID PLANKTIC FORAMINIFERA (LATE ALBIAN-MAASTRICHTIAN, CRETACEOUS)

M. Dan Georgescu[*]
Department of Geosciences, University of Calgary, Calgary, Alberta, Canada

ABSTRACT

The taxonomic revision of the late Albian-Maastrichtian planktic foraminifera traditionally assigned to the genus *Schackoina* reveals the existence of five directional lineages. Five genera are defined to accommodate these lineages; four of them are new (*Groshenyia*, *Asymetria*, *Pseudohastigerinoides* and *Neoschackoina*) and one is emended (*Schackoina*). *Groshenyia* is of late Albian age, and includes two species: *G. groshenyae* - new species and *G. pentagonalis*. *Asymetria* includes the new species *A. asymetrica* of late Albian-earliest Cenomanian age and *A.* ? sp., which is known only from lower Albian sediments. The only schackoinid species with distal bulbous projection that passes the Albian/Cenomanian boundary is *A. asymetrica*. *Schackoina* is of late Albian-Maastrichtian age, and includes *S. trituberculata*, *S. cenomana*, *S. bicornis* and *S. multispinata*. Two directional lineages evolved from *Schackoina*: *Pseudohastigerinoides* in the early Cenomanian and *Neoschackoina* in the late Santonian. *Pseudohastigerinoides* is of Cenomanian-early Turonian age and includes *P. gandolfii* and *P. moliniensis*. *Neoschackoina* of the late Santonian-early Maastrichtian consists of *N. tappanae* and *N. sellaeforma*. Test coiling, chamber shape, radially elongate termination and ornamentation development are the morphological features of paramount importance in the new schackoinid classification.

Keywords: Planktic foraminifera; Cretaceous; taxonomy; new species; new genera; evolution

[*] E-mail: dgeorge@ucalgary.ca

INTRODUCTION

Schackoinid planktics represent a relatively rare foraminiferal group, which mostly occurs as a minor component in the Cretaceous (late Barremian-Maastrichtian) assemblages. There are three characteristics of this group formalized at the family level by Pokorný (1958): small size, tests with an early low trochospiral coil and planispiral in the adult stage, and last-formed chambers with tubulospines. Schackoinid tests occur in two distinct stratigraphical intervals: upper Barremian-Aptian and upper Albian-Maastrichtian; no such tests are known in the middle and upper Albian, and apparently this indicates that the schackoinid general test architecture evolved iteratively in the Cretaceous.

Small test size was apparently a major impediment in developing an accurate taxonomical framework for the representatives of this group. Therefore, the number of chambers in the final whorl and number of tubulospines on each chamber are the most used morphological features on which the group taxonomy was based; such features are readily observable with an optical stereomicroscope. Group scarceness in the fossil record raised an additional major difficulty in its study, and the vast majority of the schackoinid reports consist of one or maximum few species collected from a relatively narrow stratigraphical interval.

A high-resolution taxonomical revision of the late Albian-Maastrichtian schackoinids is presented in this study. The test morphology, including wall ultrastructure, ornamentation and porosity characteristics, are studied with the aid of the scanning electron microscope (SEM) on tests from a variety of worldwide occurrences. The new dataset helps in reassessing species variability, leading towards a better understanding of the schackoinid taxonomy, evolution and paleobathymetry.

HISTORY OF CONCEPTS

Thalmann (1932) erected the genus *Schackoina* to accommodate the two species with tubular processes: *S. cenomana* (Schacko, 1897) and *S. multispinata* (Cushman and Wickenden, 1930); the term `tubulospine` was later proposed by Montanaro Gallitelli (1955, p. 142) for the tubular processes of the schackoinid tests. *Schackoina* received a wide and immediate recognition among taxonomists (Cushman, 1933; Morrow, 1934; Loetterle, 1937, etc). A milestone in understanding schackoinid taxonomy was the study of Reichel (1948), who used thin sections to demonstrate the tubulospine pointed terminations; in addition, *Schackoina* diversity was shown by using combinations of features such as chamber shape, chamber number, tubulospine number, umbilical diameter, etc. Reichel (1948) also first observed an evolutionary trend in the schackoinid evolution, namely the reduction of the chamber number in the final whorl.

Bolli (1957) described the double and multiple radial extensions of the chamber, a taxonomically significant morphological feature of the schackoinid tests. He described the new genus *Leupoldina* Bolli, to include the tests in which the last-formed chambers present two or more radial extensions. *Schackoina* was emended to include tests with only one radial extension per chamber irrespective of the extension characteristics: tubulospine, which is pointed, or bulbous projection. This taxonomical framework was difficult to apply from the beginning, since Bolli (1957, p. 276) hesitated to reassign to *Leupoldina* two previously described species with two tubulospines on the last-formed chambers, namely *S. cenomana bicornis* Reichel, 1948 and *S. moliniensis* Reichel, 1948. Another inconsistency was revealed when Bolli (1959, p. 264) considered that a species which has, in general, three tubulospines on the last-formed one or two chambers should be included within the genus *Schackoina*. Another merit of the Bolli's (1957) study is that the occurrence of the distal bulbous projection was for the first time mentioned in the schackoinid group; notably, this feature was known in the non-schackoinid taxa with radially elongate chambers (Cushman, 1931), but was overlooked in the description of another schackoinid species, namely *S. primitiva* Tappan, 1940.

A new taxonomical framework, which considers the radially elongate terminations and main apertural characteristics as primary taxonomical features, was developed by Loeblich and Tappan (1964, p. C658). According to this scheme *Leupoldina*, which was considered of Aptian age, includes the tests with distal bulbous projection and two symmetrically developed apertures in the final chamber of the mature specimens, whereas *Schackoina* (Aptian-Campanian, ?Maastrichtian) is restricted to the tests with hollow tubulospines and main aperture in peripheral position.

Discovery of schackoinid taxa with distal bulbous projection of late Albian-earliest Cenomanian age (Luterbacher and Premoli Silva, 1962) complicated this taxonomical framework. Initially they were assigned to *Clavihedbergella* Banner and Blow, 1959 (Luterbacher and Premoli Silva, 1962; Pflaumann and Krasheninnikov, 1977; Miles and Orr, 1980; Leckie, 1984). Caron (1985) assigned them to *Schackoina* and this framework was followed by Moullade et al. (2002), Bellier et al. (2003) and Petrizzo and Huber (2006). Loeblich and Tappan (1988, p. 461) apparently included these taxa within *Leupoldina*, as inferred from the expanded stratigraphical range of the genus.

Test ornamentation and porosity characteristics of the schackoinid were studied by Banner and Desai (1988, p. 170), who concluded that the representatives of the family Schackoinidae lack ornamentation and have a microperforate test wall. A direct phylogenetic relationship between the two genera was defined, in which *Leupoldina* is the ancestor and *Schackoina* the descendant. However, Banner and Desai (1988, p. 179) mentioned a gap spanning the uppermost Albian-Albian interval between the stratigraphical ranges of the two genera. The ancestor-descendant relationship between *Leupoldina* and *Schackoina* was first mentioned by Banner and Blow (1959), and subsequently used in other studies (van Hinte, 1963; BouDagher-Fadel, 1996; Hart, 1999; Hart et al., 2002).

The schackoinid tests of late Barremian-Aptian age were restudied by Verga and Premoli Silva (2002, 2005), who confirmed the occurrence microperforate test wall in *Leupoldina*. *Pseudoschackoina* Verga and Premoli Silva, 2005 was described to accommodate the tubulospine-bearing tests of Aptian age. Despite the name this genus appears closer to *Lilliputianella* Banner and Desai, 1988 of the family Praehedbergellidae Banner and Desai, 1988.

The extensive use of the SEM significantly changed the observation resolution and led to the beginnings of the evolutionary classification in the Cretaceous planktic foraminiferal group. Georgescu (2009) demonstrated that the non-schackoinid tests with radially elongate chambers developed several times during the late Albian-Santonian, a well-known pattern in the planktic foraminiferal evolution (Frerichs, 1971; Steineck and Fleisher, 1978). Such species evolved from globular-chambered taxa, an evolutionary process that occurred in trochospiral (Georgescu, 2009) and planispiral taxa (Georgescu and Huber, 2008). A taxonomic review of the late Albian-Maastrichtian schackoinids is a necessary step forward for a better understanding of the Cretaceous planktic foraminiferal evolution.

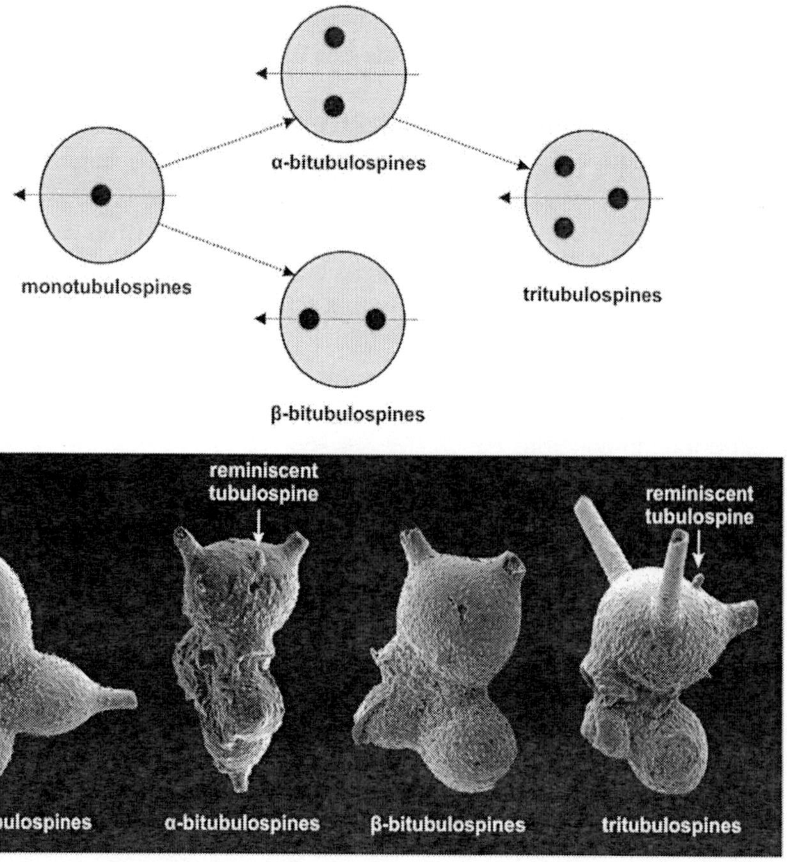

Figure 1. Tubulospine new terminology in schackoinid planktic foraminifera. The arrows in the figure above point towards the chamber anterior; tubulospines are given in green. No scale implied.

NEW MORPHOLOGICAL TERMS

The increased observation resolution generated by extensive use of the SEM created the necessity to re-evaluate schackoinid test morphological features. Tubulospine distribution, chamber shape and ornamentation development over the two test sides are features of paramount importance in the new schackoinid taxonomy, and their observed diversity requires a new and improved terminology.

A new terminology is developed to describe the tubulospine number and position (Figure 1). The tubulospine number and distribution in each schackoinid test can be described as a combination of four types defined for one chamber: monotubulospines, α-bitubulospines, β-bitubulospines and tritubulospines.

- monotubulospines: there is one tubulospine per chamber, and is situated in the test equatorial plane.
- α-bitubulospines: there are two symmetrically arranged tubulospines with respect to the test equatorial plane; the two tubulospines are situated towards the chamber anterior part.
- β-bitubulospines: there are two tubulospines in the test equatorial plane.
- tritubulospines: there are three tubulospines per chamber, with two tubulospines symmetrically arranged with respect to the equatorial plane, and shifted towards the chamber anterior part; the third is situated in the test equatorial plane and towards chamber posterior part.

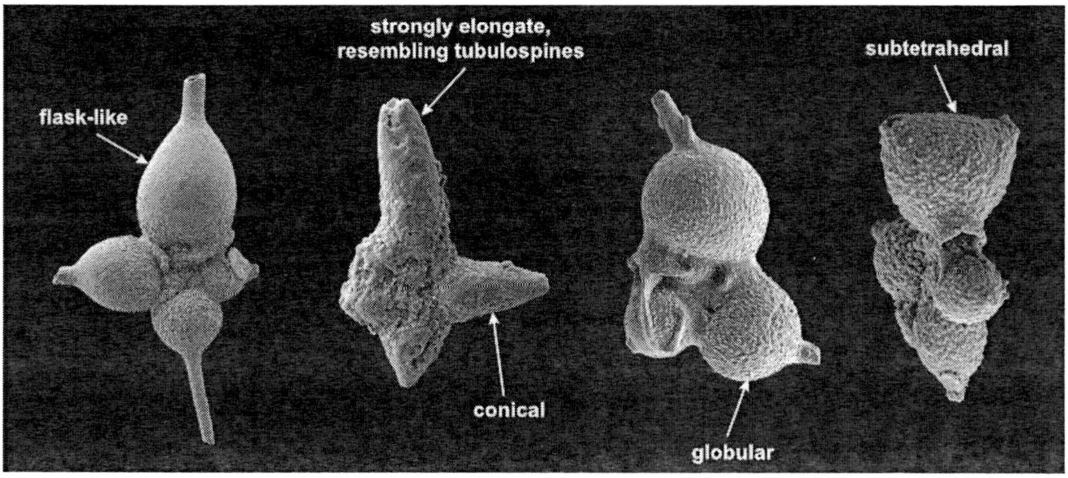

Figure 2. Chamber shape new terminology used for the schackoinid planktic foraminifera. No scale implied.

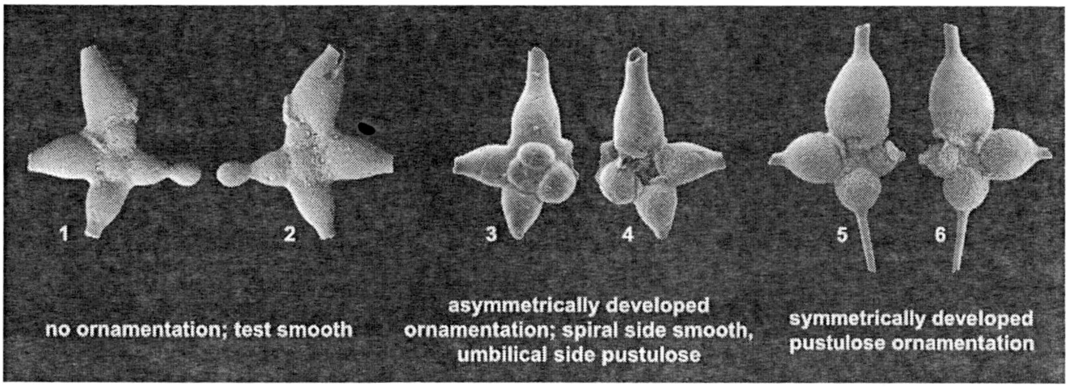

Figure 3. The new terminology used for the chamber ornamentation as a function of its development on the two test sides. No scale is implied.

The tubulospines from the chambers of the previous whorl, which penetrate through the younger chambers, are considered residual tubulospines. No taxonomical significance is conferred to the residual tubulospines.

Chamber shape is a major feature in schackoinid foraminifer taxonomy both at the species and genus/lineage level. Five types of chamber shape are recognized: flask-like, globular, conical, subtetrahedral and strongly elongate, resembling tubulospines (Figure 2).

Schackoinid test ornamentation is a feature that can be studied only with the aid of the SEM or ESEM. Three types are recognized function of the ornamentation development on the two test sides: smooth, asymmetrically developed and symmetrically developed ornamenttation (Figure 3). The asymmetrically developed ornamentation, which is smooth on the spiral side and ornamented on the umbilical side, is for the first time observed in a Cretaceous planktic foraminiferal test. Notably, different ornamentation patterns on the two test sides, for example parallel to the periphery in the spiral side and meridional on the umbilical side, can occur in the Cretaceous planktic foraminifers (Robaszynski et al., 1984; Georgescu and Huber, 2006), but they should not be confused for the newly defined asymmetrically developed ornamentation, in which one test side (i.e., spiral side) is completely smooth.

Material Studied

Collection specimens and new material were available for this study, covering the whole upper Albian-Maastrichtian range of the group considered. More than 600 SEM photographs, including detailed images, were made during the study.

Table 1. Collection specimens examined for this study

Species (original designation)	Provenance	Age	Specimen type	Collection
Schackoina primitiva	Grayson Bluff, Texas, USA	late Albian	holotype	Cushman Collection, NMNH
Schackoina primitiva			paratypes	
Schackoina multispinata	Gober Chalk, Texas, USA	Campanian	hypotypes	Loeblich and Tappan Topotype Collection, NMNH
Schackoina multispinata	Smoky Hill Shale, Niobrara Formation, Upper Missouri basin, USA	late Santonian	hypotypes	Ehrenberg Collection, NMB
Schackoina multispinata	Smoky Hill Shale, Niobrara Formation, Upper Missouri basin, USA	late Santonian	hypotypes	Bailey Collection, FH
Schackoina sellaeforma	Mooreville Chalk, Alabama, USA	early Campanian	holotype	Cushman Collection, NMNH
Hastigerinoides rohri	Gautier Formation, Trinidad, Caribbean region	late Albian	holotype	Cushman Collection, NMNH

Collection material includes specimens from the Cushman Collection (National Museum of Natural History, Washington, D. C. - NMNH), Ehrenberg Collection (Naturkundemuseum, Berlin - NMB), Jacob Whitman Bailey Collection (Farlow Herbarium, Harvard University - FH) and Loeblich and Tappan Topotype Collection (NMNH). The labelling system of each collection of provenance is used (Table 1). The collection specimens are illustrated using a variety of techniques. The specimens in the Cushman Collection were illustrated using a Leica environmental scanning electron microscope (ESEM) (Figure 4.1-2, 6-9). The specimen in the Jacob Whitman Bailey Collection is mounted in Canada balsam and is illustrated using a transmitted light microscope (Figure 4.5). SEM photographs could be taken upon the specimens in the Loeblich and Tappan Topotype Collection (Figure 4: 3) and newly collected ones from the bulk samples from the Ehrenberg Collection (Figure 4.4).

One species could not be reviewed during this study, namely *S. primitiva* (Figure 4.1-2). The holotype is in a good state of preservation, with the radially elongate chambers intact; by contrast, the test wall is heavily recrystallized and the pores completely obliterated. The two paratypes are strongly deformed.

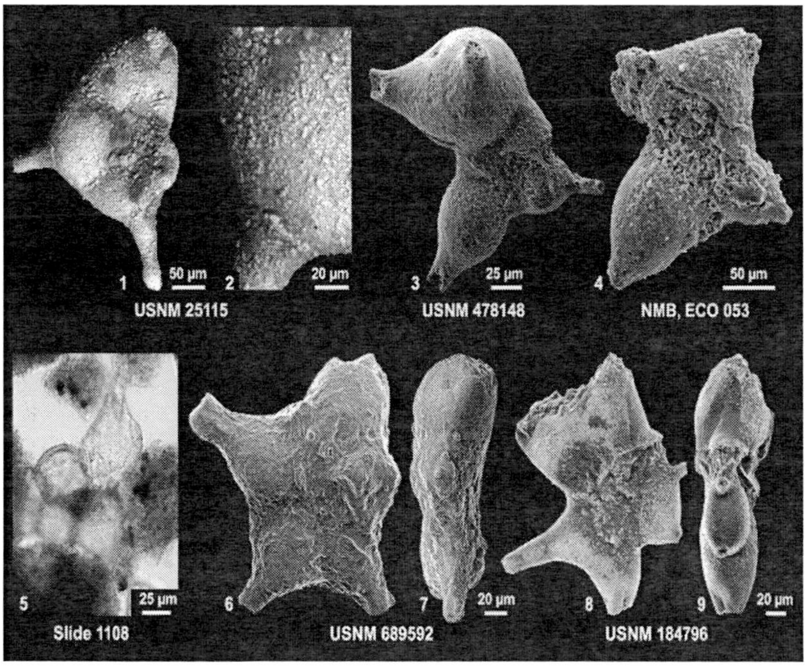

Figure 4. Collection specimen used in this study. 1-2 Holotype of *Schackoina primitiva* from the Cushman Collection (NMNH); specimen originally figured by Tappan (1943, pl. 18, Figure 14) from the Grayson Bluff of northern Texas (Denton County). 3 Hypotype of *S. multispinata* from the Gober Chalk of Texas; specimen from the Loeblich and Tappan topotype Collection (NMNH). 4 Hypotype of *S. multispinata* from the Niobrara Formation of the Upper Missouri; specimen from the Ehrenberg Collection (NMB), Sample 1595f. 5 Hypotype of *S. multispinata* from the Niobrara Formation of the Upper Missouri; specimen from the Jacob Whitman Bailey Collection (FH). 6-7 Holotype of *Hastigerinoides rohri* from the upper Albian sediments of the Gautier Formation of Trinidad, and deposited in the Cushman Collection (NMNH); specimen originally figured by Brönnimann (1952, pl. 1, Figures 8-9). 8-9 Holotype of *S. sellaeforma* from the Mooreville Chalk of Alabama, deposited in the Cushman Collection (NMNH); specimen originally figured by Masters (1976, pl. 2, Figure 14).

Table 2. Schackoinid tests yielded by the spot sample sets that were used in this study

Species	Provenance	Age	Number of specimens
Schackoina cenomana	DSDP Site 551, offshore Cobh, North Atlantic Ocean	late Cenomanian	3
S. bicornis			2
Schackoina cenomana	Valea Voievozi, Eastern Carpathians, Romania	early Turonian	23
S. bicornis			5
Paeusohastigerinoides gandolfii			21
P. moliniensis			18
Neoschackoina tappanae	ODP Hole 689C, Maud Rise, Weddell Sea	early Maastrichtian	4
Neoschackoina tappanae	ODP Hole 747A, Kerguelen Plateau	early Maastrichtian	3

A relatively small number of specimens were observed in four sets of spot samples (Table 2). Three of them are from Deep Sea Drilling Project (DSDP)/Ocean Drilling program (ODP) locations in the Atlantic and Indian Oceans, and were examined at the NMNH. The fourth set came from central Romania being also the only sample set used in this study, which was collected from an offshore location.

Most of the material was collected from eight DSDP/ODP sites worldwide (Figure 5). Sample labelling system is that currently used by the DSDP/ODP: leg number-site or hole number-core number-section number, sample depth in centimetres. The biostratigraphical framework at each location was evaluated and redone if necessary in order to assure a precise stratigraphical position for the schackoinid planktic foraminiferal occurrences. Each biozone recognized is defined and its type is mentioned using the following abbreviations: TRZ-taxon range zone, IZ-interval zone and PTRZ-partial taxon range zone. The references for the species used in the biozone definitions are given in the Appendix.

The type specimens of the new species are deposited in the Willi Karl Braun Micropaleontology Collection (University of Calgary). Specimen number is formed by the acronym WKB followed by the inventory number.

DSDP Leg 15, Site 150 (Venezuelan Basin, Caribbean Region)

A succession of upper Turonian-lower Campanian sediments was recovered in two cores, namely 9 and 10. Lithologically the upper bathyal sediments recovered in the two cores consist of calcareous clay and nannofossil marls in core 9, and dominant nannoplankton chalks in core 10. Cretaceous biostratigraphy at this site was studied for the complete succession by Premoli Silva and Bolli (1973), who assigned it a Turonian-Santonian age. Rich foraminiferal assemblages were recovered from throughout the stratigraphical succession, which is reinterpreted herein. Three biozones are recognized, and are presented from the base towards the top (Georgescu, 2010; Georgescu et al., 2011) (Table 3).

- *Marginotruncana schneegansi* Biozone (PTRZ) includes the sediments from the lowermost part of the succession, and the FO of *Dicarinella concavata* marks this biozone upper limit. Age: late Turonian.
- *D. concavata* Biozone (IZ) includes the sediments above the FO of the index species; the upper limit is marked by an unconformity, which encompasses at least most of the Coniacian and lower Santonian stratigraphical interval. Age: latest Turonian – (?) early Coniacian.
- *Ventilabrella eggeri* Biozone (TRZ) is herein defined for the sediments situated between the FO and LO of *V. eggeri*; the index species stratigraphical range is that revised by Georgescu (2010). This biozone is recognized only in a sample at DSDP Site 150 and is separated by two distinct unconformities from the adjacent successions. Age: late Santonian.

Table 3. Schackoinid stratigraphical distribution and frequency at DSDP Site 150. Frequency key: R-rare (< 5 specimens). The local stratigraphical ranges of the schackoinid species are given on grey background

DSDP Leg 15, Site 150 (core number-core section, sample depth)	Stages	Biozonation	*Scahackoina cenomana*	*S. bicornis*
9-1 46-47 cm	upper Santonian	*V. eggeri*		
9-2 94.5-95.5 cm	upper Turonian-(?) lower Coniacian	*D. concavata*	R	
10-1 25-26 cm				
10-1 58-73 cm*			R	R
10-1 75-76 cm		*M. schneegansi*	R	R
10-2 24-25 cm				
10-2 46-62 cm				
10-2 75-76 cm			R	
10 CC				

The schackoinids yielded by the upper Turonian-? Lower Coniacian sediments are very well preserved; the tests are recrystallized, but the test high resolution features, such as test wall ultrastructure, ornamentation and porosity characteristics, can be accurately studied.

DSDP Leg 32, Site 305 (Shatsky Rise, Central Pacific Ocean)

The Campanian-Maastrichtian stratigraphical succession at this site consists of white and pale orange soft chalk accumulated under bathyal conditions. They yielded rich foraminiferal assemblages, which allowed reassessing the biozonation framework originally given by Caron (1975), therefore providing an improved stratigraphical position for the occurrences of schackoinid planktic foraminifera. Eight biozones are herein recognized in the Campanian-Maastrichtian sediments, and their definition is succinctly presented below from the base towards the succession top (Table 4).

- *Hendersonites pacificus* Biozone (IZ) includes the lower Campanian sediments in the lowermost part of the succession, and only its upper part occurs at DSDP Site 305; biozone upper limit is defined by the LO of *H. pacificus*. Age: early Campanian.
- *Globotruncanita insignis* Biozone (IZ) is defined as the stratigraphical interval between the FO of *G. insignis* and FO of *Radotruncana calcarata*. Age: late early-middle Campanian.
- *R. calcarata* Biozone (TRZ) is defined between the FO and LO of the index species. Age: late Campanian.

Table 4. Schackoinid stratigraphical distribution and frequency at DSDP Site 305. Frequency key: R-rare (< 5 specimens). The local stratigraphical ranges of the schackoinid species are given on grey background

DSDP Leg 32, Site 305 (core number-core section, sample depth in centimeters)	Stages	Biozonation	*Schackoina cenomana*	*S. multispinata*	*Neoschackoina tappanae*
16-1 100-102 cm	Maastrichtian	*P. hariaensis*		R	
16-2 100-102 cm					
16-3 100-102 cm					
16-4 100-102 cm		*A. mayaroensis*		R	
16-5 100-102 cm		*R. fructicosa*		R	
17-1 100-102 cm					
17-2 100-102 cm		*G. stuarti*			
17-4 100-102 cm					
17-5 100-102 cm					
17-6 100-102 cm					
18-1 100-102 cm	Campanian				
18-2 100-102 cm					
18-3 100-102 cm					
18-4 100-102 cm					
18-5 100-102 cm		*G. subpetaloidea*			
18-6 100-102 cm					
19-1 100-102 cm					
19-2 100-102 cm					
19-3 100-102 cm					
19-4 100-102 cm					
19-5 100-102 cm				R	
19-6 100-102 cm					
20-1 100-102 cm					
20-2 100-102 cm					
21-1 50-52 cm		*R. calcarata*			R
21-2 50-52 cm					
21-2 100-102 cm					
21-3 50-52 cm					
21-4 50-52 cm					R
21-5 50-52 cm					
23-1 100-102 cm					
23-2 100-102 cm		*G. insignis*			
23-3 100-102 cm					
23-4 100-102 cm					
23-5 100-102 cm					
23-6 100-102 cm					
24-1 100-102 cm					
24-2 100-102 cm					R
24-3 100-102 cm					
24-4 100-102 cm				R	
24-5 100-102 cm					
25-1 100-102 cm					
25-2 100-102 cm					
25-3 100-102 cm					
25-4 100-102 cm					
25-5 100-102 cm					
25-6 100-102 cm					
26-1 100-102 cm		*H. pacificus*			
26-2 100-102 cm					
26-3 100-102 cm					
26-4 100-102 cm					
26-5 100-102 cm					
27-1 100-102 cm					
27-2 100-102 cm			R	R	
28-1 100-102 cm			R		

- *Globotruncanella subpetaloidea* Biozone (PTRZ) is the stratigraphical interval with *G. subpetaloidea* between the LO of *R. calcarata* and the FO of *Globotruncanita stuarti*. Age: late Campanian. The index species is keeled, and therefore cannot be considered a junior synonym of *Globotruncanella petaloidea* as considered in the taxonomical revisions by Pessagno (1967) and Robaszynski et al. (1984).
- *G. stuarti* Biozone (IZ) is the stratigraphical interval between the FO of the *G. stuarti* and the FO of *Racemiguembelina fructicosa*. Age: latest Campanian-early Maastrichtian.
- *R. fructicosa* Biozone (IZ) is the stratigraphical interval between the FO of *R. fructicosa* and FO of *Abathomphalus mayaroensis*. Age: middle Maastrichtian.
- *A. mayaroensis* Biozone (IZ) is defined between the FO of *A. mayaroensis* and the FO of *Pseudoguembelina hariaensis*. Age: late Maastrichtian.
- *P. hariaensis* Biozone (IZ) includes the stratigraphical interval from the FO of *P. hariaensis* and the Maastrichtian/Paleocene boundary. Age: latest Maastrichtian.

Foraminiferal test preservation is very good throughout the stratigraphical succession; the tests are affected by diagenesis and recrystallized, but the low magnitude of recrystallization permits accurate observations on the test wall ultrastructure and ornamentation, porosity characteristics, etc. Schackoinid species are minor component in the planktic foraminiferal assemblages in all the samples where they occur.

DSDP Leg 41, Site 370 (Deep Basin off Morocco, North Atlantic Ocean)

The succession of claystones, shales, marlstones and argillaceous limestones herein assigned to the Valanginian (?)-lower Cenomanian stratigraphical interval accumulated for the most part under reducing conditions; barren intervals are frequent especially in the lower portion of the succession. Moreover, sedimentation under lysocline resulted in thick intervals in which agglutinant forams dominate and calcareous ones occur sporadically and frequently show traces of dissolution. The first biostratigraphical framework was given by Pflaumann and Krasheninnikov (1977), and it is improved herein to accommodate the new dataset. Two planktonic foraminiferal biozones are recognized in the upper portion of this section, and combined they cover the upper Albian-lower Cenomanian stratigraphical interval (Table 5).

- *Parathalmanninella appenninica* Biozone (IZ) is the stratigraphical interval between the FO of the index species and FO of *Thalmanninella globotruncanoides*. Age: late Albian.
- *T. globotruncanoides* Biozone (IZ) is the stratigraphical interval above the FO of the index species, and only the lower part of this biozone occurs at this site. Age: early Cenomanian.

Pristine schackoinids occur sporadically in this interval. Most of them are known only as single specimen occurrences from upper Albian sediments.

Table 5. Schackoinid stratigraphical distribution and frequency at DSDP Site 370. Frequency key: R-rare (< 5 specimens); C-common (6-10 specimens). The local stratigraphical ranges of the schackoinid species are given on light grey background. The stratigraphical intervals with sedimentation below lysocline are given on dark grey background

DSDP Leg 41, Site 370 (core number-core section, sample depth)	Stages	Biozonation
38-3 101-102 cm	upper Barremian-lower Aptian	*Arenobulimina flandrini*
38-4 100-101 cm		
38-5 101-102 cm		
39-1 99-100 cm		
39-2 100-101 cm		
39-3 101-102 cm		
39-4 100-101 cm		
40-1 100-101 cm		
40-2 100-101 cm		
40-3 100-101 cm		
40-4 100-101 cm		
41-1 99-100 cm		
41-2 101-102 cm		
41-3 100-101 cm		
41-4 100-101 cm		
41-5 100-101 cm		
42-2 100-101 cm		
42-3 101-102 cm		
42-4 100-101 cm		
43-1 100-101 cm		
43-2 100-101 cm		
43-3 100-101 cm		
43-4 100-101 cm		
44-1 101-102 cm		
44-2 101-102 cm		
44-3 100-101 cm		
45-1 100-101 cm		
45-2 101-102 cm		
45-3 100-101 cm	? Valanginian-lower Barremian	NO INDEX FOSSIL
45-4 100-101 cm		
46-1 142-143 cm		
46-2 100-101 cm		
46-3 100-101 cm		
46-4 102-103 cm		
46-5 100-101 cm		
47-1 100-101 cm		
47-2 101-102 cm		
47-3 100-101 cm		
48-1 100-101 cm		
48-2 102-103 cm		
49-1 100-101 cm		
49-2 100-101 cm		
50-1 100-101 cm		
50-2 100-101 cm		
51-1 117-118 cm		
51-2 99-100 cm		

DSDP Leg 41, Site 370 (core number-core section, sample depth)	Stages	Biozonation	*Groshenyia pentagonalis*	*Asymetria* sp.	*Schackoina trituberculata*	*S. cenomana*	*S. bicornis*
20-1 100-101 cm	l. Cenomanian	*T. globotruncanoides*					
20-2 99-100 cm							
20-3 100-101 cm							
21-1 100-101 cm							
21-2 100-101 cm						R	
22-1 101-102 cm							
22-2 100-101 cm							
22-3 100-101 cm	upper Albian	*P. appenninica*					
22-4 100-101 cm							
23-1 102-103 cm							
23-2 103-104 cm						C	
23-3 100-101 cm							
23-4 100-101 cm							
24-1 100-101 cm			R			C	R
24-2 100-101 cm							
24-3 101-102 cm							
24-4 100-101 cm							
24-5 100-101 cm					R	C	
25-1 100-101 cm	middle-upper Albian	*Recurvoides imperfectus*					
25-2 100-101 cm							
26-1 100-101 cm							
26-2 100-101 cm							
26-3 102-103 cm							
26-4 101-102 cm			R				
27-1 100-101 cm							
27-2 100-101 cm							
27-3 100-101 cm							
28-1 99-100 cm							
28-2 100-101 cm							
28-3 100-101 cm							
28-4 100-101 cm							
29-1 100-101 cm							
30-1 99-100 cm							
30-2 100-101 cm							
30-3 100-101 cm							
30-4 100-101 cm							
31-1 100-101 cm							
31-2 100-101 cm	m. Al.	*M. rischi*		R			
31-3 101-102 cm	lower Albian	NO INDEX FOSSIL					
31-4 100-101 cm							
32-1 100-101 cm							
32-2 100-101 cm							
32-3 100-101 cm							
32-4 99-100 cm							
34-1 101-102 cm							
34-2 100-101 cm	u. Ap.	*L. cabri* Biozone equivalent					
34-3 100-101 cm							
34-4 101-102 cm	middle-upper Aptian	NO INDEX FOSSIL					
34-5 100-101 cm							
35-1 100-101 cm							
35-2 101-102 cm							
35-3 100-101 cm							
35-4 100-101 cm							
35-5 100-101 cm							
38-1 100-101 cm							
38-2 99-100 cm							

The upper Barremian-lower Albian interval yielded rare calcareous foraminifera. The foraminiferal assemblages are dominated by agglutinated species; two of them, *Recurvoides imperfectus* and *Arenobulimina flandrini* are significant from a biostratigraphical point of view. The *R. imperfectus* Biozone (TRZ) and *A. flandrini* Biozone (TRZ) of early-middle Albian and late Barremian-early Aptian age respectively are recognized; they are separated by a stratigraphical interval with rare foraminifera. Two levels with planktic foraminifera of middle Albian and late Aptian age respectively are recognized.

The lowermost part of the succession yielded relatively rich foraminiferal assemblages dominated by lagenids and agglutinated species; these sediments are provisory assigned to the Valanginian-lower Barremian stratigraphical interval. A detailed biostratigraphical study, which is beyond the purpose of this article, is necessary for a precise age assignment of these sediments.

DSDP Leg 47B, Site 398 (Vigo Seamount, Eastern North Atlantic Ocean)

The middle Albian-lower Cenomanian sediment succession at this site consists of calcareous mudstone at the base (cores 69 and 68) and marly nannofossil chalk in the upper part (the interval between cores 67 and 57). First stratigraphical framework was given by Sigal (1979). A major difficulty in developing a planktic foraminiferal biostratigraphy is the occurrence of reducing environments in the middle and upper Albian sediments; calcareous tests (e.g., planktic foraminifers, etc) are frequently dissolved in these environments, and only the siliceous organic debris are preserved (e.g., radiolarians, diatoms, etc). A first planktic foraminiferal zonation is herein provided for the middle Albian-lower Cenomanian sediments at the Site 398; the zonation consists of six biozones, which are defined starting from the base towards the succession top (Table 6).

- *Ticinella primula* Biozone (IZ) includes the sediments from the lower part of the section, from the section base to the FO of *Ticinella praeticinensis*. Age: middle Albian.
- *T. praeticinensis* Biozone (IZ) is the stratigraphical interval from the FO of the index species to the FO of *Pseudothalmanninella subticinensis*. Age: latest middle Albian-earliest late Albian.
- *P. subticinensis* Biozone (IZ) is the stratigraphical interval from the FO of *P. subticinensis* to the FO of *Pseudothalmanninella ticinensis*. Age: late Albian.
- *P. ticinensis* Biozone (IZ) is the stratigraphical interval between the FO of the index species and the FO of *P. appenninica*. Age: middle late Albian.
- *P. appenninica* Biozone (IZ) is the stratigraphical interval between the FO of *P. appenninica* and FO of *T. globotruncanoides*. Age: latest Albian.
- *T. globotruncanoides* Biozone (IZ) includes the uppermost part of the section, from the FO of the index species to the section top. Age: early Cenomanian.

Planktic foraminiferal preservation is excellent, most of the tests being in pristine condition. Schackoinid maximum frequency is above the Albian/Cenomanian boundary.

Table 6. Schackoinid stratigraphical distribution and frequency at DSDP Site 398. Frequency key: R-rare (< 5 specimens); F-frequent (11-25 specimens). The local stratigraphical ranges of the schackoinid species are given on light grey background. The stratigraphical interval where the micropaleontological assemblages consist only of radiolarians are given on dark grey background

DSDP Leg 47B, Site 398 (core number-core section, sample depth in centimeters)	Stages	Biozonation	*Groshenyia pentagonalis*	*Asymetria asymetrica*	*Schackoina trituberculata*	*S. cenomana*
57-1 50-52 cm	l. Cen.	*T. globotruncanoides*			R	R
57-2 50-52 cm					R	
57-3 50-52 cm					F	F
57-4 50-52 cm						
58-1 50-52 cm						
58-2 50-52 cm	upper Albian	*P. appenninica*				
59-1 50-52 cm						
59-2 50-52 cm						
59-3 50-52 cm						
59-4 50-52 cm			R	R		
59-5 50-52 cm						
60-1 50-52 cm						
60-2 50-52 cm						
60-3 50-52 cm						
61-1 50-52 cm						
61-2 50-52 cm						
62-1 50-52 cm						
62-2 50-52 cm						
62-3 50-52 cm						
62-4 50-52 cm						
62-5 50-52 cm						
62-6 50-52 cm						
63-1 50-52 cm						
63-2 50-52 cm		*P. ticinensis*				
63-3 50-52 cm						
63-4 50-52 cm						
63-5 50-52 cm						
63-6 50-52 cm						
65-1 50-52 cm						
65-2 50-52 cm						
65-3 50-52 cm		*P. subticinensis*				
65-4 50-52 cm				?R		
65-5 50-52 cm						
65-6 50-52 cm		*T. praeticinensis*				
66-1 50-52 cm						
66-2 50-52 cm	middle Albian					
67-1 50-52 cm						
67-2 50-52 cm						
67-3 50-52 cm		*T. primula*				
68-1 50-52 cm						
68-2 50-52 cm						
68-3 50-52 cm						
68-4 50-52 cm						
68-5 50-52 cm						
69-1 50-52 cm						
69-2 50-52 cm						
69-3 50-52 cm						
69-4 50-52 cm						
69-5 50-52 cm						

Table 7. Schackoinid stratigraphical distribution and frequency at DSDP Site 463. Frequency key: R-rare (< 5 specimens); C-common (6-10 specimens). The local stratigraphical ranges of the schackoinid species are given on light grey background

DSDP Leg 62, Site 463 (core number-core section, sample depth)	Stages	Biozonation	*Schackoina trituberculata*	*S. cenomana*	*S. bicornis*
21-1 52-54 cm	Campanian	*G. subpetaloidea*			
21-2 50-52 cm		*R. calcarata*			
21-3 52-54 cm					
21-4 50-54 cm					
21-5 38-40 cm					
21-5 78-80 cm					
21-6 57-59 cm					
22-1 52-54 cm					
22-2 51-53 cm					
22-3 50-52 cm					
22-4 52-54 cm		*G. insignis*			
22-4 10-12 cm					
23-1 50-52 cm					
24-1 50-52 cm					
24-2 50-52 cm					
24-3 50-52 cm					
25-1 51-53 cm					
25-2 50-52 cm		*H. pacificus*			
26-1 50-52 cm					
26-2 52-54 cm					
26-3 52-54 cm					
26-4 52-54 cm	u. Santonian	*D. asymetrica*		R	R
26-5 53-58 cm	Coniacian-l. Santonian	*D. concavata*		R	R
26-6 53-55 cm					
27-1 50-52 cm					
27-2 20-22 cm	Turonian				
29-1 50-52 cm		*M. schneegansi*			
30-1 50-53 cm					
30-2 50-52 cm					
31-1 11-13 cm					
33-1 50-52 cm					
33-2 52-54 cm					
34-1 50-52 cm		*H. helvetica*			
34-2 53-55 cm					
34-3 50-53 cm					
35-1 17-19 cm					
38-1 17-20 cm	lower Cenomanian	*T. globotruncanoides*			
43-1 49-52 cm					
43-2 50-52 cm					
43-3 50-52 cm					
43-4 48-50 cm			R		
43-5 19-22 cm					
48-1 50-52 cm			R		
48-2 50-52 cm					
48-3 53-57 cm	upper Albian	*R. appenninica*	C	R	
50-1 50-51 cm					
51-CC					
53-1 50-52 cm		*B. breggiensis*			

DSDP Leg 62, Site 463 (core number-core section, sample depth)	Stages	Biozonation	*S. multispinata*
7-4 20-22 cm	upper Campanian-lower Maastrichtian	*G. gansseri*	
8-1 50-52 cm			
8-2 50-52 cm			
8-3 15-17 cm			
9-1 50-52 cm			
9-2 50-52 cm			
9-3 47-49 cm			
10-1 50-52 cm			
10-2 50-52 cm			
10-3 50-52 cm			R
10-4 50-52 cm			
10-5 50-52 cm			
10-6 50-52 cm			
11-1 50-52 cm			
11-2 53-55 cm			
11-3 34-36 cm			
11-4 50-55 cm			
11-5 50-52 cm			
12-2 52-54 cm			
12-3 50-52 cm			
12-4 50-52 cm			
12-5 50-52 cm			
12-6 53-55 cm			
13-1 52-54 cm			
13-2 47-49 cm			
13-4 52-54 cm			
13-5 50-52 cm			
13-6 58-60 cm			
14-1 50-52 cm			
14-3 50-52 cm			
14-4 50-52 cm			
15-1 50-52 cm			
15-2 45-47 cm			
15-3 52-54 cm			
15-4 49-53 cm			
15-5 50-52 cm			
16-1 50-52 cm			
16-2 50-52 cm			
16-3 50-52 cm			
16-4 56-57 cm			
16-5 50-52 cm			R
16-6 50-52 cm			
17-1 54-57 cm			R
17-2 50-53 cm			
17-3 52-54 cm			
17-4 52-54 cm			
17-6 50-52 cm			
18-1 50-52 cm			
19-1 50-55 cm			R
19-2 52-54 cm			
19-3 47-49 cm			
19-4 49-51 cm			
19-5 52-54 cm			
19-6 51-53 cm			
20-1 51-53 cm			

Table 8. Schackoinid stratigraphical distribution and frequency at DSDP Site 511. Frequency key: R-rare (< 5 specimens); C-common (6-10 specimens). The local stratigraphical ranges of the schackoinid species are given on light grey background

DSDP Leg 73, Site 511 (core number-core section, sample depth)	Stages	Biozonation	*Schackoina cenomana*	*Neoschackoina tappanae*	*N. sellaeforma*
23-1 4-6 cm	u. Campan.	*G. havanensis*			
22-4 22-25 cm		*G. impensus*			
24-5 21-24 cm					
24-6 20-24 cm				C	R
24-7 20-23 cm				C	
28-2 21-25 cm	lower Campanian	*A. cretacea*			
28-6 18-22 cm					
29-1 12-15 cm					
30-1 23-27 cm					
30-5 12-16 cm					
31-5 8-11 cm					
32-1 19-22 cm					
32-4 23-27 cm					
33-1 20-24 cm					
34-2 20-24 cm					
35-1 20-24 cm					
36-1 24-28 cm		*M. marginata*			
36-7 23-27 cm					
37-1 23-27 cm					
38-1 24-28 cm					
39-2 13-17 cm					
40-1 38-40 cm					
40-4 26-29 cm					
40-6 20-23 cm					
41-2 21-23 cm					
41-3 55-57 cm					
42-1 25-30 cm	upper Coniacian-Santonian				
42-3 21-23 cm					
42-4 21-23 cm					
42-5 16-20 cm			R		
43-4 23-29 cm					
44-1 24-26 cm			R		
44-3 16-20 cm					
45-1 20-23 cm					
46-1 102-104 cm					
46-2 23-29 cm					
46-2 102-104 cm					
46-3 100-102 cm					
46-4 50-52 cm					
46-5 65.5-67.5 cm			R		
46 CC			R		
47-1 22-24 cm	upper Turonian	*Whiteinella* sp.			
47-1 25-30 cm					
47-1 50-52 cm					
47-1 100-102 cm					
47-2 22-24 cm					
47-2 50-52 cm					
47-2 100-102 cm					
47-3 20-25 cm					
47-3 25-31 cm					
47-3 50-52 cm			R		
47-3 100-102 cm			R		
47-4 50-52 cm					
47-4 100-102 cm					
47-5 13-15 cm					
47-5 50-52 cm			R		
47-5 102-104 cm					
47-6 18-25 cm					

DSDP Leg 62, Site 463 (Mid-Pacific Mountains, Central Pacific Ocean)

The upper Albian-lower Maastrichtian sediments at this site are dominated by nannofossil chalks. The biostratigraphy at this site was studied at various stratigraphical intervals by Boersma (1981), Ando et al. (2009), Georgescu and Huber (2009) and Georgescu (2010, 2011a); the biostratigraphical framework is redone herein. A succession of twelve planktic foraminiferal biozones, which are defined from the base towards the top, is recognized (Table 7).

- *Biticinella breggiensis* Biozone (TRZ) is the stratigraphical interval between the FO and LO of *B. breggiensis*; only the upper part of this biozone occurs at Site 463. Age: late Albian.
- *P. appenninica* Biozone (IZ) is the stratigraphical interval between the FO of *P. appenninica* and the FO of *T. globotruncanoides*. Age: latest Albian.
- *T. globotruncanoides* Biozone (IZ) is the stratigraphical interval between the FO of the index species and the unconformity at the upper part of the lower Cenomanian; therefore, only the lower part of this biozone occurs at this site. Age: early Cenomanian.
- *Helvetoglobotruncana helvetica* Biozone (TRZ) is the stratigraphical interval between the FO and LO of *H. helvetica*. There is an unconformity separating the *T. globotruncanoides* Biozone and *H. helvetica* Biozone, which spans the middle Cenomanian-earliest Turonian time interval. Age: early Turonian.
- *M. schneegansi* Biozone (PTRZ) is the stratigraphical interval with *M. schneegansi* between the LO of *H. helvetica* and FO of *D. concavata*. Age: late Turonian.
- *D. concavata* Biozone (IZ) is the stratigraphical interval between the FO of the index species and FO of *Dicarinella asymetrica*. Age: latest Turonian-early Santonian.
- *D. asymetrica* Biozone (TRZ) is the stratigraphical interval between the FO and LO of *D. asymetrica*. Age: late Santonian.
- *H. pacificus* Biozone (IZ) is the stratigraphical interval between the LO of *D. asymetrica* and LO of *H. pacificus*. Age: early Campanian.
- *G. insignis* Biozone (IZ) is the stratigraphical interval between the FO of the index species and FO of *R. calcarata*. Age: late early-middle Campanian.
- *R. calcarata* Biozone (TRZ) is the stratigraphical interval between the FO and LO of the index species. Age: late Campanian.
- *G. subpetaloidea* Biozone (PTRZ) is the stratigraphical interval between the LO of *R. calcarata* and FO of *Gansserina gansseri*. Age: late Campanian.
- *G. gansseri* Biozone (IZ) is the stratigraphical interval between the FO of *G. gansseri* and the top of the Cretaceous section; only the lower part of this biozone occurs at Site 463. Age: latest Campanian-early Maastrichtian.

The schackoinid specimens are relatively very well preserved. However, they are affected by low magnitude recrystallization.

Table 9. Schackoinid stratigraphical distribution and frequency at ODP Site 762C. Frequency key: R-rare (< 5 specimens); C-common (6-10 specimens). The local stratigraphical ranges of the schackoinid species are given on light grey background

ODP Leg 122, Hole 762C (core number-core section, sample depth)	**Stages**	**Biozonation**	*Schackoina cenomana*	*Pseudohastigerinoides gandolfii*	*P. moliniensis*
66-4 110-111 cm	Conaician-lower Santonian	*G. vescicarinata*			
67-1 20-21 cm					
67 CC 30-31 cm					
68-1 20-21 cm					
68-1 120-121 cm					
69-1 20-21 cm					
69-1 120-121 cm					
69-2 20-21 cm					
70-1 20-21 cm					
70 CC 25-27 cm					
71-1 25.5-27 cm	Turonian	*F. maslakovae*			
71-1 118.5-119.5 cm					
71-2 19.5-20.5 cm					
71-2 123.5-124.5 cm					
72-1 20-21 cm					
72-1 112-113 cm					
72-2 20-21 cm					
72-2 110.5-112 cm		*H. hoelzli*			
73-1 19.5-20.5 cm			F		
73-1 119.5-121 cm					
73-2 22-24 cm			R		
73-2 120-121 cm		*H. helvetica*			
73-3 12.5-13.5 cm					
74-1 19.5-21 cm					
74-1 115-116 cm					
74-2 20-21 cm				R	R
74-2 119-120.5 cm			C		
75-1 26-27 cm		*W. archaeo-cretacea*		C	R
75-1 113-114 cm			R	R	R
75-2 20-21 cm	u. Cen.	*R. cushmani*	C	C	R
75-2 120-121 cm				R	

ODP Leg 122, Hole 762C (core number-core section, sample depth)	**Stages**	**Biozonation**	*Schackoina cenomana*	*S. multispinata*
60-1 20-21 cm	lower-middle Campanian	*G. ventricosa*		
60-1 120-121 cm				
60-2 20-21 cm				
60-2 120-121 cm				
60-3 20-21 cm				
60-3 120-121 cm				
61-1 20-21 cm				
61-1 120-121 cm				
61-2 20-21 cm		*G. elevata* equivalent	R	R
61-2 120-121 cm			R	
62-1 20-21 cm				
62-1 120-121 cm				
62-2 20-21 cm				
62-2 120-121 cm				
62-3 20-21 cm				
62-3 120-121 cm				
62-4 20-21 cm				
62-4 120-121 cm				
63-1 20-21 cm				
63-1 120-121 cm				
63-2 20-21 cm				
63-2 120-121 cm				
64-1 18.5-19.5 cm				
64-1 125-126 cm				
64-2 20-21 cm				
64-2 120-121 cm				
64-3 20-21 cm				
64-3 120-121 cm				
65-1 20-21 cm				
66-1 21.5-22.5 cm				
66-1 118-119 cm	u. Santon.	*D. asymetrica*		
66-2 20-21 cm				
66-2 120-121 cm				
66-3 20-21 cm				
66-3 120-121 cm				
66-4 20-21 cm				

Table 10. Schackoinid stratigraphical distribution and frequency at ODP Site 1050C. Frequency key: R-rare (< 5 specimens); C-common (6-10 specimens); F-frequent (11-25 specimens). The local stratigraphical ranges of the schackoinid species are given on light grey background

ODP Leg 171B, Hole 1050C (core number-core section, sample depth)	Stages	Biozonation	*Groshenyia groshenyae*	*G. pentagonalis*	*Asymetria asymetrica*	*Schackoina tritubercolata*	*S. cenomana*
26-CC 9-10 cm	upper Albian	*P. appenninica*					R
27-1 5-6 cm						R	C
27-1 19-21 cm					C		
27-1 75-76 cm					C		
27-1 125-127 cm				C	C		R
27-2 25-26 cm							
27-2 85-88 cm							
27-2 125-126 cm							
27-3 5-6 cm							
27-3 30-32 cm							
27-3 79-84 cm							
27-3 139-141 cm							
27-4 34-36 cm				R			
27-4 80-83 cm				C		R	
27-4 134-136 cm			R				
27-5 32-34 cm							
27-5 90-94 cm				C			
27-5 123-125 cm							
27-6 29-31 cm							
27-6 74-78 cm							
27-7 20-22 cm					R		
27-CC							
28-1 20-21 cm				R			
28-1 70-73 cm				F			
28-3 60-63 cm		*P. ticinensis*	R		F		
28-4 66-69 cm						C	
28-4 80-83 cm						C	
28-4 146-150 cm							
28-5 52-53 cm					R	R	
28-6 66-69 cm				C	F	R	
29-1 57-60 cm							
29-2 133-135 cm						R	
29-3 83-86 cm							
29-4 80-83 cm							
29-5 72-75 cm							
29-6 21-24 cm							
29-6 62-65 cm			R	R			
30-1 102-105 cm							
30-2 72-75 cm							
30-1 92-97 cm				R			
30-3 72-75 cm				C	C	R	
30-4 76-81 cm						R	
30-5 82-85 cm							
30-6 80-84 cm							
31-1 81-85 cm				R	F		
31-1 132-134 cm					R		
31-2 80-84 cm					R		
31-2 130-134 cm				R	R		
31-3 80-83 cm				R			
31-4 22-27 cm							
31-4 87-90 cm							
31-5 81-84 cm			R	C			
31-6 79-83 cm							
31-6 88-91 cm			R				
31 CC			R	C			

ODP Leg 171B, Hole 1050C (core number-core section, sample depth)	Stages	Biozonation	*Asymetria asymetrica*	*Schackoina tritubercolata*	*S. cenomana*
23-1 4-5 cm	middle Cenomanian	*T. reicheli*			R
23-1 50-51 cm					C
23-1 126.5-128.5 cm					
23-2 19-21 cm					
23-2 98.5-99.5 cm					
23-2 109-111 cm					
23-3 22-24 cm					C
23-3 49.5-50.5 cm					C
23-3 122-124 cm					R
24-3 127-130 cm					C
24-3 69-71 cm					
25-1 27-30 cm					
25-1 82-85 cm					R
25-1 133-136 cm					R
25-2 70-72 cm				R	
25-2 83-87 cm					
25-2 87-88 cm					
25-2 130-132 cm				F	
25-3 37-38 cm					R
25-3 60-64 cm	lower Cenomanian	*T. globotruncanoides*		R	
25-3 142-146 cm					
25-4 16-20 cm					
26-1 7-9 cm					
26-1 53-56 cm					R
26-1 139-141 cm					
26-2 24-26 cm					
26-2 56-59 cm				R	R
26-2 114-116					R
26-2 115-117 cm					
26-3 33-36 cm					
26-3 73-76 cm					
26-3 120-122 cm					
26-4 34-37 cm					
26-4 83-86 cm					
26-4 140-142 cm					
26-5 28-30 cm			R		R
26-5 83-86 cm			R	R	

DSDP Leg 71, Site 511 (Falkland Plateau, South Atlantic Ocean)

Pristine planktic foraminiferal tests occur throughout the zeolitic claystones of upper Turonian-Campanian age. The biostratigraphical framework used in this study is that given by Huber (1992) and Huber et al. (1995). Five planktic foraminiferal biozones are recognized, and they are briefly presented from the succession base upwards (Table 8).

- *Whiteinella baltica* Biozone (PRTZ) is the stratigraphical interval with *W. baltica* from the last occurrence of *Praeglobotruncana stephani* to the first occurrence of *Archaeoglobigerina cretacea*. Only the upper part of this biozone was studied at this site. Age: late Turonian.
- *Marginotruncana marginata* Biozone (IZ) is the stratigraphical interval between the FO of *A. cretacea* and LO of *M. marginata*. Age: late Coniacian-early Campanian.
- *A. cretacea* Biozone (PTRZ) is the stratigraphical interval with *A. cretacea* between the LO of *M. marginata* and FO of *Globigerinelloides impensus*. Age: middle Campanian.
- *G. impensus* Biozone (TRZ) is the stratigraphical interval between the FO and LO of the index species. Age: late Campanian.
- *G. havanensis* Biozone (IZ) is the stratigraphical interval between the FO of *G. havanensis* and the section top; only the lower part of this biozone occurs at Site 511. Age: late Campanian.

The planktic foraminiferal assemblages in the Upper Cretaceous sediments at Site 511 are in excellent state of preservation. Specimens that lack the test wall recrystallization are frequent.

ODP Leg 122, Hole 762C (Exmouth Plateau, Eastern Indian Ocean)

The upper Cenomanian-middle Campanian sediments at Hole 762C consist of nannofossil chalks. Although the planktic foraminiferal biostratigraphy was studied by Wonders (1992) and Petrizzo (2000), a new biostratigraphical framework is herein developed. Nine planktic foraminiferal biozones are recognized and they are presented from the succession base towards the top (Table 9).

- *Rotalipora cushmani* Biozone (TRZ) is the stratigraphical interval between the succession base to the FO of *Whiteinella archaeocretacea*; only the upper part of this biozone occurs in Hole 762C. Age: late Cenomanian.
- *W. archaeocretacea* Biozone (IZ) is the stratigraphical interval between the FO of the index species and FO of *H. helvetica*. Age: earliest Turonian.
- *H. helvetica* Biozone (TRZ) is the stratigraphical interval between the FO and LO of the index species. Age: early-middle Turonian.
- *Hedbergella hoelzli* Biozone is the stratigraphical interval between the FO of *H. hoelzli* and FO of *Falsotruncana maslakovae*. Age: late Turonian.

- *F. maslakovae* Biozone is the stratigraphical interval between the FO of *F. maslakovae* and FO of *Globotruncanita vescicarinata*. Age: latest Turonian.
- *G. vescicarinata* Biozone is the stratigraphical interval between the FO of the index species and FO of *D. asymetrica*. Age: Coniacian-early Santonian.
- *D. asymetrica* Biozone is the stratigraphical interval between the FO and LO of the index species. Age: late Santonian.
- *Globotruncanita elevata* Biozone equivalent is the stratigraphical interval between the LO of *D. asymetrica* and FO of *Globotruncana ventricosa*; *G. elevata* was not recognized in the studied samples. Age: early Campanian.
- *G. ventricosa* Biozone is the stratigraphical interval between the FO of *G. ventricosa* and the section top; only the biozone lower part is included in this study. Age: middle Campanian.

The schackoinid tests are very well-preserved in general, affected by low magnitude recrystallization that allows accurate observations on the test ultrastructure, ornamentation and porosity.

ODP Leg 171B, Hole 1050C (Blake Plateau, Western North Atlantic Ocean)

The schackoinid planktic foraminifers occur in upper Albian-middle Cenomanian sediments. Nannofossil claystones and nannofossil chalks are the dominant lithological types in the upper Albian and lower-middle Cenomanian respectively. The biostratigraphical framework for these sediments used in this study was provided by Huber et al. (1999), Bellier and Moullade (2002) and Petrizzo and Huber (2006). Four planktic foraminiferal biozones are recognized, and they are briefly presented below, from the succession's base upwards (Table 10).

- *P. ticinensis* Biozone (IZ) is defined from the succession base to the FO of *P. appenninica*. Age: middle late Albian.
- *P. appenninica* Biozone (IZ) comprises the stratigraphical interval between the FO of P. appenninica and FO of *T. globotruncanoides*. Age: latest Albian.
- *T. globotruncanoides* Biozone (IZ) is the stratigraphical interval between the FO of *T. globotruncanoides* and FO of *Thalmanninella. reicheli*. Age: early Cenomanian.
- *T. reicheli* Biozone (IZ) is the stratigraphical interval between the FO of the index species and FO of *R. cushmani*. Age: middle Cenomanian.

Test preservation is excellent in the upper Albian nannofossil claystone, where pristine tests are frequent. Preservation is very good in the lower-middle Cenomanian nannofossil chalk: accurate observations on the test wall ultrastructure, ornamentation and porosity characteristics can be made despite the low magnitude recrystallization.

Systematic Classification

Higher classification units are after Loeblich and Tappan (1988). Evolutionary classification units are after Georgescu (2009, 2010, 2012).

Order FORAMINIFERIDA Eichwald, 1830
Suborder GLOBIGERININA Delage and Hérouard, 1896
Superfamily ROTALIPORACEA Sigal, 1958
Family SCHACKOINIDAE Pokorný, 1958

Remarks. Family Schackoinidae is herein reassigned to superfamily Rotaliporacea based on the early low trochospiral stage, which occurs at least in the groups oldest representatives and is herein removed from superfamily Planomalinacea Bolli et al., 1957.

Genus/Directional Lineage *Groshenyia* – n. gen./new directional lineage

Type species.- *Schackoina pentagonalis* Reichel, 1948, by original designation.

Species included. Initiating species (IS): *Groshenyia groshenyae* – new species and first descendant species (FDS): *G. pentagonalis* (Reichel, 1948).

Diagnosis. Smooth late Albian schackoinids, which exhibit a transition from very low trochospiral tests with petaloid or clavate chambers and extraumbilical-peripheral aperture to tests with adult planispiral coiling, radially elongate chambers bearing one distal bulbous projection and equatorial aperture.

Description. Test is very low trochospiral in *G. groshenyae* and with adult planispirally coiled stage in *G. pentagonalis*. Earlier chambers are globular or subglobular, those in the adult stage tangentially elongate, petaloid, more rarely clavate in *G. groshenyae* and radially elongate, bearing one distal bulbous projection in *G. pentagonalis*. Sutures are distinct, depressed and radial on both test sides. Test slightly asymmetrical in edge view in *G. groshenyae* and symmetrical in *G. pentagonalis*. Aperture exhibits a gradual transition from a low arch in extraumbilical-peripheral position in the *G. groshenyae* to a low to medium high arch in equatorial position in *G. pentagonalis*; it is bordered by a delicate imperforate lip, which is rarely preserved; relict periapertural structures occasionally occur in the umbilical region. Chamber surface is smooth. Test wall is calcitic, hyaline, simple and perforate; pores are circular to subcircular.

Remarks. *Groshenyia* differs from *Leupoldina* of the late Barremian-Aptian by having elongate chambers of *G. pentagonalis* with one distal bulbous projection without the tendency to bifurcate and test wall with larger pores. Tests with elongate chambers bearing distal bulbous projections of late Aptian-early late Albian are not known, thereby leading to the conclusion that the two resulted through iterative evolution. *Lilliputianella* Banner and Desai, 1988 and *Lilliputianelloides* BouDagher-Fadel, Banner and Whittaker, 1997 of the Aptian present radially elongate chambers, which can be either petaloid or pointed, but do not develop distal bulbous projections as *Groshenyia*.

Derivation. The genus name honours Dr D. Grosheny (Université de Strasbourg) as appreciation for her outstanding contributions in foraminiferal studies.

Stratigraphical range. Upper Albian; from the *P. ticinensis* Biozone to the lower part of *P. appenninica* Biozone).

Geographic distribution. Western Tethys (Switzerland, Italy), North Atlantic Ocean (Blake Plateau, Mazagan Plateau, Bermuda Rise, offshore Spanish Sahara and Vigo Seamount) and Caribbean region (Trinidad, Puerto Rico Trench).

IS *Groshenyia groshenyae* – new species
Figures 6.1-6.9

1980 *Clavihedbergella simplex* (Morrow); Miles and Orr, pl. 4, figures 8-10.
2003 *Praehedbergella* sp.; Bellier et al., pl. 3, figures 13-14.

Holotype. Specimen WKB 010110.

Holotype dimensions. Maximum diameter: D_{max}=0.159 mm; minimum diameter: D_{min}=0.134 mm; D_{min}/D_{max}=0.842; thickness: t=0.057 mm; t/D_{max}=0.358; umbilical diameter: U_D=0.057 mm; U_D/D_{max}=0.358.

Paratypes. Five specimens, WKB 010111-010115.

Dimensions. D_{max}=0.130-0.181 mm; D_{min}=0.112-0.144 mm; D_{min}/D_{max}=0.791-0.861; t=0.052-0.065 mm; t/D_{max}=0.332-0.392; UD=0.038-0.057 mm; UD/D_{max}=0.209-0.358. The ranges are based on the average measurements of six specimens (holotype and paratypes).

Material. 16 Specimens

Type locality. Western North Atlantic Ocean (Blake Plateau), ODP Hole 1050C; geographical coordinates: 30° 05' N, 76° 14' W.

Type level. Upper Albian, Sample 171B-1050C-28-4, 80-83 cm (*P. ticinensis* Biozone).

Derivation. As for the genus.

Diagnosis. *Groshenyia* with very low trochospiral coil and petaloid to, more rarely clavate last-formed chambers.

Description. Test is very low trochospiral with the distinct tendency to become planispiral in the adult stage, unequally biumbilicate, and consisting of 7-8 chambers arranged in 1.5-2 whorls; there are 4-4.5 (commonly 4.5) chambers in the final whorl. Earlier chambers are globular, with low chamber size increase; chambers of the last whorl are tangentially elongate, petaloid, more rarely clavate, often backward oriented, and with higher size increase rate. Sutures are distinct, depressed and radial on both sides of the test; sutures between the chambers of the last whorl are deeply incised, resulting in a strongly lobate outline. Test is slightly asymmetrical in edge view due to the very low trochospire and the tendency to develop planispirally coiled tests. Aperture is a low arch, extraumbilical-peripheral in position and bordered by an imperforate lip; relict periapertural structures can be occasionally observed in the umbilical region. Umbilici are shallow and wide, with a diameter of circa one half of the maximum test diameter. Chamber surface is smooth. Test wall is calcitic, hyaline, simple and perforate; pores are simple, circular or subcircular, and with a diameter of 0.9-1.3 μm.

Remarks. *Groshenyia groshenyae* differs from all the species of the genus *Leupoldina* mainly by lacking last-formed chamber distal bulbous projections. It differs from *Clavihedbergella subcretacea* (Tappan, 1943) by having petaloid to, more rarely clavate rather than subcylindrical last formed chambers, lower trochospire resulting in an unequally biumbilicate test, simple periphery rather than with a peripheral band of low pore density and smaller pore size and (iv) smaller pores with a diameter of 0.9-1.3 μm rather than 1.5-3.2 μm. It differs from *Pseudoclavihedbergella simplicissima* (Magné and Sigal in Cheylan et al., 1954) mainly by having very low trochospire resulting in an unequally biumbilicate test rather than being mono-umbilicate, main aperture shifted more towards the periphery, smooth chamber surface rather than ornamented with dome-like pustules and smaller pores (0.9-1.3 μm rather than 0.9-2.2 μm in diameter). It differs from *P. amabilis* (Loeblich and Tappan, 1961) mainly by having simple test wall rather than incipiently reticulate and smaller pores, with a diameter of 0.9-1.3 μm rather than 1.8-5.0 μm. The gross test architecture of *G. groshenyae* resembles that of the Aptian "*Globigerinelloides*" *sigali* Longoria, 1974, but the porosity is different, with larger pores in the former; apparently the two species evolved iteratively, and there is a gap comprising the late Aptian-early late Albian time interval between the LO of "*G.*" *sigali* and FO of *G. groshenyae*.

Stratigraphical range. Upper Albian; from the *P. ticinensis* Biozone to the *P. appenninica* Biozone).

Geographic distribution. Western North Atlantic Ocean (Blake Plateau).

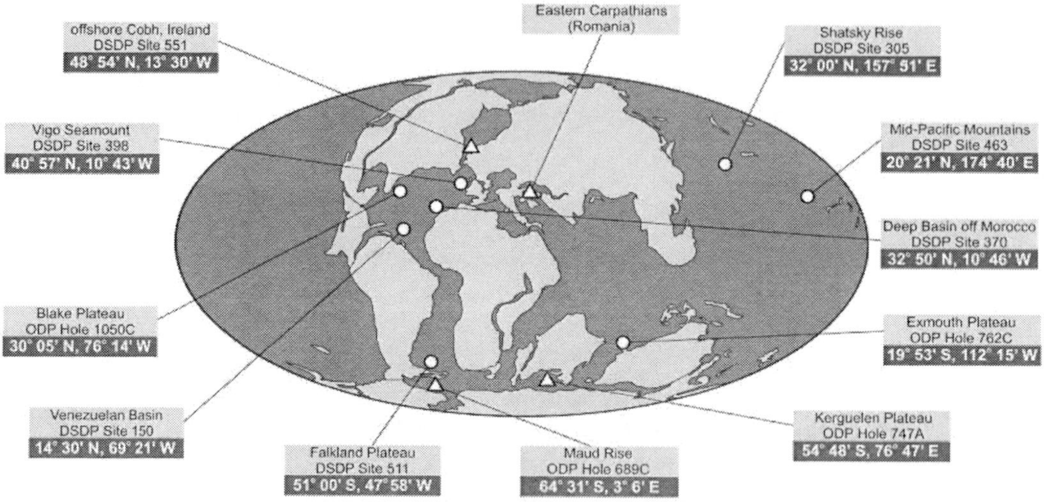

Figure 5. Geographical locations of the DSDP/ODP sites/holes (red circles) and spot sample sets (red triangles) used in this study. Base map after Hay et al. (1999), with modifications.

FDS *Groshenyia pentagonalis* (Reichel, 1948)
Figures 6.10-6.18, 7.1-7.12

1948 *Schackoina pentagonalis*; Reichel, p. 395, figures 1, 6: 1, 7: 1.

1948 *Schackoina pentagonalis* var. *aperta*; Reichel, p. 397, figures 2, 6: 2.

1952 *Hastigerinoides rohri*; Brönnimann, p. 55, pl. 1, figures 8-9, text-figure 29: a-f.

1959 *Schackoina gandolfii* Reichel; Bolli, p. 263, pl. 20, figures 13, 17-18.

1960 *Schackoina cenomana* (Schacko) *pentagonalis* Reichel; Masella, p. 26, pl. 4, figures 5, 9-14, pl. 5, figures 1-5.

1962 *Clavihedbergella pentagonalis* (Reichel); Luterbacher and Premoli Silva, pl. 22, figure A: 1-4.

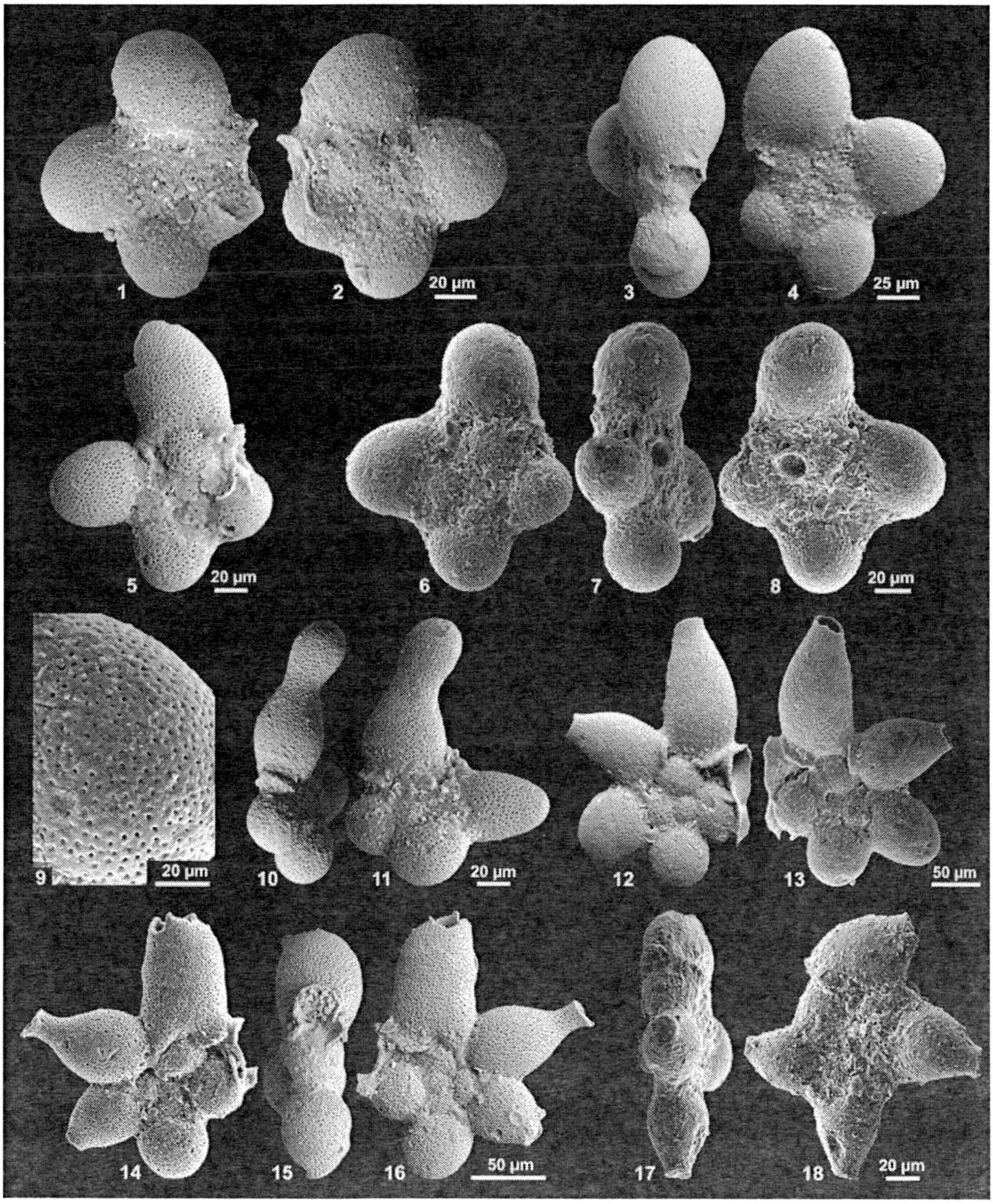

Figure 6. Specimens of *Groshenyia groshenyae* new genus/directional lineage and new species/initiating species and *G. pentagonalis* (Reichel, 1948) from the western North Atlantic Ocean (Blake Plateau). 1-2 Paratype of *G. groshenyae*, Sample 171B-1050C-27-4, 134-136 cm (late Albian, *P. appenninica* Biozone). 3-4 Paratype of *G. groshenyae*, Sample 171B-1050C-29-6, 62-65 cm (late Albian, *P. ticinensis* Biozone). 5 Paratype of *G. groshenyae*, Sample 171B-1050C-28-3, 62-65 cm (late Albian, *P. ticinensis* Biozone). 6-9 Holotype of *G. groshenyae*, Sample 171B-1050C-28-4, 80-83 cm (late Albian, *P. ticinensis* Biozone). 10-11 Hypotype of *G. pentagonalis*, Sample 171B-1050C-28-6, 66-69 cm (late Albian, *P. ticinensis* Biozone). 12-13 Hypotype of *G. pentagonalis*, Sample 171B-1050C-31-core catcher (late Albian, *P. ticinensis* Biozone). 14-16 Hypotype of *G. pentagonalis*, Sample 171B-1050C-31-core catcher (late Albian, *P. ticinensis* Biozone). 17-18 Hypotype of *G. pentagonalis*, Sample 171B-1050C-30-3, 72-75 cm (late Albian, *P. ticinensis* Biozone).

1964 *Schackoina pustulans* Bolli; Todd and Low, p. 407, pl. 1, figure 7.
1965 *Schackoina* sp.; Neagu, p. 36, pl. 10, figure 5.
1966 *Schackoina cenomana pentagonalis* Reichel; Barbieri, figure 2: 12.
1977 *Leupoldina pentagonalis* (Reichel); Masters, p. 425, pl. 14, figures 2-3.
1977 *Clavihedbergella moremani* (Cushman); Pflaumann and Krasheninnikov, p. 547, pl. 7, figures 2-3.
1979 *Schackoina pentagonalis* Reichel; Sigal, pl. 5, figure 4.
1980 *Clavihedbergella moremani* (Cushman); Miles and Orr, pl. 4, figures 6-7.
1984 *Clavihedbergella moremani* (Cushman); Leckie, p. 599, pl. 8, figures 11-12.
2003 *Schackoina leckiei*; Bellier et al., p. 137, pl. 1, figures 1-2, 10-13, pl. 2, figures 1-12.
2005 *Schackoina primitiva* Tappan; Neagu, p. 316, pl. 1, figure 31.

Material. 65 Specimens.

Diagnosis. *Groshenyia* with planispiral adult stage, equatorial main aperture, and last-formed chambers with distal bulbous projection.

Description. Test is evolute-involute, very low trochospiral in the early stage and planispirally coiled in the adult. It consists of 8-10 chambers arranged in 1.5-2 whorls; there are 4-5.5 (commonly 4.5-5) chambers in the final whorl. Early chambers are globular, with low chamber size increase. Chambers in the adult stage are often backward oriented and with variable shape: petaloid, more rarely clavate, and up to four of the last-formed ones are radially elongate, bearing one distal bulbous projection; elongation axis is tangential to the previous whorl or intersects it at an angle of 65-80°. Chambers in the last whorl have higher size increase rate, when compared to the earlier ones. Sutures are distinct, depressed and radial on both test sides. Test is symmetrical in edge view. Aperture is a low to medium high arch in equatorial position and is bordered by a delicate imperforate lip, which is rarely preserved; relict periapertural structures can occasionally occur in the umbilical region. Umbilici are shallow and wide, with a diameter of circa one half of the maximum test diameter at the level of the last-formed clavate chamber. Chamber surface is smooth. Test wall is calcitic, hyaline, simple and perforate; pore simple, circular or subcircular, and with a diameter of 0.5-1.6 μm.

Remarks. Chamber distal bulbous projections are absent in the original material illustrated by Reichel (1948) but, even in the absence of this important taxonomical feature the species can be readily identified by the evolute-involute tests and elongate chamber backward orientation. However, Luterbacher and Premoli Silva (1962) illustrated for the first time the distal bulbous projections in *E. pentagonalis*, assigning it to the genus *Clavihedbergella* Banner and Blow, 1959; this taxonomical solution was followed by Miles and Orr (1980) and Leckie (1984). Georgescu (2009) reviewed *Clavihedbergella* and showed that the type species *C. subcretacea* (Tappan, 1943) has subcylindrical elongate chambers rather than with distal bulbous projection and much larger pores (1.5-3.2 μm), more than twice the pore size in *G. pentagonalis* (0.5-1.6 μm); in addition, *C. subcretacea* lacks the general schackoinid test appearance and therefore cannot be considered congeneric with *G. pentagonalis*. Bellier et al. (2003) selected the specimen illustrated by Leckie (1984, pl. 8, Figures 11-12) as holotype for their new species *Schackoina leckiei*; this specimen clearly belongs to *G. pentagonalis*, and therefore *S. leckiei* is the latter species junior synonym. *Groshenyia pentagonalis* differs from *G. groshenyae* by having evolute-involute tests, adult planispiral stage rather than being completely trochospiral, last-formed chambers with distal bulbous projection rather than petaloid or, more rarely, clavate and larger pores, with a diameter of 0.5-1.6 μm rather than 0.9-1.3 μm; the similarities between the early stage of *G. pentagonalis* and the tests of *G. groshenyae*, and test ultrastructure indicate that *G. pentagonalis* evolved from *G. groshenyae*. *Leupoldina pustulans* (Bolli, 1957) of the Aptian differs from *G. pentagonalis* by having smaller, submicronic pores and occasionally bifurcate chamber extensions bearing distal bulbous projections; apparently the two species are homeomorphs that resulted through iterative evolution and therefore are not congeneric. *Pessagnoina moremani* (Cushman, 1931) of the late Cenomanian was reviewed by Georgescu (2009); it cannot be confused with *E. pentagonalis* for it has (i) trochospiral tests rather than with adult planispiral stage, (ii) umbilical-peripheral aperture rather than equatorial in the adult stage and (iii) larger pores (2.7-4.1 μm rather than 0.5-1.6 μm).

Stratigraphical range. Upper Albian (from the *P. ticinensis* Biozone to *P. appenninica* Biozone).

Geographic distribution. Europe (Switzerland, Italy), North Atlantic Ocean (Blake Plateau, Mazagan Plateau, Bermuda Rise, offshore Spanish Sahara and Vigo Seamount) and Caribbean region (Trinidad, Puerto Rico Trench).

Genus/Directional Lineage *Asymetria* – n. gen./new directional lineage

Type species.- *Asymetria asymetrica* – new species.

Species included. IS or FDS: *Asymetria asymetrica* – new.

Diagnosis. Schackoinid planktic foraminifera with low trochospiral test, last-formed chambers with one distal bulbous projection, and asymmetrically developed ornamentation: smooth on the spiral side and pustulose on the umbilical one.

Description. Test is low trochospiral; earlier chambers are globular, then reniform; the last-formed one to three is radially elongate and bears one distal bulbous projection. Sutures are distinct, depressed, and straight to slightly curved on both test sides. Test is convex-concave, slightly asymmetrical in edge view. Aperture is a medium high arch, extraumbilical-peripheral in position and bordered by a wide, tunnel-shaped lip; relict periapertural structures occur in the umbilical region. Chamber surface exhibits asymmetrically developed ornamentation; pustulose ornamentation is developed only on the umbilical side, whereas the spiral side is smooth. Test wall is calcitic, hyaline, simple and perforate; pores are simple, and with circular outline.

Remarks. *Asymetria* differs from any other Cretaceous planktic foraminifer by the asymmetrically developed ornamentation, which is pustulose on the umbilical side, and smooth on the spiral side.

Derivation. The name invokes the asymmetrical ornamentation.

Stratigraphical range. Upper Albian-lowermost Cenomanian (from the *P. ticinensis* Biozone to the lowermost part of the *T. globotruncanoides* Biozone). An uncertain occurrence in the early late Albian *P. subticinensis* Biozone at DSDP Site 398 is based on one poorly preserved specimen.

Geographic distribution. North Atlantic Ocean (Blake Plateau, Bermuda Rise and Vigo Seamount).

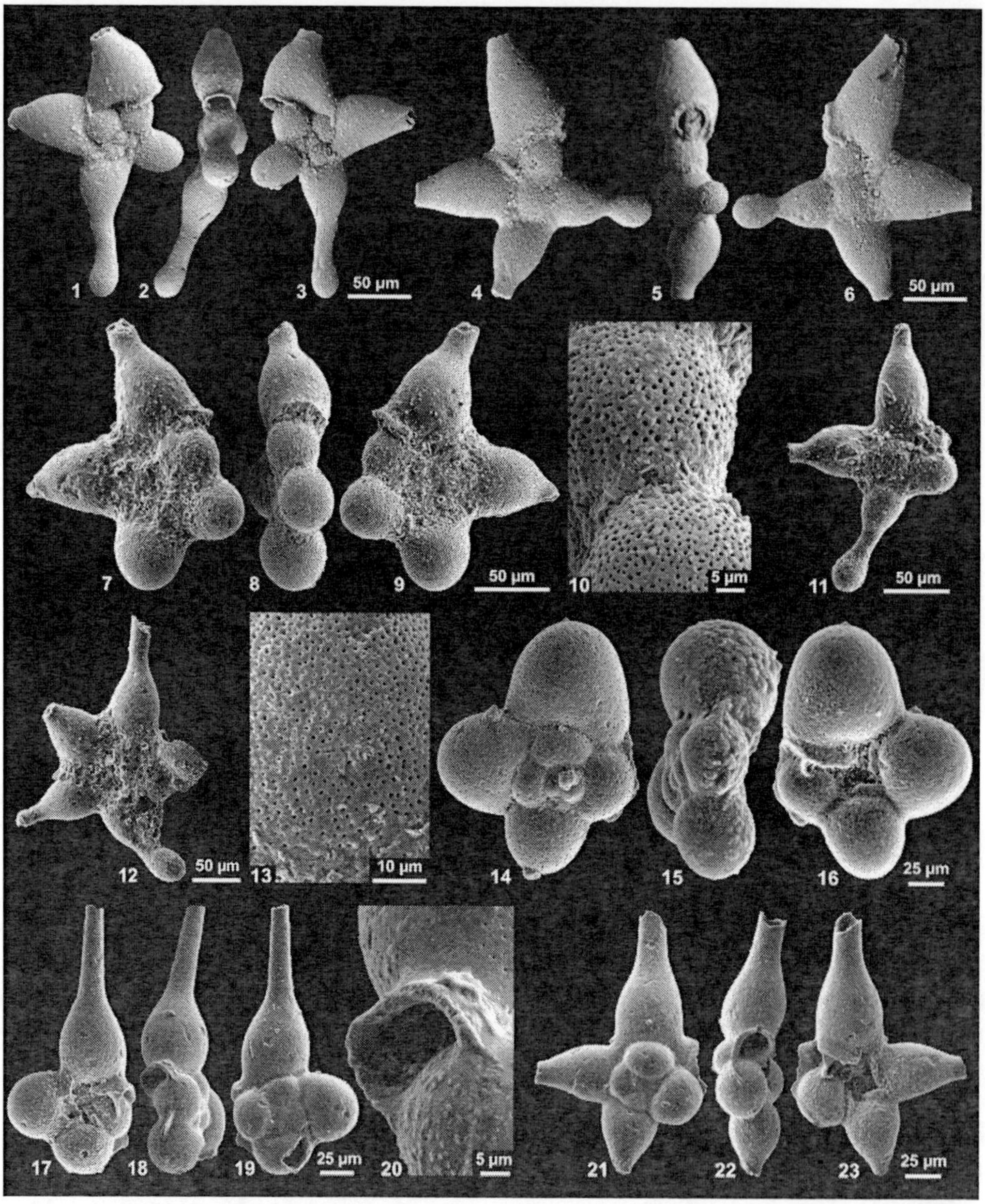

Figure 7. Specimens of *Groshenyia pentagonalis* (Reichel, 1948), *Asymetria*? sp. and *A. asymetrica* new genus/undifferentiated lineage and new species/undifferentiated species from the western North Atlantic Ocean (Blake Plateau). 1-3 Hypotype of *G. pentagonalis*, Sample 171B-1050C-27-1, 125-127 cm (late Albian, *P. appenninica* Biozone). 4-6 Hypotype of *G. pentagonalis*, Sample 171B-1050C-28-6, 66-69 cm (late Albian, *P. ticinensis* Biozone). 7-10 Hypotype of *G. pentagonalis*, Sample 171B-1050C-31-1, 81-85 cm (late Albian, *P. ticinensis* Biozone). 11 Hypotype of *G. pentagonalis*, Sample 171B-1050C-28-1, 70-73 cm (late Albian, *P. appenninica* Biozone). 12-13 Hypotype of *G. pentagonalis*, Sample 171B-1050C-28-1, 70-73 cm (late Albian, *P. appenninica* Biozone). 14-16 Hypotype of *A.*? sp., Sample 171B-1049B-12-3 (early Albian, *M. rischi* Biozone). 17-20 Paratype of *A. asymetrica*, Sample 171B-1050C-30-3, 72-75 cm (late Albian, *P. ticinensis* Biozone). 21-23 Paratype of *A. asymetrica*, Sample 171B-1050C-31-1, 81-85 cm (late Albian, *P. ticinensis* Biozone).

IS or FDS *Asymetria asymetrica* – new species
Figures 7.17-7.23, 8.1-8.17

1979 *Schackoina* cf. *cenomana* (Schacko); Sigal, pl. 4, figure 18.
1980 *Clavihedbergella moremani* (Cushman); Miles and Orr, pl. 4, figures 4-5.
2002 *Schackoina* sp. 2; Moullade et al., p. 135, figure 6: A-C.
2003 *Schackoina leckiei*; Bellier et al., p. 137, pl. 1, figures 3-9, pl. 3, figures 1-10.
2006 *Schackoina leckiei* Bellier, Moullade and Tronchetti; Petrizzo and Huber, pl. 1, figure 3.

Holotype. Specimen WKB 010116.

Holotype dimensions. D_{max}=0.164 mm; D_{min}=0.146 mm; D_{min}/D_{max}=0.890; t=0.074 mm; t/D_{max}=0.451; U_D=0.046 mm; UD/D_{max}=0.280.

Paratypes. Five specimens, WKB 010117-010121.

Dimensions. D_{max}=0.135-0.164 mm; D_{min}=0.096-0.146 mm; D_{min}/D_{max}=0.671-0.890; t=0.043-0.074 mm; t/D_{max}=0.294-0.451; U_D=0.040-0.051 mm; U_D/D_{max}=0.280-0.328. The ranges are based on the average measurements of ten specimens (holotype, paratypes and topotypes).

Material. 95 Specimens.

Type locality. Western North Atlantic Ocean (Blake Plateau), ODP Hole 1050C; geographical coordinates: 30° 05' N, 76° 14' W.

Type level. Upper Albian, Sample 171B-1050C-27-1, 125-127 cm (*P. appenninica* Biozone).

Derivation. As for the genus.

Description. Test is low trochospiral and consists of 8-10 chambers arranged in 2-2.5 whorls; there are 4-5 (commonly 4-4.5) chambers in the final whorl. Earlier chambers are globular, then reniform in shape, gradually increasing in size. Chambers of the last whorl are radially elongate, with the elongation axis perpendicular or at a high angle to the previous whorl; up to three of the last formed ones bear one distal bulbous projection, which is rarely preserved. Sutures are distinct, depressed, straight to slightly curved, and radial on both test sides. Test shape is convex-concave, slightly asymmetrical in edge view. Aperture is a medium high arch extraumbilical-peripheral in position, and bordered by a wide tunnel-shaped lip; relict periapertural structures occur in the umbilical region. Umbilici are shallow and medium-sized, with a diameter of circa one third to one half of the maximum test diameter at the level of the first radially elongate chamber. Chamber ornamentation is asymmetrical: spiral side smooth and umbilical side ornamented with small pustules (0.6-1.1 μm diameter), which are concentrated around the umbilicus; pustules occur also on the distal bulbous projections. Test wall is calcitic, hyaline, simple, and perforate; pores are simple, circular, and with a diameter of 0.4-1.0 μm.

Remarks. *Asymetria asymetrica* differs from any other Cretaceous planktic foraminiferal species by the asymmetrically developed ornamentation, which is completely absent on the spiral side and pustulose on the umbilical one. Pustule occurrence on the distal bulbous projections is unique among the leupoldinid and schackoinid foraminifera, which further ads to this species uniqueness. The origins of *A. asymetrica* are unknown. Its early stage exhibits similarities with the late Aptian-early Albian *Microhedbergella miniglobularis* Huber and Leckie, 2011 and *M. renilaevis* Huber and Leckie, 2011; however, these two species are smooth and there is a time gap spanning the latest early Albian-early late Albian between their last occurrence and the first occurrence of *A. asymetrica*. Therefore, a direct ancestor-descendant relationship cannot be demonstrated at this time, but specimens resembling the *M. miniglobularis-M. renilaevis* group, and with petaloid last-formed chambers are known in the lower Albian (*M. rischi* Biozone) sediments of the Blake Plateau (Figure 7.14-16); these rare specimens herein assigned to *A.*? sp. document the beginnings of the chamber elongation in this group, and may indicate the *Asymetria* lineage initiation, which would eventually lead to the evolution of *A. asymetrica* in the late Albian. Additional material is necessary to confirm or reject this presumed phylogenetic relationship.

Stratigraphical range. Upper Albian-lowermost Cenomanian (from the *P. ticinensis* Biozone to the lowermost part of the *T. globotruncanoides* Biozone).

Geographic distribution. North Atlantic Ocean (Blake Plateau, Bermuda Rise and Vigo Seamount).

Genus/Directional Lineage *Schackoina* Thalmann, 1932 – emended

Type species.- *Siderolina cenomana* Schacko, 1897, by original designation.

1932 *Hantkenina* (*Schackoina*); Thalmann, p. 288.
1933 *Schackoina* Thalmann; Cushman, p. 267.
1959 *Schackoina* (*Schackoina*) Thalmann; Banner and Blow, p. 9.
1964 *Schackoina* Thalmann; Loeblich and Tappan, p. C658.
1971 *Schackoina* Thalmann; Subbotina, p. 138.
1977 *Schackoina* Thalmann; Masters, p. 426.
1979 *Schackoina* Thalmann; Neagu, p. 310.
1985 *Schackoina* Thalmann; Caron, p. 25 (partly).
1988 *Schackoina* Thalmann; Loeblich and Tappan, p. 461.
2002 *Schackoina* Thalmann; Korchagin, p. 28.

Species included. IS: *Schackoina trituberculata* (Morrow, 1934), FDS: *S. cenomana* (Schacko, 1897), second descendant species (SDS): *S. bicornis* Reichel, 1948 and third descendant species (TDS): *S. multispinata* (Cushman and Wickenden, 1930).

Emended diagnosis. Schackoinid planktic foraminifera with early trochospiral coil and planispiral adult stage, last-formed chambers with tubulospines, and pustulose ornamentation symmetrically developed on both test sides.

Emended description. Test is very low trochospiral in the early stage and planispirally coiled in the adult stage. Earlier chambers are globular or subglobular, those of the adult stage are flask-like, radially elongate, and terminate with one tubulospine; the early chambers of the final whorl of the *S. trituberculata* lack tubulospines, which occur in *S. cenomana*, *S. bicornis* and *S. multispinata* developed on all the chambers of the final whorl; there are monotubulospines over the last-formed chambers of *S. trituberculata* and on all the chambers of the final whorl of *S. cenomana*, α-bitubulospines on the last-formed chambers of the *S. bicornis* and tritubulospines on the last-formed chambers of *S. multispinata*. Sutures are distinct, depressed, straight and radial on both test sides. Test is symmetrical in edge view. Aperture is an arch in peripheral position, and bordered by an imperforate lip; relict periapertural structures occasionally occur in the umbilical region. Chamber surface is symmetrically ornamented with scattered dome-like pustules on both test sides; pustules are concentrated around the umbilici. Test wall is calcitic, hyaline, simple and perforate; pores are simple and circular.

Remarks. *Schackoina* is emended to accommodate a directional lineage of four species, which consistently present flask-like chambers in the adult stage. The most obvious evolutionary trend in the Schackoina directional lineage is the gradual increase of the tubulospine number in the last-formed one or two chambers. *Schackoina* differs from *Leupoldina*, *Groshenyia* and *Asymetria* mainly by having symmetrically developed ornamentation and chamber termination(s) consisting of tubulospines rather than having distal bulbous projections.

Stratigraphical range. Upper Albian-Maastrichtian (from the upper part of the *P. ticinensis* Biozone to the *P. hariaensis* Biozone).

Geographic distribution. Cosmopolitan.

IS *Schackoina trituberculata* (Morrow, 1934) – emended
Figures 8.18-8.21, 9.4-9.15

1934 *Hantkenina trituberculata*; Morrow, p. 195, pl. 29, figures 26-27.
1946 *Schackoina trituberculata* (Morrow); Cushman, p. 148, pl. 61, figures 13-14.
1954 *Schackoina* cf. *trituberculata* (Morrow); Montanaro Gallitelli, pl. 1, upper five figures.
1966 *Schackoina cenomana cenomana* (Schacko); Neagu, pl. 1, figures 9-17.
1970 *Schackoina cenomana cenomana* (Schacko); Neagu, p. 63, pl. 14, figures 15-23.
1964 *Schackoina cenomana tappanae* Montanaro Gallitelli; Boccaletti and Pirini, p. 57, pl. 1, Figures 19-20, p. 2, figures 1-7, text-figure e.
1966 *Schackoina* cf. *cenomana tappanae* Montanaro Gallitelli; Barbieri, figure 2: 7-11.
1974 *Schackoina cenomana* (Schacko); Herb, p. 751, pl. 1, figure 6.
1975 *Schackoina cenomana* (Schacko); Playford et al., p. 338, figure 2: 2-3.
1977 *Schackoina tappanae* Montanaro Gallitelli; Masters, p. 437, pl. 17, figure 5.
1978 *Schackoina pentagonalis aperta* Reichel; Caron, p. 658, pl. 8, figure 7.
1979 *Schackoina cenomana* (Schacko); Sigal, pl. 5, figures 2, 5.
1980 *Schackoina cenomana* (Schacko); Miles and Orr, p. 798, pl. 1, figures 10-11.
1984 *Schackoina cenomana* (Schacko); Premoli Silva and McNulty, pl. 2, figures 8, 12.
1990 *Schackoina hermi* Weidich, p. 164, pl. 57, figures 13-27.
1997 *Schackoina cenomana* (Schacko); BouDagher-Fadel et al., p. 184, pl. 10.7, figures 4-5.
2003 *Schackoina cenomana* (Schacko); Bellier et al., pl. 3, figures 11-12.
2006 *Schackoina cenomana* (Schacko); Coccioni et al., pl. 1, figure 12.

Material. 110 Specimens.

Diagnosis. *Schackoina* with monotubulospines only on the last-formed chambers of the final whorl.

Description. Test is very low trochospiral in the early stage and planispiral, evolute-involute and biumbilicate in the adult, consisting of 6 to 7 chambers arranged in 1.5-2 whorls; there are 3.5-4 (commonly 4) chambers in the final whorl. Earlier chambers are globular, with low size increase. Up to three of the last-formed chambers are flask-like, radially elongate and with monotubulospines, which have the long axis perpendicular to the previous whorl. Sutures are distinct, straight and radial on both test sides. Test is symmetrical in edge view. Aperture is a low to medium high arch in peripheral position, and bordered by an imperforate lip; no relict periapertural structures occur in the umbilical region. Umbilici are shallow and medium-sized, its diameter being of one fourth to one half of the maximum test diameter at the tubulospine base level. Chamber surface is symmetrically ornamented on the two test sides with dome-like or irregular pustules (3.0-5.9 μm maximum dimensions), concentrated around the umbilicus and very rare or absent towards the chamber periphery; the ornamented zones are symmetrically developed on both test sides. Test wall is simple, calcitic, hyaline and perforate; pores are simple, with circular outline, and a diameter of 0.2-0.5 μm.

Remarks. *Schackoina trituberculata* differs from *G. pentagonalis* and *A. asymetrica* mainly by having the radially elongate chambers with monotubulospine rather than strongly elongate and with one distal bulbous projection. This is the oldest *Schackoina* species and the globular chambers in the earlier portion of the final whorl apparently indicates a hedbergellid ancestry; this is also supported by the earlier trochospiral stage that occurs in all the species of the *Schackoina* directional lineage. A hedbergellid species, herein assigned to *Praehedbergella* sp. based on its smooth surface (Gorbachik and Moullade, 1973; Moullade et al., 2002), and recorded in the *P. ticinensis* Biozone in ODP Hole 1050C (Figure 9.1-3), can be considered transitional between the hedbergellid and schackoinid groups and possibly the ancestor of *Schackoina*; the very low trochospire and small number of chambers (7 to 8) are features that also occur in *S. trituberculata*; additional material is necessary to clarify the taxonomic status of *P.* sp.

Stratigraphical range. Upper Albian-Cenomanian (from the *P. ticinensis* Biozone to the *R. cushmani* Biozone).

Geographic distribution. Cosmopolitan.

FDS *Schackoina cenomana* (Schacko, 1897)
Figures 9.16-9.18, 10.1-10.7

1897 *Siderolina cenomana*; Schacko, 1897, p. 166, pl., 4, figures 3-5.
1902 *Siderolina cenomana* Schacko; Egger, p. 174, pl. 21, figure 42.
1928 *Siderolina cenomana* Schacko; Franke, p. 193, pl. 18, figure 11.
1930 *Hantkenina cenomana* (Schackoina); Cushman and Wickenden, p. 40, pl. 6, figures 1-3.
1934 *Hantkenina trituberculata*; Morrow, pl. 29, figure 24.
1946 *Schackoina trituberculata* (Morrow); Cushman, pl. 61, figure 16.
1948 *Schackoina cenomana* (Schacko); Reichel, p. 400.
1951 *Schackoina cenomana* (Schacko); Noth, p. 74, pl. 5, figures 9-10.
1952 *Schackoina cenomana* (Schacko); Bermudez, p. 108, pl. 20, figures 1-2.
1957 *Schackoina cenomana* (Schacko); Bolli et al., pl. 2, figures 1-2.
1959 *Schackoina gandolfii* Reichel; Bolli, p. 263, pl. 20, figures 12, 14-16.
1960 *Schackoina cenomana* (Schacko) *tappanae* Montanaro Gallitelli; Masella, pl. 7, figures 1-15, pl. 8, figures 6-15, pl. 10, figures 8-12.
1961 *Schackoina cenomana* (Schacko); Loeblich and Tappan, p. 270, pl. 1, figures 2-7.
1962 *Schackoina cenomana gandolfii* Reichel; Luterbacher and Premoli Silva, pl. 22, figures 1-3.

1962 *Schackoina cenomana* (Schacko); Samuel, p. 181, pl. 8, figures 2-3.
1964 *Schackoina cenomana gandolfii* Reichel; Boccaletti and Pirini, pl. 1, figures 4-9, text-figure b.
1965 *Schackoina cenomana* (Schacko); Takayanagi, p. 202, pl. 21, figure 1.
1965 *Schackoina cenomana* (Schacko); Viterbo, pl. 10, figure 2: 5.
1966 *Schackoina cenomana gandolfii* Reichel; Barbieri, figure 2: 1.
1966 *Schackoina cenomana cenomana* (Schacko); Barbieri, figure 2: 2-6.
1966 *Schackoina cenomana* (Schacko); Marianos and Zingula, p. 334, pl. 37, figure 2.
1966 *Schackoina cenomana cenomana* (Schacko); Neagu, p. 365, pl. 1, figures 1-2.
1966 *Schackoina cenomana* (Schacko); Salaj and Samuel, p. 165, pl. 7, figure 8.
1968 *Schackoina cenomana* (Schacko); Scheibnerová, p. 57, pl. 7, figures 5-7.
1968 *Schackoina trituberculata* (Morrow); Ansary and Tewfik, p. 46, pl. 5, figure 9.
1969 *Schackoina cenomana* (Schacko); Douglas, p. 162, pl. 6, figure 5.
1969 *Schackoina cenomana* (Schacko); Porthault, p. 529, pl. 1, figure 2.
1970 *Schackoina cenomana* (Schacko); Eicher and Worstell, p. 298, pl. 9, figures 1-2, 4.
1971 *Schackoina cenomana* (Schacko); Subbotina, p. 138, pl. 1, figures 1-3.
1974 *Schackoina cenomana* (Schacko); Herb, p. 751, pl. 5, figures 14-15.
1975 *Schackoina cenomana* (Schacko); Luterbacher, pl. 1, figures 6-7.
1977 *Schackoina cenomana* (Schacko); Carter and Hart, p. 28, pl. 1, figure 10.
1977 *Schackoina cenomana* (Schacko); Masters, p. 430, pl. 16, figures 1-2.
1977 *Schackoina tappanae* Montanaro Gallitelli; Masters, p. 437, pl. 17, figure 6.
1977 *Schackoina cenomana* (Schacko); Pflaumann and Krasheninnikov, p. 548, pl. 7, figures 5-7.
1977 *Schackoina cenomana* (Schacko); Premoli Silva and Boersma, pl. 2, figure 2.
1978 *Schackoina cenomana* (Schacko); Caron, p. 658, pl. 8, figures 8-9.
1980 *Schackoina cenomana* (Schacko); Miles and Orr, pl. 1, figures 12-13.
1980 *Schackoina cenomana* (Schacko); Sliter, pl. 17, figures 4, 8.
1983 *Schackoina cenomana* (Schacko): McNeil and Caldwell, p. 248, pl. 19, figure 17.
1983 *Schackoina cenomana* (Schacko); Krasheninnikov and Basov, p. 803, pl. 1, figures 1-3.
1984 *Schackoina cenomana* (Schacko); McNeil, figure 11: 1.
1984 *Schackoina cenomana* (Schacko); Leckie, p. 598, pl. 9, figures 5, 9.
1985 *Schackoina cenomana* (Schacko); Caron, p. 76, figure 35: 5-9.
1989 *Schackoina cenomana* (Schacko); Frerichs, p. 11.
1992 *Schackoina cenomana* (Schacko); Huber, pl. 1, figure 11.
1997 *Schackoina cenomana* (Schacko); BouDagher-Fadel et al., p. 184, pl. 10.7, figures 1-3.
1998 *Schackoina* cf. *cenomana* (Schacko); Bellier, p. 339, pl. 3. figure 11.
2000 *Schackoina cenomana* (Schacko); Petrizzo, figure 13: 5.
2009 *Schackoina cenomana* (Schacko); Dalby et al., pl. 1, figures 15-16.
2010 *Schackoina cenomana* (Schacko); Gebhardt et al., figure 6: 28-29, 34.
2010 *Schackoina cenomana* (Schacko): Patterson et al., p. 21, figure 8: 27-28.

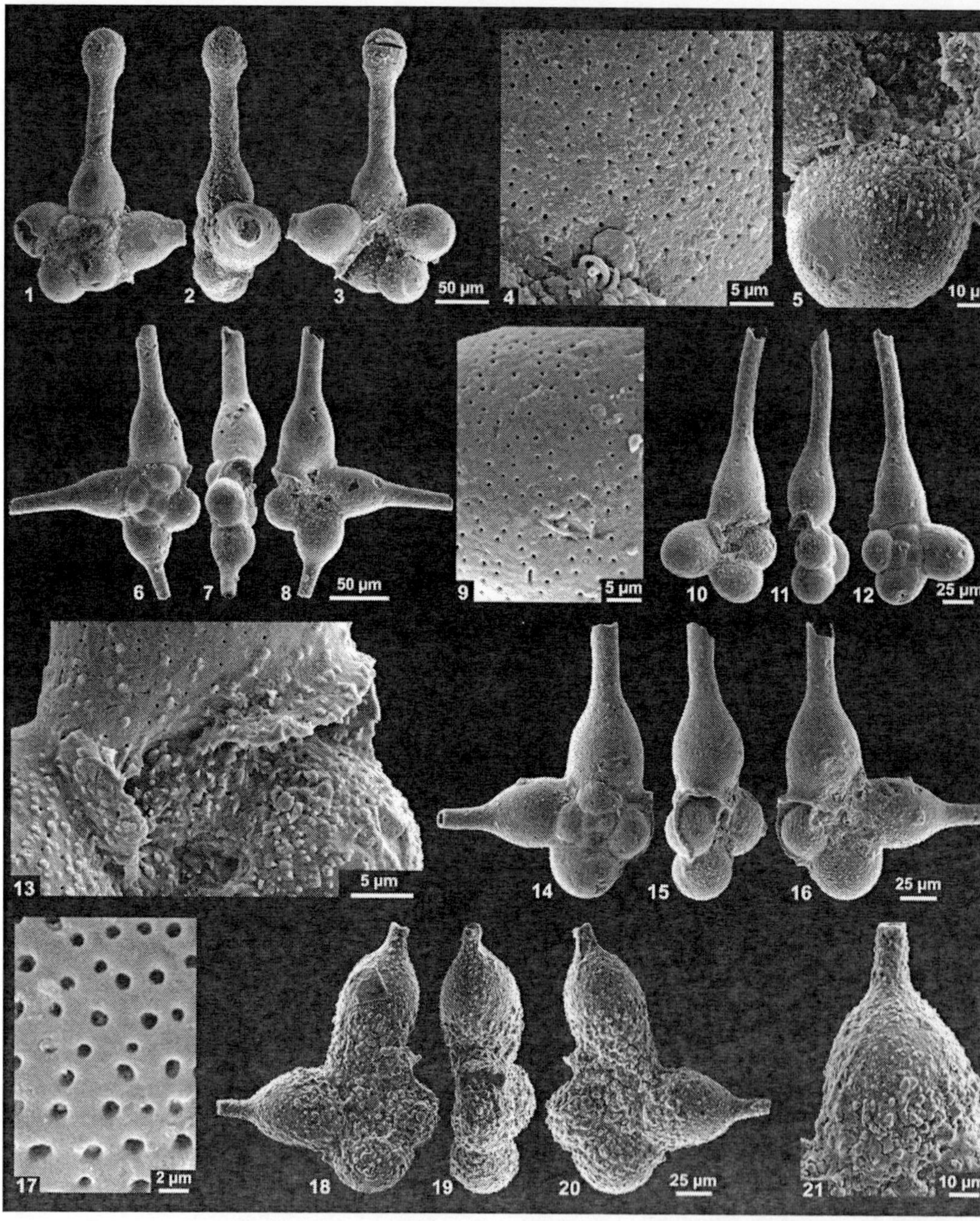

Figure 8. Specimens of *Asymetria asymetrica* new genus/directional lineage and new species/initiating species and *Schackoina trituberculata* (Morrow, 1934) from the western North Atlantic Ocean (Blake Plateau). 1-5 Holotype of *A. asymetrica*, Sample 171B-1050C-27-1, 125-127 cm (late Albian, *P. appenninica* Biozone). 6-9 Paratype of *A. asymetrica*, Sample 171B-1050C-30-3, 72-75 cm (late Albian, *P. ticinensis* Biozone). 10-13 Paratype of *A. asymetrica*, Sample 171B-1050C-31-1, 81-85 cm (late Albian, *P. ticinensis* Biozone). 14-17 Paratype of *A. asymetrica*, Sample 171B-1050C-27-1, 125-127 cm (late Albian, *P. appenninica* Biozone). 18-21 Hypotype of *S. trituberculata*, Sample 171B-1050C-28-4, 80-83 cm (late Albian, *P. ticinensis* Biozone).

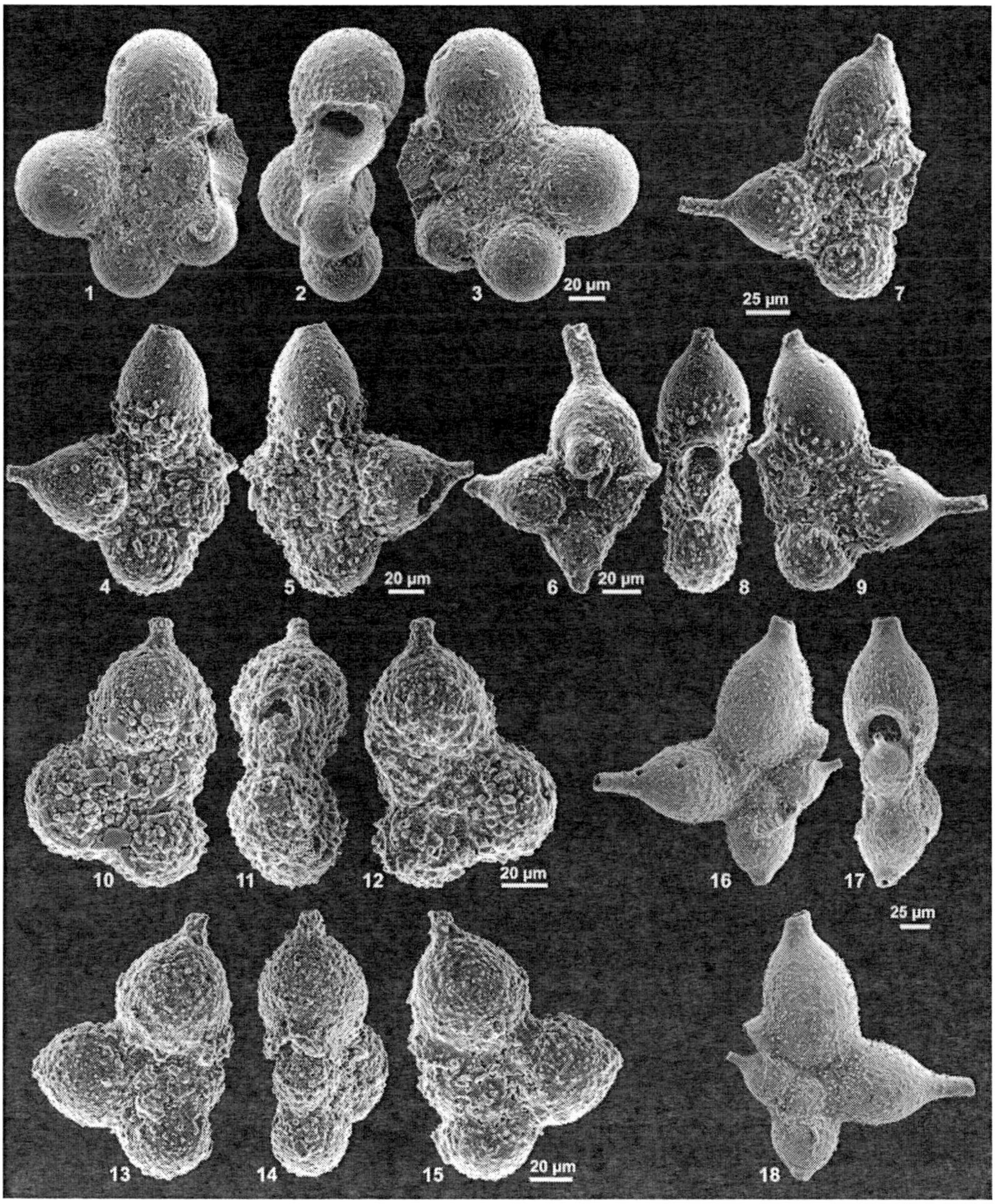

Figure 9. Specimens of *Praehedbergella* sp., *Schackoina trituberculata* (Morrow, 1934) and *S. cenomana* (Schacko, 1897) from the western North Atlantic Ocean (Blake Plateau) (1-15) and South Atlantic Ocean (Blake Plateau) (16-18). 1-3 Hypotype of *P.* sp., Sample 171B-1050C-31-2, 80-84 cm (late Albian, *P. ticinensis* Biozone). 4-5 Hypotype of *S. trituberculata*, Sample 171B-1050C-28-4, 80-83 cm (late Albian, *P. ticinensis* Biozone). 6 Hypotype of *S. trituberculata*, Sample 171B-1050C-28-4, 80-83 cm (late Albian, *P. ticinensis* Biozone). 7-9 Hypotype of *S. trituberculata*, Sample 171B-1050C-28-4, 80-83 cm (late Albian, *P. ticinensis* Biozone). 10-12 Hypotype of *S. trituberculata*, Sample 171B-1050C-25-2, 130-132 cm (middle Cenomanian, *T. reicheli* Biozone). 13-15 Hypotype of *S. trituberculata*, Sample 171B-1050C-25-2, 130-132 cm (middle Cenomanian, *T. reicheli* Biozone). 16-18 Hypotype of *S. cenomana*, Sample 71-511-46-core catcher (Turonian, *Praeglobotruncana* sp. Biozone).

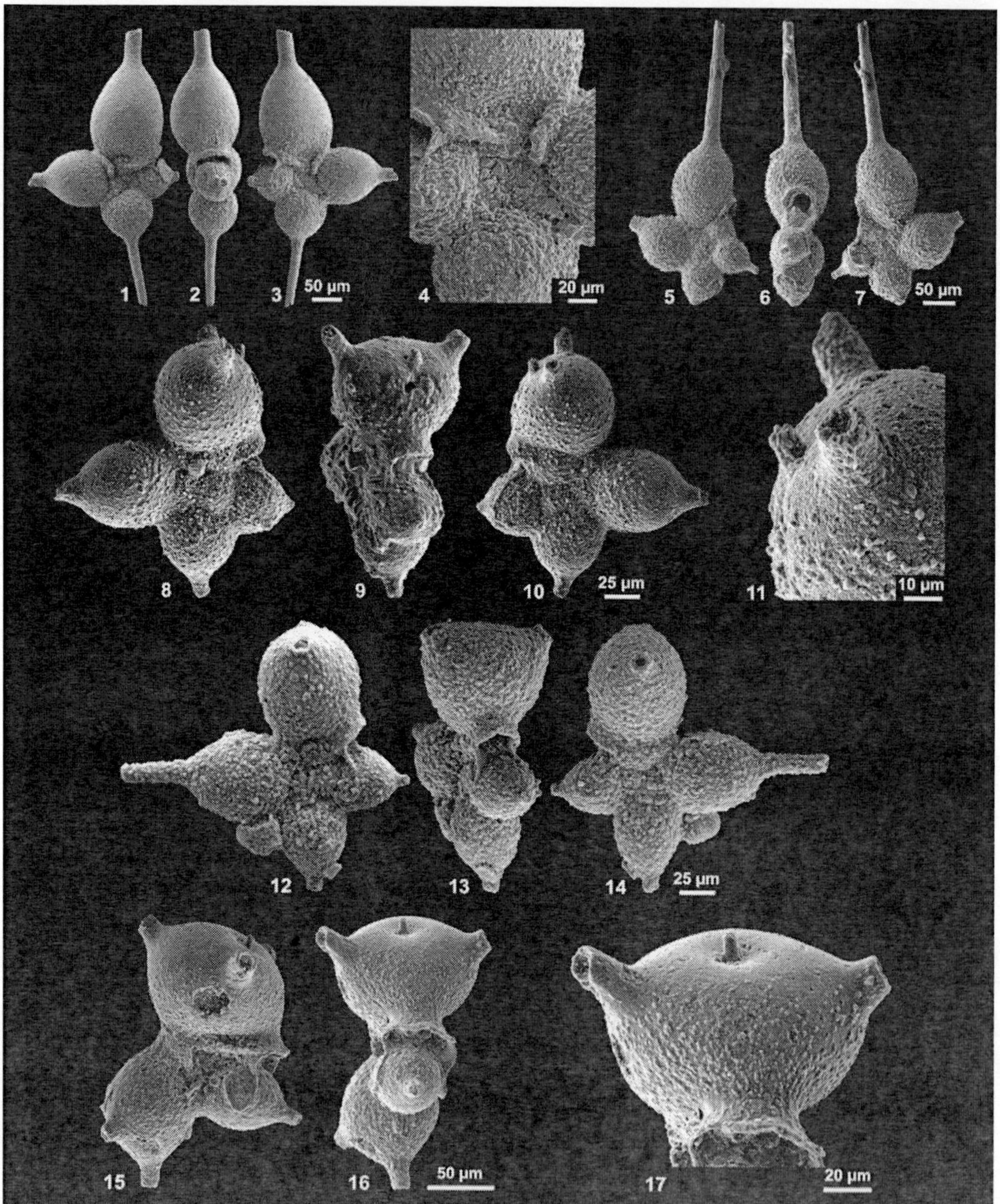

Figure 10. Specimens of *Schackoina cenomana* (Schacko, 1897), *S. bicornis* Reichel, 1948 and *S. multispinata* (Cushman and Wickenden, 1930) from South Atlantic Ocean (Falkland Plateau) (1-7), Caribbean region (Venezuelan Basin) (8-11), equatorial Central Pacific Ocean (Mid-Pacific Mountains) (12-14) and USA (Texas) (15-17). 1-4 Hypotype of *S. cenomana*, Sample 71-511-49-5, 65.5-67.5 cm (late Cenomanian, *Praeglobotruncana* sp. Biozone). 5-7 Hypotype of *S. cenomana*, Sample 71-511-46-core catcher (Turonian, *Praeglobotruncana* sp. Biozone). 8-11 Hypotype of *S. bicornis*, Sample 10-150-10-1, 58-73 cm (late Turonian, *M. schneegansi* Biozone). 12-14 Hypotype of *S. bicornis*, Sample 62-463-26-5, 53-58 cm (Coniacian, *D. concavata* Biozone). 15-17 Hypotype of *S. multispinata* from the Gober Chalk (Fennin County, Texas, USA) deposited in the Loeblich and Tappan Topotype Collection, NMNH, (USNM 478148).

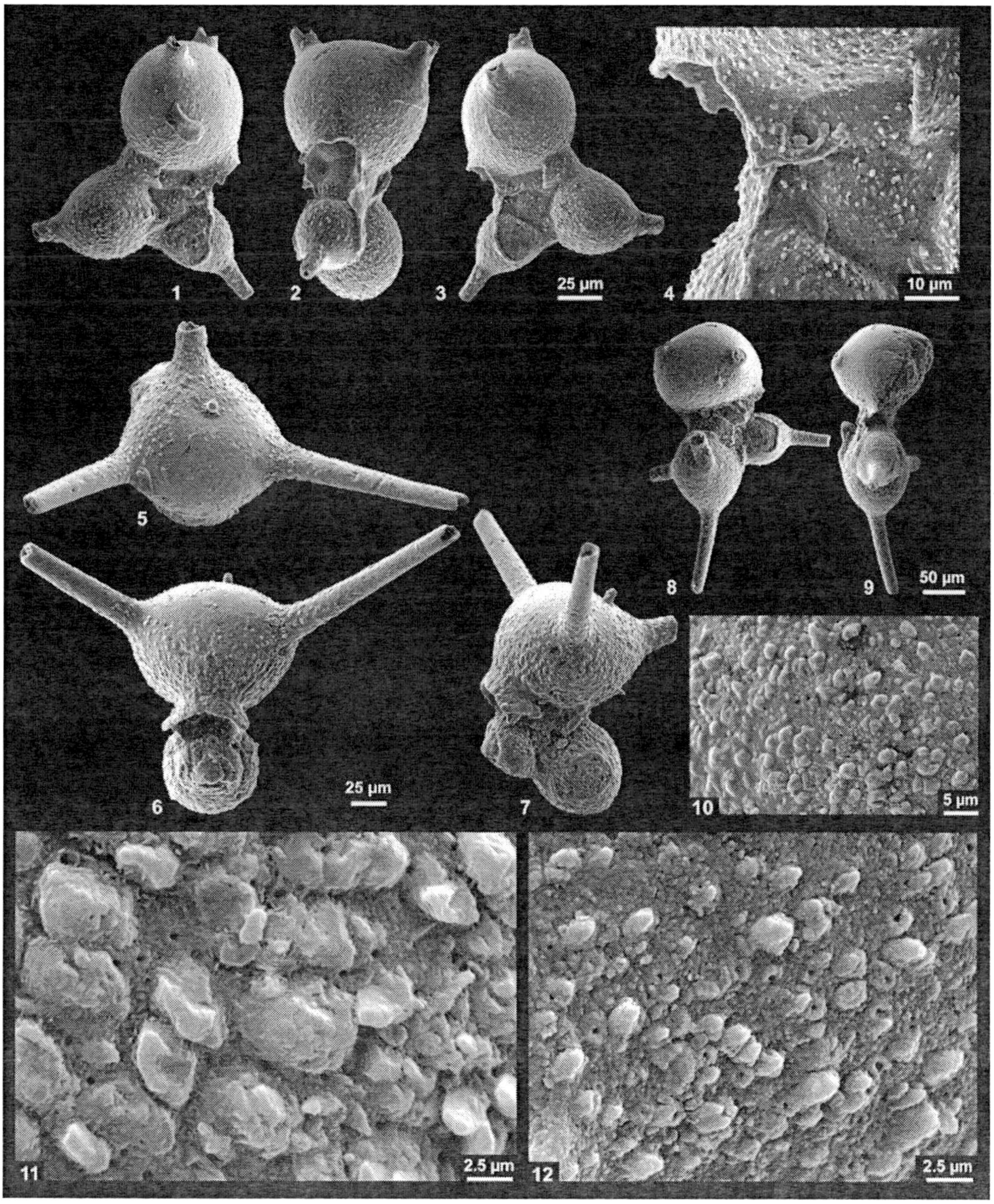

Figure 11. Specimens of *Schackoina multispinata* (Cushman and Wickenden, 1930) from the USA, Alabama (1-4, 11) and Texas (5-10, 12). 1-4, 11 Hypotype from the Arkadelphia Marl (Clark County, Alabama) deposited in the Loeblich and Tappan Topotype Collection, NMNH. 5-7, 12 Hypotype from the Gober Chalk (Fennin County, Texas) deposited in the Loeblich and Tappan Topotype Collection, NMNH, (USNM 478148). 8-10 Hypotype from the Gober Chalk (Fennin County, Texas) deposited in the Loeblich and Tappan Topotype Collection, NMNH, (USNM 478148).

Material. 205 Specimens.

Diagnosis. *Schackoina* with monotubulospines on all chambers of the final whorl.

Description. Test with early very low trochospiral stage and planispiral, evolute-involute and biumbilicate in the adult, consisting of 6-8 chambers arranged in 1.5-2 whorls; there are 3-4 chambers, mostly 4, rarely 4.5 or 5, in the final whorl. Earlier chambers are globular or subglobular and increase slowly in size. All the chambers in the final whorl are flask-like, and with monotubulospines. Chamber elongation axis is perpendicular to the previous whorl. Sutures are distinct, depressed, straight and radial on both test sides. Test is symmetrical in edge view. Aperture is a low to medium high arch in peripheral position, and is bordered by a narrow imperforate lip; relict periapertural structures of the last-formed chambers occur occasionally in the umbilical region. Umbilici are shallow and medium-sized, with a diameter of circa one fourth to one half of the maximum test diameter at the tubulospine base level. Chamber surface is symmetrically ornamented with scattered dome-like pustules (1.9-4.8 μm), which are denser around the umbilici and rarer towards the periphery. Test wall is simple, calcitic, hyaline and perforate; pores are simple, circular, and with a diameter of 0.2-0.5 μm.

Remarks. *Schackoina cenomana* differs from *S. trituberculata*, its descendant mainly by having monotubulospines developed on all the chambers of the final whorl rather than only over the last-formed chambers of the final whorl; it differs from *G. pentagonalis* by having the chambers of the final whorl with monotubulospines rather than radially elongate and with one distal bulbous projection and symmetrical ornamentation consisting of dome-like pustules rather than being completely smooth; it differs from *A. asymetrica* by having tests with adult planispiral stage rather than completely trochospiral, the chambers of the final whorl with monotubulospines rather than radially elongate and with one distal bulbous projection, main aperture in equatorial rather extraumbilical-peripheral position and symmetrically developed ornamentation rather than only on the umbilical side.

Stratigraphical range. Upper Albian-lower Campanian (from the *P. appenninica* Biozone to the *G. elevata* Biozone).

Geographic distribution. Cosmopolitan.

SDS *Schackoina bicornis* Reichel, 1948
Figures 10.8-10.14

1934 *Hantkenina trituberculata*; Morrow, pl. 29, figure 28 (only).

1946 *Schackoina trituberculata* (Morrow); Cushman, pl. 61, figure 15.

1948 *Schackoina cenomana bicornis*; Reichel, p. 401, figures 4: a-g, 6: 4, 7: 4, 8: b, 10: 8, 15.

1948 *Schackoina moliniensis*; Reichel, p. 402, figure 10: 13.

1948 *Schackoina cenomana bicornis* ? ou *S. moliniensis* ?; Reichel, figure 10: 9.

1955 *Schackoina cenomana bicornis* Reichel; Zanzucchi, p. 373, pl. 28, figure b: 5-7.

1960 *Schackoina cenomana* (Schacko) *cenomana* (Schacko); Masella, p. 24, pl. 1, figures 1-9, pl. 2, figures 1-4, pl. 10, figures 3-4.

1960 *Schackoina cenomana* (Schacko) *multispinata* (Cushman and Wickenden); Masella, p. 28, pl. 8, figures 16-21, pl. 9, figures 1-17, pl. 10, figures 13-15.

1960 *Schackoina cenomana* (Schacko) *utriculus*; Masella, p. 27, pl. 5, figures 6-8.

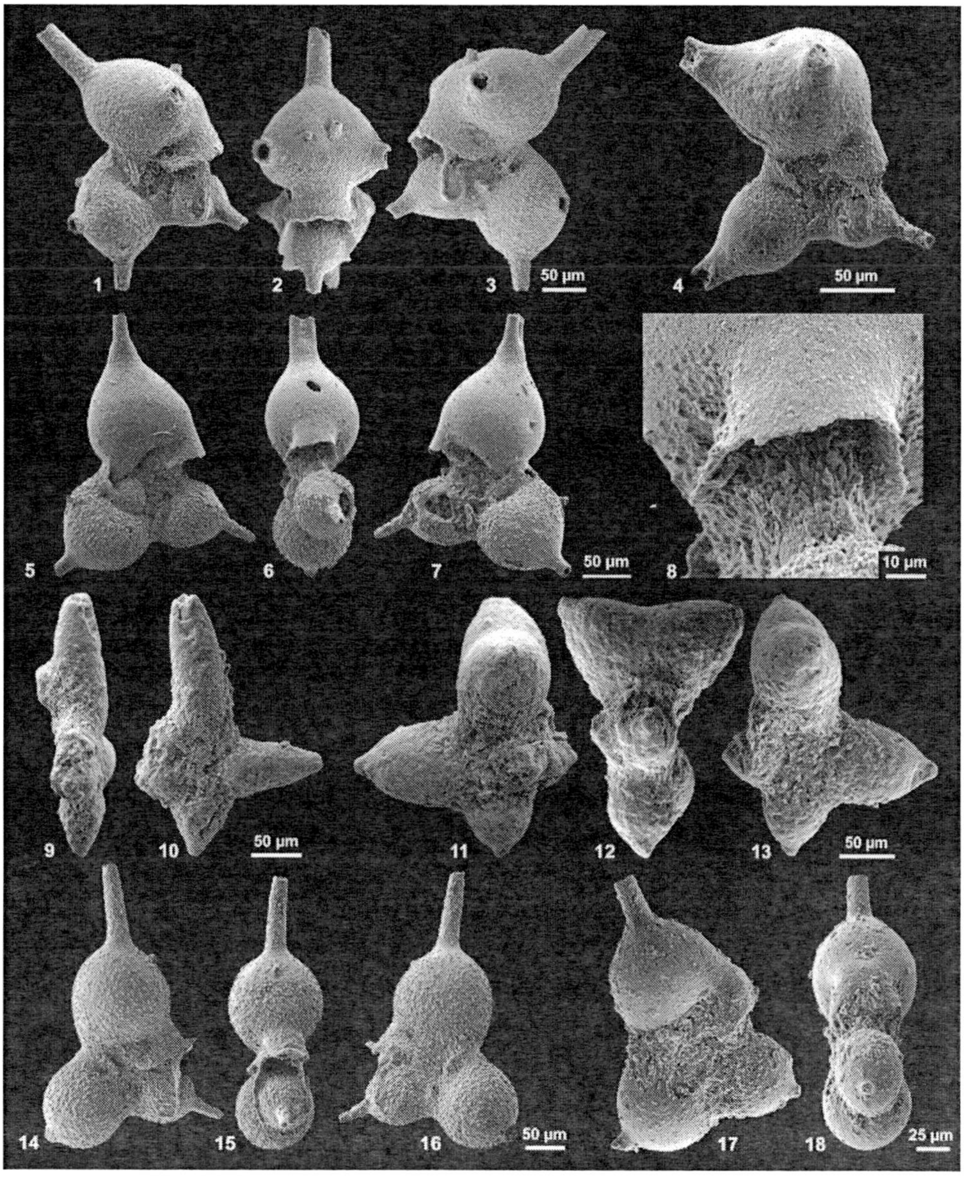

Figure 12. Specimens of *Schackoina multispinata* (Cushman and Wickenden, 1930), *Pseudohastigerinoides gandolfii* (Reichel, 1948), *P. moliniensis* (Reichel, 1948) and *Neoschackoina tappanae* (Montanaro Gallitelli, 1955) from the Central Pacific Ocean (Shatsky Rise) (1-3), USA (Texas) (4), Indian Ocean (Kerguelen Plateau) (5-8, 14-16), Valea Voievozi (Eastern Carpathians, Romania) (9-13) and South Atlantic Ocean (Maud Rise) (17-18). 1-3 Hypotype of *S. multispinata*, Sample 32-305-16-5, 60-76 cm (late Maastrichtian, *P. hariaensis* Biozone). 4 Hypotype of *S. multispinata* from the Gober Chalk (Fennin County, Texas) deposited in the Loeblich and Tappan Topotype Collection, NMNH, (USNM 478148). 5-8 Hypotype of *S. multispinata*, Sample 120-747A-23-6, 80-84 cm (early Maastrichtian, upper part of the *G. gansseri* Biozone). 9-10 Hypotype of *R. gandolfii*, Sample Neagu T. 144/67 (early Turonian, *H. helvetica* Biozone). 11-12 Hypotype of *R. moliniensis*, Sample Neagu T. 144/67 (early Turonian, *H. helvetica* Biozone). 14-16 Hypotype of *N. tappanae*, Sample 120-747A-24-1, 79-83 cm (early Maastrichtian, upper part of the *G. gansseri* Biozone). 17-18 Hypotype of *N. tappanae*, Sample 113-689B-28-1, 77-80 cm (early Maastrichtian, *G. havanensis* Biozone).

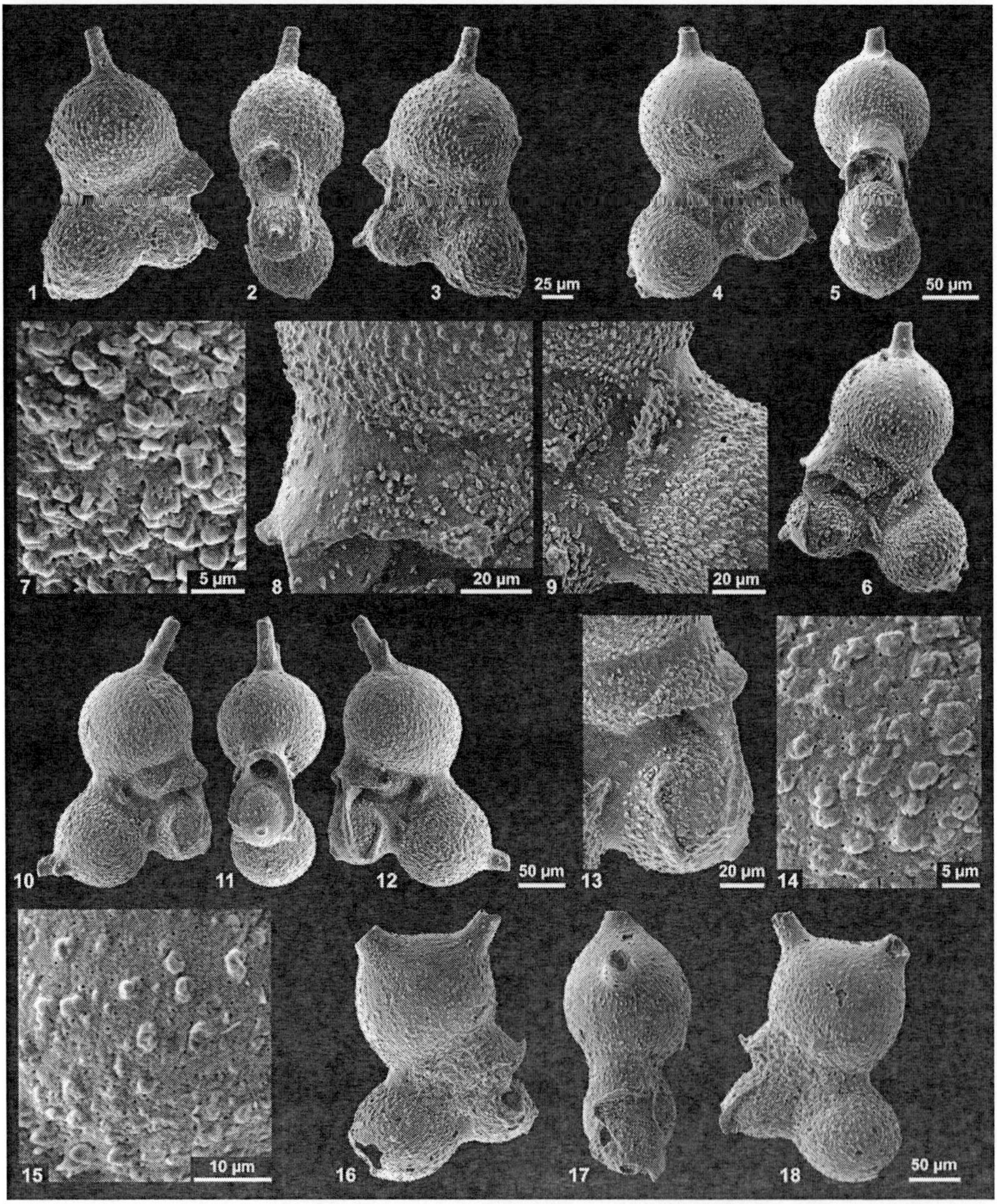

Figure 13. Specimens of *Neoschackoina tappanae* (Montanaro Gallitelli, 1955) and *N. sellaeforma* (Masters, 1976) from the South Atlantic Ocean (Falkland Plateau) (1-16, 18) and USA (Alabama) (17). 1-3 Hypotype of *N. tappanae*, Sample 71-511-24-7, 20-23 cm (late Campanian, *G. impensus* Biozone). 4-9 Hypotype of *N. tappanae*, Sample 71-511-24-6, 20-24 cm (late Campanian, *G. impensus* Biozone). 10-15 Hypotype of *N. tappanae*, Sample 71-511-24-6, 20-24 cm (late Campanian, *G. impensus* Biozone). 16-18 Hypotype of *N. sellaeforma*, Sample 71-511-24-6, 20-24 cm (late Campanian, *G. impensus* Biozone).

1961 *Schackoina multispinata* (Cushman and Wickenden); Loeblich and Tappan, p. 271, pl. 1, figures 8-10.
1962 *Schackoina cenomana* cf. *bicornis* Reichel; Luterbacher and Premoli Silva, pl. 22, figure B: 3.
1962 *Schackoina cenomana bicornis* Reichel; Samuel, p. 181, pl. 8, figures 5-6.
1964 *Schackoina moliniensis* Reichel; Boccaletti and Pirini, p. 54, pl. 1, figures 1-3, text-figure a.
1964 *Schackoina cenomana bicornis* Reichel; Boccaletti and Pirini, p. 56, pl. 1, figures 12-18, text-figure d.
1966 *Schackoina multispina bicornis* Reichel; Neagu, p. 366, pl. 1, figures 5-6, pl. 2, figures 3-7.
1967 *Schackoina multispina* (Cushman and Wickenden); Burckle et al., figure 2: 5.
1969 *Schackoina multispinata* (Cushman and Wickenden); Porthault, p. 529, pl. 1, figure 3.
1970 *Schackoina multispinata* (Cushman and Wickenden); Eicher and Worstell, p. 300, pl. 9, figures 5, 8.
1970 *Schackoina cenomana bicornis* Reichel; Neagu, p. 63, pl. 15, figures 7-25.
1977 *Schackoina bicornis* Reichel; Masters, p. 429, pl. 15, figure 3-4.
1977 *Schackoina multispinata* (Cushman and Wickenden); Masters, pl. 16, figure 4.
1979 *Schackoina multispina bicornis* Reichel; Sigal, pl. 5, figure 3.
1980 *Schackoina multispinata* (Cushman and Wickenden); Sliter, 1980, p. p. 369, figures 5, 6 (only).
1981 *Schackoina* sp.; Belford, p. 51, pl. A6, figures 23-26.
2000 *Schackoina multispinata* (Cushman and Wickenden); Petrizzo, figure 13: 5.
2005 *Schackoina cenomana* (Schacko); Verga and Premoli Silva, figure 1: 2.
2010 *Schackoina cenomana* (Schacko); Gebhardt et al., figure 6: 33.

Material. 17 Specimens.

Diagnosis. *Schackoina* with the last-formed one or two chambers with α-bitubulospines.

Description. Test is very low trochospiral in the early stage and planispiral, evolute-involute and biumbilicate when adult, consisting of 7-8 chambers arranged in 1.5-2 whorls; there are 4 chambers in the final whorl, rarely 4.5. Earlier chambers are globular or subglobular, with low size increase, whereas earlier chambers of the final whorl are flask-like, and with monotubulospines; the last-formed one or two chambers have α-bitubulospines; the chambers with α-bitubulospines are subtetrahedral in shape. Sutures are distinct, depressed, straight and radial on both test sides. Test is symmetrical in edge view. Aperture is a low to medium high arch in equatorial position, and bordered by an imperforate lip; no relict periapertural structures occur in the umbilical region. Umbilici are shallow and medium-sized, with the diameter of circa one half of the maximum test diameter at the level of the tubulospine base. Chamber surface is symmetrically ornamented on the two sides with dome-like pustules (2.0-2.6 μm), which are denser in the periumbilical zones. Test wall is simple, calcitic, hyaline and perforate; pores are simple, circular, and with a diameter of 0.2-1.5 μm.

Remarks. *Schackoina bicornis* differs from *S. cenomana* and *S. trituberculata* mainly by having α-bitubulospines on the last-formed one or two chambers rather than monotubulospines; it differs from *G. pentagonalis* and *A. asymetrica* mainly by having the chambers of the final whorl with monotubulospines and then with α-bitubulospines rather than radially elongate and with one distal bulbous projection, and chamber surface symmetrically ornamented with dome-like pustules rather than smooth or with asymmetrically developed ornamentation respectively.

Stratigraphical range. Upper Albian-lower Santonian (from the *P. appenninica* Biozone to the *D. concavata* Biozone).

Geographic distribution. Cosmopolitan.

TDS *Schackoina multispinata* (Cushman and Wickenden, 1930)
Figures 10.15-10.17, 11.1-11.12, 12.1-12.8

1937 *Schackoina trituberculata* (Morrow); Loetterle, p. 47, pl. 7, figure 7.
1930 *Hantkenina multispinata*; Cushman and Wickenden, p. 40, pl. 6, figures 4-6.
1931 *Hantkenina multispinata* Cushman and Wickenden; Cushman, p. 88, pl. 11, figures 10-11.
1946 *Schackoina multispinata* (Cushman and Wickenden); Cushman, p. 148, pl. 61, figures 11-12.
1952 *Schackoina multispinata* (Cushman and Wickenden); Bermudez, p. 108, pl. 20, figure 3.
1959 *Schackoina multispinata* (Cushman and Wickenden); Bolli, p. 264, pl. 20, figure 19.
1962 *Schackoina cushmani*; Barr, p. 565, pl. 69, figure 3, text-figure 5 a-h.
1962 *Schackoina multispinata* (Cushman and Wickenden); Luterbacher and Premoli Silva, figure C: 1-3.
1963 *Schackoina multispinata* (Cushman and Wickenden); Graham and Church, p. 61, pl. 7, figure 9.
1964 *Schackoina cenomana multispinata* (Cushman and Wickenden); Boccaletti and Pirini, pl. 2, figure 20, text-figure g.
1967 *Schackoina multispinata* (Cushman and Wickenden); Pessagno, p. 280, pl. 60, figure 1.
1968 *Schackoina multispinata* (Cushman and Wickenden); Sliter, p. 100, pl. 15, figure 7.
1971 *Schackoina multispinata* (Cushman and Wickenden); Subbotina, p. 139, pl. 1, figures 4-5.
1977 *Schackoina multispinata* (Cushman and Wickenden); Masters, p. 434, pl. 16, Figure 6, pl. 17, figure 2.
1977 *Schackoina multispinata* (Cushman and Wickenden); Sliter, p. 541, pl. 7, figures 8-9.
1981 *Schackoina multispinata* (Cushman and Wickenden); Frerichs and Dring, p. 68, pl. 2, figure 12.
1981 *Schackoina multispinata* (Cushman and Wickenden); McNeil and Caldwell, p. 249, pl. 20, figures 1-2.
1985 *Schackoina multispinata* (Cushman and Wickenden); Caron, p. 76, figure 35: 10-13.
1992 *Schackoina multispinata* (Cushman and Wickenden); Quilty, p. 381, pl. 2, figures 1-2.
1992 *Schackoina* sp.; Quilty, p. 381, pl. 2, figure 3.

Material. 24 Specimens.

Diagnosis. *Schackoina* with the last-formed one or two chambers with tritubulospines.

Description. Test is a very low trochospiral in early stage and planispiral, evolute-involute and biumbilicate in the adult stage, and consists of 7-8 chambers arranged in 2-2.5 whorls; there are three chambers in the final whorl. Earlier chambers are globular or subglobular, those in the final whorl flask-like or subtetrahedral; earlier chambers of the final whorl with monotubulospines, then with α-bitubulospines and the last-formed one or two with tritubulospines; the chambers with α-bitubulospines may be absent. Sutures are distinct, depressed, straight and radial on both test sides. Test is symmetrical in edge view. Umbilici are shallow, with the diameter of approximately one half of the test diameter at the tubulospine base. Aperture is a low to medium high arch equatorial in position, and bordered by an imperforate lip; relict periapertural structures occur in the umbilical regions. Chamber surface is symmetrically ornamented with scattered dome-like pustules with a diameter of 1.0-3.5 μm, which are concentrated around the umbilical regions and less dense in the peripheral region around the tubulospine base. Test wall is calcitic, hyaline, simple and perforate; pores are simple and circular, with a diameter of 0.2-0.5 μm.

Remarks. *Schackoina multispinata* differs from *S. trituberculata*, *S. cenomana* and *S. bicornis* mainly by having the last-formed one or two chambers with tritubulospines. It differs from the species of *Groshenyia* and *Asymetria* by having the chambers with tubulospines rather than radially elongate and with one bulbous distal projection.

Stratigraphical range. Upper Santonian-Maastrichtian (from the *D. asymetrica* Biozone throughout *P. hariaensis* Biozone).

Geographic distribution. Cosmopolitan.

Genus/Directional Lineage *Pseudohastigerinoides* – n. gen./new directional lineage

Type species.- *Schackoina moliniensis* Reichel, 1948, by original designation.

Species included. IS: *Pseudohastigerinoides gandolfii* (Reichel, 1948) and FDS: *P. moliniensis* (Reichel, 1948).

Diagnosis. Schackoinid planktic foraminifera with strongly elongate chambers resembling tubulospines in the adult stage.

Description. Test is very low trochospiral in the early stage and planispiral in the adult stage. Early chambers are globular and subglobular, those of the final whorl conical or highly conical, and occasionally flask-like, the last formed one or two strongly radially elongate, resembling tubulospines; there is one chamber extension on each chamber in *P. gandolfii* and two or three elongate extensions, the two anterior ones symmetrically arranged with respect to the equatorial plane in *P. moliniensis*. Sutures are distinct, depressed, straight and radial on both test sides. Test is symmetrical in edge view; an equatorial groove developed on the chambers with symmetrically developed chamber extensions occurs in *P. moliniensis*. Aperture is a low arch at the base of the last-formed chamber, equatorial in position, and bordered by a narrow imperforate lip; relict periapertural structures occasionally occur in the umbilici. Chamber ornamentation consists of scattered dome-like pustules, which are concentrated around the umbilici, and are symmetrically developed over the two test sides. Test wall is calcitic, hyaline, simple and perforate; pores are simple and with circular outline.

Remarks. *Pseudohastigerinoides* differs from any other schackoinid planktic foraminifer mainly by the last-formed chambers, which are strongly elongate resembling tubulospines, rather than having radially elongate chambers with one distal projection or tubulospines.

Derivation. The Greek prefix "*pseudo-*" (=false) is added to the pre-existing genus name *Hastigerinoides*, invoking the similarities in the general test architecture between the two genera.

Stratigraphical range. Cenomanian-lower Turonian (from the upper part of the *T. globotruncanoides* Biozone to the lower part of *H. helvetica* Biozone).

Geographic distribution. Europe (Switzerland, France, Italy, Romania), Caribbean region (Puerto Rico Trench), northern Africa (Tunisia), Australia and eastern Indian Ocean (Exmouth Plateau).

IS *Pseudohastigerinoides gandolfii* (Reichel, 1948)
Figures 12.9-10

1948 *Schackoina gandolfii*; Reichel, p. 397, text-figures 3: a-g, 6: 3, 7: 3, 8: a, 10: 1, 3, 4, pl. 8, figure 1.
1952 *Schackoina gandolfii* Reichel; Bermudez, p. 108, pl. 20, figure 4.
1954 *Schackoina gandolfii* Reichel; Aurouze and de Klasz, figure 1c.
1955 *Schackoina gandolfii* Reichel; Zanzucchi, p. 373, pl. 28, figure c: 1-4.
1960 *Schackoina cenomana gandolfii* Reichel; Masella, pl. 2, figures 5-12, pl. 3, Figures 1-13, pl. 10, Figures 6-7.
1960 *Schackoina cenomana cenomana* (Schacko); Masella, pl. 10, figures 1-2.
1962 *Schackoina cenomana gandolfii* Reichel; Luterbacher and Premoli Silva, figure B: 1-2.
1964 *Schackoina gandolfii* Reichel; Todd and Low, p. 406, pl. 1, figures 5-6, 8.
1966 *Schackoina cenomana gandolfii* Reichel; Neagu, p. 366, pl. 1, figures 3-4, 18-21, pl. 2, figures 1-2.
1966 *Schackoina gandolfii* Reichel; Salaj and Samuel, p. 166, pl. 7, figure 6.
1970 *Schackoina cenomana gandolfii* Reichel; Neagu, p. 63, pl. 15, figures 1-6.
1972 *Schackoina cenomana cenomana* (Schacko); Gawor-Biedowa, p. 64, pl. 6, figure 1.
1973 *Schackoina cenomana* (Schacko); Quilty, pl. 1, figure 6.
1975 *Schackoina cenomana* (Schacko); Playford et al., p. 338, figure 2: 1, 4.
1977 *Schackoina cenomana gandolfii* Reichel; Premoli Silva and Boersma, p. 637, pl. 2, figure 1.

Material. 43 Specimens.

Diagnosis. *Pseudohastigerinoides* with one or two highly conical chambers in the final whorl and last-formed one or two chambers strongly radially elongate, resembling tubulospines.

Description. Test with early very low trochospiral stage and planispiral, evolute-involute and biumbilicate adult stage; it consists of 7-8 chambers arranged in 1.5-2 whorls; there are 4, rarely 4.5 chambers in the final whorl. Earlier chambers are globular or subglobular and with a slow size increase. The early one or two of the early chambers of the final whorl are conical or highly conical, and occasionally flask-like, the last formed one or two strongly radially elongate, resembling tubulospines; the first chamber in the final whorl can occasionally be globular; chamber elongation axis is perpendicular to the previous whorl; chambers increase in size at high rate in the last whorl. Sutures are distinct, depressed, straight and radial on both test sides. Test symmetrical in edge view; periphery is rounded. Aperture is a low arch at the base of the last-formed chamber, equatorial in position, and is bordered by a narrow imperforate lip; relict periapertural structures occasionally occur in the umbilici. Chamber ornamentation is symmetrically developed on the two test sides, and consists of scattered dome-like pustules (1.5-4.3 μm), which are concentrated around the umbilical regions. Test wall is simple, calcitic, hyaline and perforate; pores are circular, with a diameter of 0.2-0.5 μm.

Remarks. *Pseudohastigerinoides gandolfii* differs from species of *Groshenyia* and *Asymetria* mainly by lacking chamber distal bulbous projection; moreover, *P. gandolfii* is a younger species, which is recorded only in the proximity of the Cenomanian/Turonian boundary and in the lower Turonian. It differs from the species of the genus *Schackoina* mainly by lacking flask-like chambers in the last whorl and tubulospines. *Hastigerinoides alexanderi* (Cushman, 1931) of the late Santonian differs from *P. gandolfii* by its medium high to high aperture rather than a low arch and larger pore size (1.6-1.8 μm rather than 0.5-1.0 μm); however, the two species are separated by the middle Turonian-lower Santonian Stratigraphical interval.

Stratigraphical range. Cenomanian-lower Turonian (from the upper part of the *T. globotruncanoides* Biozone to the lower part of *H. helvetica* Biozone).

Geographic distribution. Europe (Switzerland, France, Italy, Romania), Caribbean region (Puerto Rico Trench), northern Africa (Tunisia), Australia and Indian Ocean (Kerguelen Ridge, Exmouth Plateau).

FDS *Pseudohastigerinoides moliniensis* (Reichel, 1948)
Figures 12.11-12.13

1948 *Schackoina moliniensis*; Reichel, p. 402, text-figures 5: a-d, 6: 5, 7: 5, 8: c, 10: 10, 13.
1952 *Schackoina bicornis* Reichel; Bermudez, p. 108, pl. 20, figure 5.
1955 *Schackoina moliniensis* Reichel; Zanzucchi, p. 373, pl. 28, figure c: 8-10.
1960 *Schackoina cenomana alberti*; Masella, p. 26, pl. 5, figures 11-16, pl. 6, figures 1-9, pl. 10, figure 5.
1960 *Schackoina cenomana trinacriae*; Masella, p. 27, pl. 5, figures 9-10.
1962 *Schackoina moliniensis* Reichel; Samuel, p. 183, pl. 8, figures 7-8.
1966 *Schackoina multispina bicornis* Reichel; Neagu, p. 366, pl. 1, figures 7-8, pl. 2, figures 8-22.
1966 *Schackoina multispinata* (Cushman and Wickenden); Salaj and Samuel, 1966, p. 166, pl. 7, figure 7.
1969 *Schackoina bicornis* Reichel; Porthault, p. 529, pl. 1, figure 1.
1972 *Schackoina cenomana bicornis* Reichel; Gawor-Biedowa, p. 65, pl. 6, figure 2.
1972 *Schackoina moliniensis* Reichel; Gawor-Biedowa, p. 66, pl. 6, figure 3.
1973 *Schackoina bicornis* Reichel; Quilty, pl. 1, figure 7.
1977 *Schackoina alberti* Masella; Masters, p. 428, pl. 15, figure 2.
1978 *Schackoina moliniensis* Reichel; Caron, p. 658, pl. 8, figures 10-11.
1980 *Schackoina multispinata* (Cushman and Wickenden); Sliter, p. 369, pl. 17, figures 7, 9 (only).

Material. 40 Specimens.

Diagnosis. *Pseudohastigerinoides* with the last one or two chambers with two or three elongate extensions, two of them symmetrically arranged with respect to the test equatorial plane.

Description. Test planispiral, involute-evolute and biumbilicate at least in the adult stage. The test consists of 7-8 chambers arranged in 1.5-2 whorls; there are commonly 4 chambers in the final whorl, rarely 4.5. Earlier chambers of the final whorl are conical in shape, occasionally flask-like; the last one or two have two or three elongate extensions, two of them symmetrically arranged with respect to the test equatorial plane. Sutures are distinct, depressed, straight and radial on both test sides. Test symmetrical in edge view; periphery with an equatorial groove developed on the chambers with symmetrically developed chamber extensions. Aperture is a low arch in equatorial position at the base of the last-formed chamber; it is bordered by an imperforate lip, which is rarely preserved. Umbilici are shallow and without relict periapertural structures. Chamber surface is ornamented with scattered dome-like pustules (1.8-3.4 μm in diameter), which are concentrated around the umbilical regions; ornamentation is symmetrically developed on the two test sides. Test wall is simple, calcitic, hyaline and perforate; pores are circular, with a diameter of 0.2-0.5 μm.

Remarks. *Pseudohastigerinoides moliniensis* differs from *P. gandolfii* by having two or three extensions resembling tubulospines over the last one or two chambers rather than one.

Stratigraphical range. Uppermost Cenomanian-lower Turonian (from the upper part of the *R. cushmani* Biozone to the lower part of *H. helvetica* Biozone).

Geographic distribution. Europe (Switzerland, Italy, Romania) and eastern Indian Ocean (Exmouth Plateau).

Genus/Directional Lineage *Neoschackoina* – n. gen./new directional lineage

Type species.- Schackoina sellaeforma Masters, 1976, by original designation.

Species included. IS: *Neoschackoina tappanae* (Montanaro Gallitelli, 1955) and FDS: *N. sellaeforma* (Masters, 1976).

Diagnosis. Schackoinid planktic foraminifera with globular chambers with monotubulospines or β-bitubulospines.

Description. Test planispiral, evolute-involute and biumbilicate consisting of globular chambers. Chambers are globular throughout and terminate with monotubulospines in *N. tappanae* and subrectangular to trapezoidal terminating with β-bitubulospines in *N. sellaeforma*. Sutures are distinct, depressed, straight and radial on both test sides. Test is symmetrical in edge view, slightly compressed in *N. sellaeforma*. Aperture is a medium high to high arch in equatorial position, and is bordered by a sparsely perforate lip; relict periapertural structures occur in the umbilici. Chamber surface is symmetrically ornamented with irregular pustules; less dense pustules occur occasionally around the tubulospine base. Test wall is calcitic, hyaline, simple and perforate; pores are simple, and circular.

Remarks. *Neoschackoina* differs from *Groshenyia* and *Asymetria* mainly by having the chambers terminating with tubulospines, rather that radially elongate and with one distal bulbous projection. It differs from *Schackoina* mainly by the test consisting of globular rather than flask-like chambers at least in the adult stage, and chambers terminating with monotubulospines and β-bitubulospines rather than monotubulospines, α-bitubulospine and tritubulospines. It differs from *Pseudohastigerinoides* mainly by having globular chambers rather than bearing radially elongate resembling tubulospines.

Derivation. The Greek prefix *neo-* (=new) is added to the pre-existing name *Schackoina*.

Stratigraphical range. Upper Santonian-lower Maastrichtian (from the *D. asymetrica* Biozone to the upper part of *G. gansseri* Biozone).

Geographic distribution. Europe (Italy, Poland), South Atlantic Ocean (Falkland Plateau, Maud Rise), Indian Ocean (Kerguelen Plateau) and USA (Alabama, Kansas, Wyoming).

IS *Neoschackoina tappanae* (Montanaro Gallitelli, 1955)
Figures 12.14-18, 13.1-15

1955 *Schackoina tappanae*; Montanaro Gallitelli, p. 142, pl. 1, figures 1-10.
1964 *Schackoina tappanae* Montanaro Gallitelli; Cati, p. 255, pl. 41, figure 10.
1969 *Schackoina multispinata* (Cushman and Wickenden); Douglas, p. 163, pl. 11, figure 13.
1975 *Schackoina multispinata* (Cushman and Wickenden); Frerichs et al., p. 306, pl. 3, figure 12.
1977 *Schackoina cenomana* (Schacko); Sliter, p. 541, pl. 7, figure 7.
1981 *Schackoina multispinata* (Cushman and Wickenden); Frerichs and Dring, p. 68, pl. 2, figure 11 (only).
1990 *Schackoina multispinata* (Cushman and Wickenden); Huber, p. 503, pl. 2, figure 1.
1992 *Schackoina tappanae* Montanaro Gallitelli; Gawor Biedova, p. 79, pl. 13, figure 1.

Material. 26 Specimens.

Diagnosis. *Neoschackoina* with the last-formed chambers with monotubulospine.

Description. Test planispiral, evolute-involute and biumbilicate, consisting of 6-7 chambers; there are 3.5-4 chambers in the final whorl, most commonly 3.5. Chambers are globular throughout, and slowly increase in size; the last-formed chambers have monotubulospines. Sutures are distinct, depressed, straight and radial on both test sides. Test is symmetrical in edge view. Aperture is a medium high or high arch in equatorial position, and bordered by a wide perforate lip. Umbilici are shallow, with a diameter of circa one third of the test maximum diameter at the level of the tubulospine base; relict periapertural structures occur in the umbilical regions. Chamber surface is symmetrically ornamented with pustules on both test sides; pustules irregular, with a maximum dimension of 3.2-4.5 μm, and are often distributed over the entire chamber surface; specimens with lesser amount of ornamentation around the tubulospine base occasionally occur. Test wall is calcitic, hyaline, simple and perforate; pores are simple, circular and with a diameter of 0.2-0.4 μm.

Remarks. *Neoschackoina tappanae* differs from the species of the *Schackoina* directional lineage mainly by the globular rather than flask-like chambers. It differs from the species of the directional lineage *Pseudohastigerinoides* mainly by the globular chambers with well-developed monotubulospines rather than radially elongate resembling tubulospines. There is no evidence to document the existence of an early trochospiral coil in *N. tappanae*.

Stratigraphical range. Upper Santonian-lower Maastrichtian (from the *D. asymetrica* Biozone to the upper part of *G. gansseri* Biozone).

Geographic distribution. Europe (Italy, Poland), South Atlantic Ocean (Falkland Plateau, Maud Rise), Indian Ocean (Kerguelen Plateau) and USA (Kansas, Wyoming).

FDS *Neoschackoina sellaeforma* (Masters, 1976)
Figures 13.16-18

1976 *Schackoina sellaeforma*; Masters, p. 327, pl. 1, figures 13-16.
1977 *Schackoina sellaeforma* Masters; Masters, p. 437, pl. 17, figures 3-4.

Material. 3 Specimens.

Diagnosis. *Neoschackoina* with last-formed one to three chambers with β-bitubulospines.

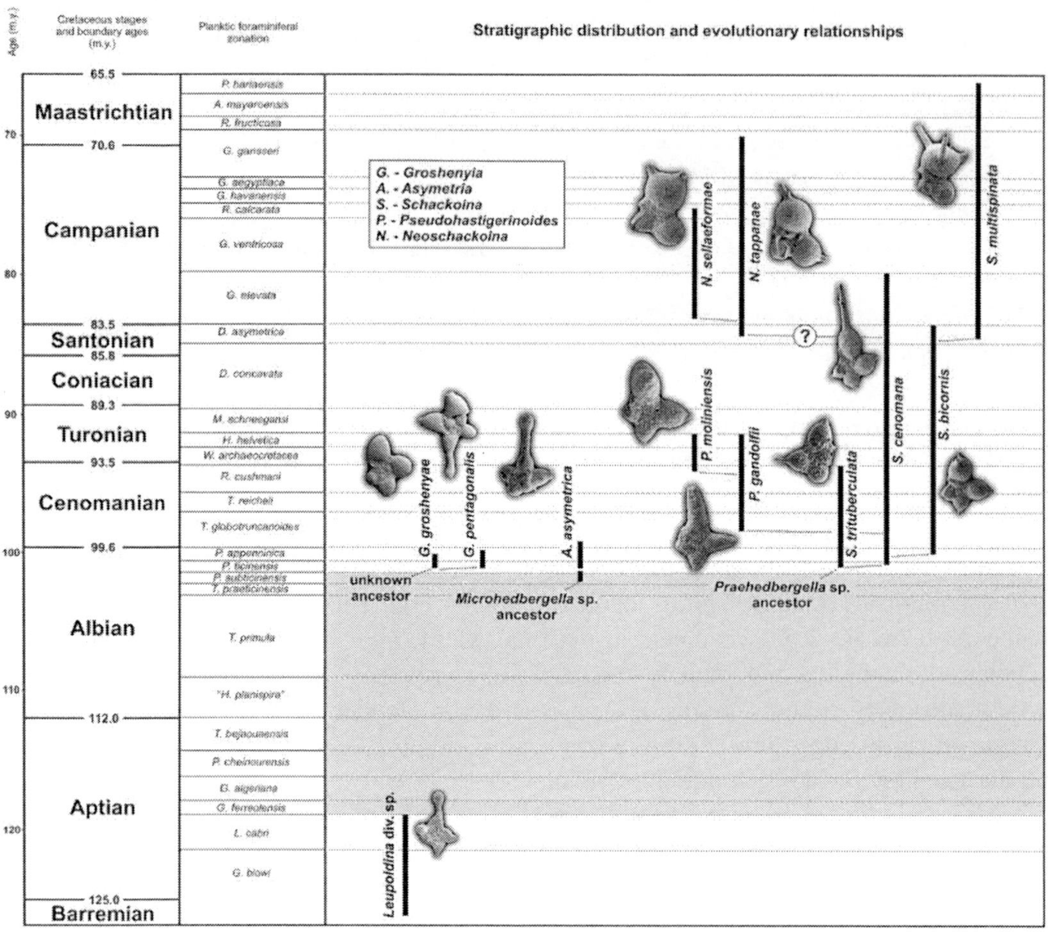

Figure 14. Stratigraphic distribution and evolutionary relationships of the schackoinid foraminifera; the stratigraphical interval without schackoinid tests is given in red. Planktic foraminiferal zonation is after Robaszynski and Caron (1995). Ages are after Gradstein et al. (2004).

Description. Test is planispiral, evolute-involute and biumbilicate and consists of 6-7 chambers; there are 3-3.5 chambers in the final whorl. Earlier chambers are globular or with one monotubulospine, the last-formed one to three are subrectangular to trapezoidal, and with β-bitubulospines. Sutures are distinct, depressed, straight and radial on both test sides. Test is symmetrical and slightly compressed in edge view. Aperture is a medium high arch in equatorial position, and bordered by a wide sparsely perforate lip. Umbilici are shallow, and with a diameter of circa one third of the test maximum diameter at the level of the tubulospine base; relict periapertural structures occur in the umbilical regions. Chamber surface is symmetrically ornamented with pustules on both test sides; pustules are irregular in shape, with the maximum dimension of 2.4-3.9 μm; less dense pustules occur occasionally around the tubulospine base. Test wall is calcitic, hyaline, simple and perforate; pores are simple, circular and with a diameter of 0.2-0.4 μm.

Remarks. *Neoschackoina sellaeforma* differs from all the other schackoinid planktic foraminiferal species by having β-bitubulospines on the last-formed one to three chambers.

Stratigraphical range. Campanian (from the *G. elevata* Biozone to the *R. calcarata* Biozone).

Geographic distribution. USA (Alabama) and South Atlantic Ocean (Falkland Plateau).

CONCLUSION

This taxonomical review of the Cretaceous (late Albian-Maastrichtian) genus *Schackoina*, which is based on the extensive use of the SEM, results in a more accurate view of the test morphology, intraspecific variability, stratigraphic distribution and evolutionary relationships within this relative small group of planktic foraminifers. A new classification is herein developed to accommodate the newly discovered morphological features and evolutionary trends. The oldest planktic foraminifera with schackoinid appearance of late Barremian-early Aptian are included within the genus *Leupoldina*. Ttest that were traditionally assigned to the genus *Schackoina* are younger and first occur in the upper Albian sediments (*P. ticinensis* Biozone), following a gap encompassing the upper Aptian-lower upper Albian stratigraphic interval without schackoinid tests (Figure 14).

Groshenyia is a new genus, which accommodates a new directional lineage, consists of two species: *G. groshenyae* - new species and *G. pentagonalis*. The former is very low trochospiral throughout and the last-formed chambers tangentially elongate, petaloid and more rarely clavate; the latter has an adult planispiral stage and the last-formed one to four chambers radially elongate and with one distal bulbous projection. The test is smooth, and with circular pores, with a diameter of 0.5-1.6 μm. The first occurrence is recorded for both species in the *P. ticinensis* Biozone at ODP hole 1050C (Blake Plateau); additional studies are necessary to document the lineage initiation timing. The gradual development of the chambers bearing one distal bulbous projection is for the first time documented in the schackoinid group. Apparently the distal bulbous projection developed in a similar morphological sequence as in the non-schackoinid lineages *Pessagnoina* Georgescu, 2009 and *Eohastigerinella* Morozova, 1957 of the Late Cretaceous (Georgescu, 2009, 2011b).

The new genus *Asymetria* is erected to accommodate the late Albian-earliest Cenomanian tests with asymmetrically developed ornamentation, pustulose on the umbilical side and smooth on the spiral one, an unique feature among the Cretaceous planktics; it includes only one species, *A. asymetrica* - new species. This new taxon is trochospiral throughout and has the last-formed one to three chambers with one distal bulbous projection. The *Asymetria* lineage cannot be defined at the present level of knowledge because its ancestry is unknown. Rare tests with incipient chamber elongation that can potentially belong to a species ancestral to *A. asymetrica* are herein reported from the lower Albian sediments from ODP Hole 1049B (Blake Plateau) and possibly DSP Site 370 (Deep Basin off Morocco); such tests are assigned to *A.*? sp. The similarities in the general test appearance of *A.* ? sp. and the early stages of *A. asymetrica*, as well as the similarities in the wall ultrastructure and porosity characteristics, indicate an evolutionary relationship between the two taxa. Additional study is necessary to document the evolution of the asymmetrically developed ornamentation, being given that *A.* ? sp. is a smooth species.

Genus *Schackoina* is reviewed to accommodate a directional lineage consisting of four species, which in evolutionary order are: *S. trituberculata*, *S. cenomana*, *S. bicornis* and *S.*

multispinata. Schackoina species are characterized by tests with early low planispiral coil and planispiral adult stage, flask-like chambers with tubulospines, and pustulose ornamentation symmetrically developed on the two test sides. This directional lineage evolved in the late Albian from the globular-chambered *Praehedbergella* in the upper part of the *P. ticinensis* Biozone. Monotubulospines are developed only on the last-formed chambers of the final whorl in the earliest species of *Schackoina*, namely *S. trituberculata*; the earlier chambers of the last whorl are globular and without tubulospines, demonstrating thereby the direct evolutionary relationships with the representatives of the hedbergellid stock. Apparently the evolution within the *Schackoina* directional lineage is towards the tubulospine development on earlier chambers and increase of tubulospine number per chamber. Monotubulospines occur on all the chambers of the final whorl in *S. cenomana*. Multiple tubulospines evolve only on the last-formed chambers of *S. bicornis* (α-bitubulospines) and *S. multispinata* (tritubulospines).

A genus that accommodates a new directional lineage is described to accommodate the Cenomanian-early Turonian schackoinids with strongly elongate chambers resembling tubulospines: *Pseudohastigerinoides*. Two species are included within the *Pseudohastigerinoides* lineage: *P. gandolfii* and *P. moliniensis*. There is one radial extension on each chamber in the former species and two or three extensions on the last-formed one or two chambers in the latter. Gross test architecture features (chamber number, development of radially elongate extensions, etc) indicate that the most likely ancestor of *P. gandolfii* is *S. trituberculata*.

The new genus *Neoschackoina* accommodates the youngest schackoinid directional lineage, which evolved in the late Santonian and became extinct in the early Maastrichtian. *Neoschackoina* consists of two species: *N. tappanae* and *N. sellaeforma*. The evolution in this directional lineage is most apparent in the chamber shape and tubulospines: monotubulospines and globular chambers in the former species and β-bitubulospines and laterally compressed chambers in the latter. The origins of *Neoschackoina* are not precisely known. It could have evolved from *S. cenomana* through the development of globular chambers with one tubulospine on all the chambers of the final whorl; however, this phylogenetic relationship is not supported by the chamber shape evolution: globular (earlier chambers of the final whorl in *S. trituberculata*), then flask-like (*S. cenomana*) and again globular (*N. tappanae*). Another possibility is that *Neoschackoina* evolved from a non-schackoinid ancestor, but this requires additional studies at ultrastructure level on the representatives of the globular-chambered hedbergellids and globigerinelloidids of Santonian age.

Eleven species with schackoinid test general appearance are recognized in the upper Albian-Maastrichtian. They can be easily identified based on features readily observable with the optical microscope (Figure 15).

The taxonomical review of the late Albian-Maastrichtian schackoinid planktic foraminifera shows that it is possible to develop an evolutionary classification framework based on high resolution data collected from throughout the stratigraphic ranges of the studied species. One of the most important advantages of this method is that lineages, which are natural evolutionary units, can be defined, thereby avoiding the lumping at supraspecific level, which is frequent in the classification based on typological principles.

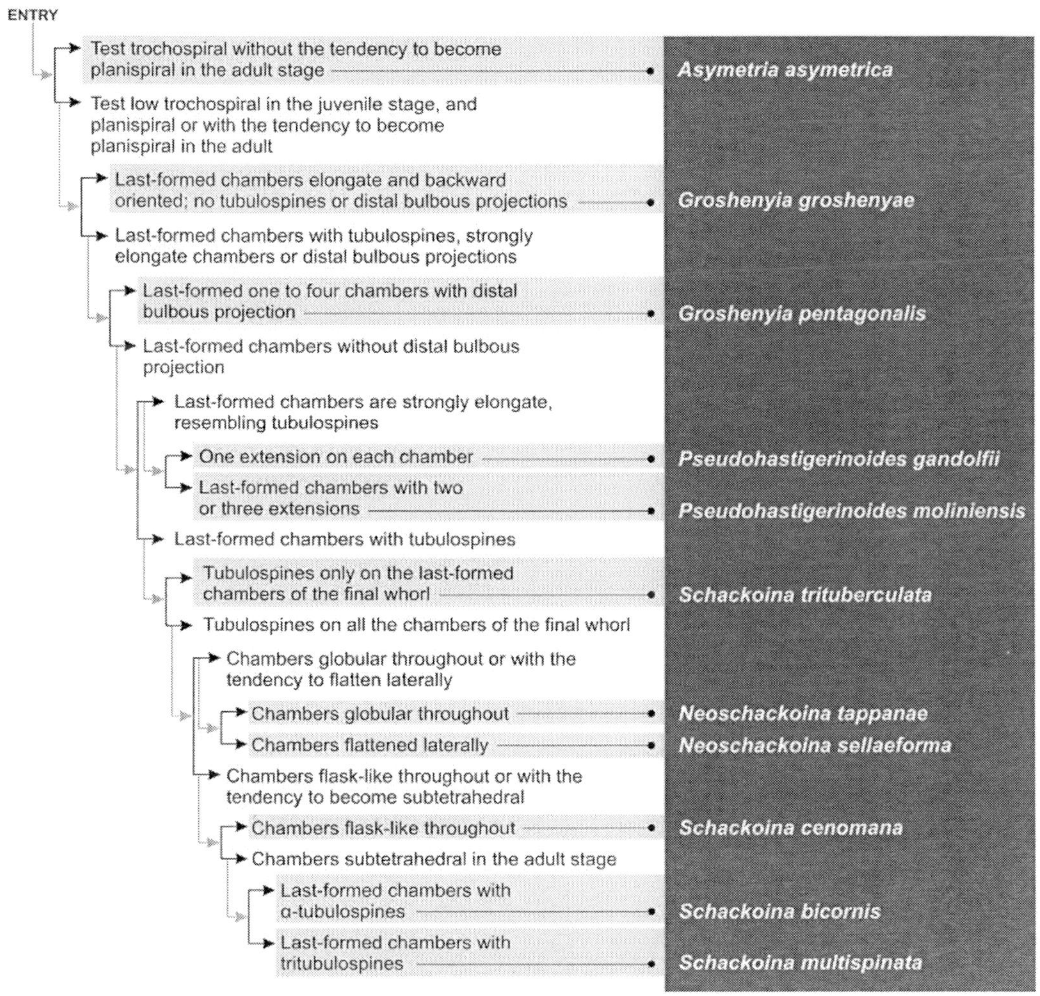

Figure 15. Practical identification key for the schackoinid species; key end features are given on yellow background.

ACKNOWLEDGMENTS

The DSDP/ODP/IODP headquarters are thanked for allowing the resampling of the sediments cored by the boreholes drilled under their auspices. Dr D. Lazarus (Naturkundemuseum, Berlin) is thanked for allowing the restudy of the Cretaceous foraminifera of the Ehrenberg Collection. Dr R. K. Edgar (Farlow Herbarium, Harvard University) is thanked for allowing the restudy of the planktic foraminifera from the Jacob Whitman Bailey Collection. Dr T. Neagu (University of Bucharest, Romania) is thanked for the spot samples from eastern Carpathians (Romania). The author thanks Dr L.V. Hills (University of Calgary) for the constructive criticism and suggestions to improve the manuscript. Dr M. Schoel (Microscope and Imaging Facility, University of Calgary) and S. Whittaker (NMNH, Washington, D.C.) are thanked for the professional and enthusiastic help during the SEM and ESEM operations.

APPENDIX

Alphabetical list of species used in the Cretaceous biostratigraphy at the eight DSDP/ODP sites or holes studied.

Abathomphalus mayaroensis (Bolli, 1951);
Archaeoglobigerina cretacea (d'Orbigny, 1840);
Arenobulimina flandrini (Moullade, 1966);
Biticinella breggiensis (Gandolfi, 1942);
Dicarinella asymetrica (Sigal, 1952);
Dicarinella concavata (Brotzen, 1934);
Falsotruncana maslakovae Caron, 1981;
Gansserina gansseri (Bolli, 1951);
Globigerinelloides impensus Sliter, 1977;
Globotruncanella petaloidea (Gandolfi, 1955);
Globotruncanella subpetaloidea (Gandolfi, 1955);
Globotruncanita elevata (Brotzen, 1934);
Globotruncanita insignis (Gandolfi, 1955);
Globotruncanita stuarti (de Lapparent, 1918);
Globotruncanita vescicarinata (Belford, 1981);
Hedbergella hoelzli (Hagn and Zeil, 1954);
Helvetoglobotruncana helvetica (Bolli, 1945);
Hendersonites pacificus Georgescu, 2011;
Marginotruncana marginata (Reuss, 1845);
Marginotruncana schneegansi (Sigal, 1952);
Parathalmanninella appenninica (Gandolfi, 1942);
Pseudoguembelina hariaensis Nederbragt, 1991;
Pseudothalmanninella subticinensis (Gandolfi, 1942);
Pseudothalmanninella ticinensis (Gandolfi, 1942);
Racemiguembelina fructicosa (Egger, 1902);
Radotruncana calcarata (Cushman, 1927);
Recurvoides imperfectus (Hanzliková, 1973);
Rotalipora cushmani (Morrow, 1934);
Thalmanninella globotruncanoides Sigal, 1948;
Thalmanninella reicheli (Mornod, 1950);
Ticinella praeticinensis Sigal, 1966;
Ticinella primula Luterbacher in Renz et al., 1963;
Ventilabrella eggeri Cushman, 1928;
Whiteinella archaeocretacea Pessagno, 1967;
Whiteinella baltica Douglas and Rankin, 1969.

References

Ando, A., Nakano, T., Kaiho, K., Kobayashi, T., Kokado, E. and B.-K. Khim 2009. Onset of seawater 87Sr/86Sr excursion prior to Cenomanian-Turonian oceanic anoxic event 2? New Late Cretaceous strontium isotope curve from the Central Pacific Ocean. *Journal of Foraminiferal Research*, 39: 322-334.

Ansary, S.E. and Tewfik, N.M. 1968. Planktonic foraminifera and some new benthonic species from the subsurface Upper Cretaceous of Ezz El Orban area, Gulf of Suez. *Journal of Geology of the United Arab Republic*, 10: 31-76.

Aurouze, G. and de Klasz, I. 1954. Sur la présence de Schackoïnes dans le Crétacé supérieur de France, de Baviére et de Tunisie. *Bulletin de la Société Géologique de France*, 4(1-3): 97-103.

Banner, F.T. and Blow, W.H. 1959. The Classification and Stratigraphical Distribution of the Globigerinaceae. *Palaeontology*, 2: 1-27.

Banner, F.T. and Desai, D. 1988. A review and revision of the Jurassic-Early Cretaceous Globigerinina, with special reference to the Aptian assemblages of Speeton (North Yorkshire, England). *Journal of Micropaleontology*, 7: 143-185.

Barbieri, F. 1966. Segnalazione dei generi *Gabonella* e *Grimsdaleinella* (foraminiferi) nel Cretacico dell'Appennino settentrionale. *Bolletino della Societá Geologica Italiana*, 85: 13-20.

Barr, F.T. 1961. Upper Cretaceous planktonic foraminifera from the Isle of Wight, England. *Palaeontology*, 4: 552-580. [published in January 1962]

Belford, D.J. 1981. Late Cretaceous planktic foraminifera in Valdivia core KL1 from the Scott Plateau, off northwestern Australia. *Bureau of Mineral Resources, Geology and Geophysics Bulletin*, 213: 48-67.

Bellier, J.-P. 1998. Cretaceous planktonic foraminifers, Eastern Equatorial Atlantic. In Mascle, J., Lohmann, G.P. and Moullade, M. (eds), *Proceedings of the Ocean Drilling Program Scientific Results,* 159: 335-345. College Station, Texas (Ocean Drilling Program).

Bellier, J.-P. and Moullade, M. 2002. Lower Cretaceous planktonic foraminiferal biostratigraphy of the western North Atlantic (ODP Leg 171B) and taxonomic clarification of key index species. *Revue de Micropaléontologie*, 45: 9-26.

Bellier, J.-P., Moullade, M. and Tronchetti, G. 2003. Les Schackoïnes (Foraminifères planctoniques) de l'Albien supérieur du Blake Nose (marge atlantique de la Floride, Leg ODP 171B). *Revue de Micropaléontologie*, 46: 131-141.

Bermúdez, P.J. 1952. Estudio sistemático de los foraminíferos rotaliformes. *Boletín de Geología, Venezuela*, 2: 1-230.

Boccaletti, M. and Pirini, C. 1964. Ritrovamento di Schackoina nel complesso argilloso-calcareo di S. Fiora. *Palaeontographia Italica*, 59: 53-61. [published in 1965]

Boersma, A. 1981. Cretaceous and early Tertiary foraminifers from Deep Sea Drilling Project Leg 62 Sites in the Central Pacific. In Stout, L.N. (ed.), *Initial Reports of the Deep Sea Drilling Project,* 62: 377-396. United States Government Printing Office, Washington, D. C.

Bolli, H.M. 1945. Zur Stratigraphie der Oberen Hreide in den höheren helvetischen Decken. *Eclogae Geologicae Helvetiae*, 37: 217-328.

Bolli, H.M. 1951. The genus *Globotruncana* in Trinidad, B. W. I. *Journal of Paleontology*, 25: 187-199.

Bolli, H.M. 1957. The foraminiferal genera *Schackoina* Thalmann, emended and *Leupoldina*, n. gen. in the Cretaceous of Trinidad B. W. I. *Eclogae Geologicae Helvetiae*, 50: 271-278.

Bolli, H.M. 1959. Planktic foraminifera from the Cretaceous of Trinidad,
B. W. I. *Bulletins of American Paleontology*, 39(179): 255-277.

Bolli, H.M., Loeblich, A.R.Jr. and Tappan, H. 1957. Planktic foraminiferal families Hantkeninidae, Orbulinidae, Globorotaliidae and Globotruncanidae. In Loeblich, A.R.Jr. (ed.), *Studies in Foraminifera*. United States National Museum Bulletin, 215: 3-50.

Bou Dagher-Fadel, M.K. 1996. Re-evaluation of the phylogeny of early Cretaceous planktonic foraminifera (Praehedbergellidae). *Cretaceous Research*, 17: 767-771.

Bou Dagher-Fadel, M.K., Banner, F.T. and Whittaker, J.E. 1997. *The Early Evolutionary History of Planktic Foraminifera*. Chapman and Hall, London.

Brotzen, F. 1934. Foraminiferen aus dem Senon Palästinas. *Zeitschrift des Deutsche-Palästinavereins*, 57: 28-72.

Brönnimann, P. 1952. Globigerinidae from the Upper Cretaceous (Cenomanian-Maestrichtian) of Trinidad, B.W.I. *Bulletins of American Paleontology*, 34(140): 5-71.

Burckle, L.H., Saito, T. and Ewing, M. 1967. A Cretaceous (Turonian) core from the Naturaliste Plateau southeast Indian Ocean. *Deep-Sea Research*, 14: 421-426.

Caron, M. 1975. Late Cretaceous planktonic foraminifera from the northwestern Pacific: Leg 32 of the Deep Sea Drilling Project. In Larson, R.L., Moberly, R. et al. (eds), *Initial Reports of the Deep Sea Drilling Project,* 32: 719-724. United States Government Printing Office, Washington, D. C.

Caron, M. 1978. Cretaceous planktonic foraminifers from DSDP Leg 40, southeastern Atlantic Ocean. In Bolli, H.M., Ryan, W.B.F. et al. (eds), *Initial Reports of the Deep Sea Drilling Project,* 40: 651-678. United States Government Printing Office, Washington, D. C.

Caron, M. 1981. Un nouveau genre de foraminifère planctonique du Crétacé: *Falsotruncana* nov. gen. *Eclogae Geologicae Helvetiae*, 74: 65-73.

Caron, M. 1985. Cretaceous planktic foraminifera. In Bolli, H.M., Saunders, J.B., and Perch-Nielsen, K. (eds), *Plankton Stratigraphy*, 17-86. Cambridge University Press, Cambridge.

Carter, D.J. and Hart, M.B. 1977. Aspects of mid-Cretaceous stratigraphical micropaleontology. *Bulletin of the British Museum (Natural History)*, 29: 1-135.

Cati, F. 1964. Una fauna campaniana dei Monti Berici (Vicenza). *Giornale di Geologia*, 32: 199-271.

Cheylan, G., Magné, J., Sigal, J. and Grekoff, N. 1954. Résultats géologiques et micropaléontologiques du sondage d'El Krachem (Haut plateau Algérois). Description de quelques espèces nouvelles. *Bulletin de la Société Géologique de France*, 3(4-6): 471-492.

Coccioni, R., Luciani, V. and Marsili, A. 2006. Cretaceous oceanic anoxic events and radially elongated chambered planktonic foraminifera: paleoecological and paleogeographic implications. *Palaeogeography, Palaeoclimatology, Palaeoecology*, 235: 66-92.

Cushman, J.A. 1927. New and interesting foraminifera from Mexico and Texas. *Contributions from the Cushman Foundation for Foraminiferal Research*, 3: 111-119.

Cushman, J.A. 1928. Additional genera of foraminifera. *Contributions from the Cushman Laboratories for Foraminiferal Research*, 4: 1-10.

Cushman, J.A. 1931. *Hastigerinella* and other interesting foraminifera from the Upper Cretaceous of Texas. *Contributions from the Cushman Foundation for Foraminiferal Research*, 7: 83-90.

Cushman, J.A. 1933. Foraminifera their classification and economic use. *Cushman Laboratory for Foraminiferal Research, Special Publications*, 4: 1-349.

Cushman, J.A. 1946. Upper Cretaceous foraminifera of the Gulf Coastal region of the United States and adjacent areas. *United States Geological Survey Professional Paper*, 206: 1-241.

Cushman, J.A. and Wickenden, R.T.D. 1930. The development of *Hantkenina* in the Cretaceous with a description of a new species. *Contribution from the Cushman Laboratory for Foraminiferal Research*, 6: 39-43.

Dalby, A.P., Patterson, R.T. and Haggart, J.W. 2009. Distribution of Albian-Cenomanian from Charlotte Islands, British Columbia, Canada: Constraints on the timing of the northward migration of the Wrangellia Terrane. *Journal of Foraminiferal Research*, 39: 231-245.

Delage, Y. and Hérouard, E. 1896. *Traité de zoologie concrete. Tome I: La cellule et les protozoaires*. Schleicher Frères, Paris.

Douglas, R.G. 1969. Upper Cretaceous planktonic foraminifera in northern California. Part 1 – Systematics. *Micropaleontology*, 15: 151-209.

Douglas, R.G. and Rankin, C. 1969. Cretaceous planktonic foraminifera from Bornholm and their zoogeographic significance. *Lethaia*, 2: 185-217.

Egger, J.G. 1902. Foraminiferen und Ostrakoden aus den Kreidemergeln der Oberbayerischen Alpen. *Abhandlungen der Matematisch–Physikalischen Klasse der Könglich Bayerischen Akademie der Wissenschaften*, 21: 1-230.

Eicher, D.L. and Worstell, P. 1970. Cenomanian and Turonian foraminifera from the Great Plains, United States. *Micropaleontology*, 16: 269-324.

Eichwald, C.E. von 1830. *Zoologia specialis II*. D.E. Eichwaldus, Vilnae.

Franke, A., 1928. Die Foraminiferen der Oberen Kreide Nord- und Mitteldeutschlands. *Preußischen Geologischen Landesanstalt*, 111: 1-207.

Frerichs, W.E. 1971. Evolution of planktonic foraminifera and paleotemperatures. *Journal of Paleontology*, 45: 963-968.

Frerichs, W.E. 1989. Distribution of the Planomalinidae and the Schackoinidae in Upper Cretaceous strata of the Western Interior. *Contributions to Geology, University of Wyoming*, 27: 7-13.

Frerichs, W.E., Atherton, C.C. and Shive, P.N. 1975. Planktonic foraminifera from the chalky members of the Niobrara Formation, Centennial Valley, Wyoming. *Journal of Foraminiferal Research*, 5: 294-312.

Frerichs, W.E. and Dring, N.B. 1981. Planktonic foraminifera from the Smoky Hill Shale of west central Kansas. *Journal of Foraminiferal Research*, 11: 47-69.

Gandolfi, R. 1942. Ricerche micropaleontologiche e stratigrafiche sulla scaglia e sul flysch Cretacici dei Dintorni di Balerna (Canton Ticino). *Rivista Italiana di Paleontologia*, 48: 1-160.

Gandolfi, R. 1955. The genus *Globotruncana* in northeastern Colombia. *Bulletin of American Paleontology*, 36 (155): 1-118.

Gawor Biedova, E. 1972. The Albian, Cenomanian and Turonian foraminifers of Poland and their Stratigraphical importance. *Acta Paleontologica Polonica*, 17: 1-155.

Gawor Biedova, E. 1992. Campanian and Maastrichtian foraminifera from the Lublin Upland, eastern Poland. *Palaeontologia Polonica*, 52: 1-187.

Gebhardt, H., Friedrich, O., Schenk, O., Fox, L., Hart, M. and Wagreich, M. 2010. Paleoceanographic changes at the northern Tethyan margin during the Cenomanian-Turonian Oceanic Anoxic Event (OAE-2). *Marine Micropaleontology*, 77: 25-45.

Georgescu, M.D. 2009. Upper Albian–lower Turonian non-schackoinid planktic foraminifera with elongate chambers: morphology reevaluation, taxonomy and evolutionary classification. *Revista Española de Micropaleontología*, 41: 255-293.

Georgescu, M.D. 2010. Origin, taxonomic revision and evolutionary classification of the late Coniacian-early Campanian planktic foraminifera with multichamber growth in the adult stage. *Revista Española de Micropaleontología*, 42: 59-118.

Georgescu, M.D. 2010. Evolutionary classification of the Upper Cretaceous (Turonian-lower Campanian) planktic foraminifera with incipient meridional ornamentation. *Journal of Micropaleontology*, 29: 149-161.

Georgescu, M.D. 2011a. A new type of test wall in the Late Cretaceous (Late Santonian-Campanian) heterohelicid planktic foraminifera. *Revue de Micropaléontologie*, 54: 105-114.

Georgescu, M.D. 2011b. New data on the evolutionary classification of the Late Cretaceous (late Coniacian–Santonian) planktic foraminifera with elongate chambers. *Revista Española de Micropaleontología*, 43: 39-54.

Georgescu, M.D. 2011c. Iterative evolution, taxonomic revision and evolutionary classification of the praeglobotruncanid planktic foraminifera, Cretaceous (late Albian-Santonian). *Revista Española de Micropaleontología*, 43: xxx-xxx. [published March 2012]

Georgescu, M.D. and Huber, B.T. 2008. Taxonomic re-evaluation and phylogeny of the stellate planktonic foraminiferal genus *Hastigerinoides* Brönnimann, 1952. *Journal of Foraminiferal Research*, 38: 52-58.

Georgescu, M.D. and Huber, B.T. 2009. Early evolution of the Cretaceous serial planktic foraminifera (late Albian–Cenomanian). *Journal of Foraminiferal Research*, 39: 335-360.

Georgescu, M.D., Quinney, A.E. and Anderson, K.D. 2011. New data on the taxonomy, evolution and biostratigraphical significance of the Turonian–Coniacian (Late Cretaceous) planktic foraminifer *Huberella* Georgescu, 2007. *Micropaleontology*, 57: 247-254.

Gorbachik, T.N. and Moullade, M. 1973. Caractères microstructuraux de la paroi du test des foraminifers planctoniques du Crétacé inférieur et leur signification sur le plan taxinomique. *Compte rendu hebdomadaire des séances de l'academie des sciences*, 277: 2661-2664.

Gradstein, F.M., Ogg, J.G. and Smith, A.G. 2004. *A Geologic Time Scale 2004*. Cambridge University Press, Cambridge.

Graham, J.J. and Church, C.C. 1963. Campanian foraminifera from the Stanford University Campus California. *Stanford University Publications, Geological Sciences*, 8: 1-106.

Hagn, H. and Zeil, W. 1954. Globotruncanen aus dem Ober-Cenoman und Unter-Turon der Bayerischen Alpen. *Eclogae Geologicae Helvetiae*, 47: 1-60.

Hanzliková, E. 1973. Foraminifers of the variegated Godula Member in Moravia (Cenomanian-Turonian). *Sbornik Geologickych Ved, Paleontologie*, 15: 119-184.

Hart, M.B. 1999. The evolution and biodiversity of Cretaceous planktic Foraminiferida. *Geobios*, 32: 247-255.

Hart, M.B., Oxford, M.J. and Hudson, W. 2002. The early evolution and palaeobiogeography of Mesozoic planktonic foraminifera. In Crame, J.A. and Owen, A.W. (eds), *Paleobiogeography and Biodiversity Change: the Ordovician and Mesozoic-Cenozoic Radiations.* Geological Society of London, Special Publications, 194: 115-125.

Hay, W.H., DeConto, R.M., Wold, C.N., Wilson, K.M., Voigt, S., Schulz, M., Wold, A.R., Dullo, W.C., Ronov, A.B., Balukhovsky, A.N. and Söding, E. 1999. Alternative global Cretaceous paleogeography. In Barrera, E. and Johnson, C.C. (eds), *Evolution of the Cretaceous Ocean-Climate System.* Geological Society of America, Special Publications, 332: 1-47.

Herb, R. 1974. Cretaceous planktonic foraminifera from the Eastern Indian Ocean. In Davies, T.A., Luyendyk, B.P. et al. (eds), *Initial Reports of the Deep Sea Drilling Project,* 24: 745-769. United States Government Printing Office, Washington, D.C.

Hinte, J.E. van 1963. Zur Stratigraphie und Mikropaläontologie der Oberkreide und des Eozäns des Krappfeldes (Kränten). *Jahrbuch der Geologischen Bundesanstalt*, 8: 1-147.

Huber, B.T. 1990. Maestrichtian planktonic foraminifer biostratigraphy of the Maud Rise (Weddell Sea, Antarctica): ODP Leg 113 Holes 689B and 690C. In Barker, P.F., Kennett, J.P. et al. (eds), *Proceedings of the Ocean Drilling Program Scientific Results,* 113: 489-513. College Station, Texas (Ocean Drilling Program).

Huber, B.T. 1992. Upper Cretaceous planktic foraminiferal biozonation for the Austral Realm. *Marine Micropaleontology*, 20: 107-128.

Huber, B.T., Hodell, D.A. and Hamilton, C.P. 1995. Middle-Late Cretaceous climate of the southern high latitudes: Stable isotopic evidence for minimal equator-to-pole thermal gradients. *Geological Society of America Bulletin*, 107: 1164-1191.

Huber, B.T. and Leckie, M.R. 2011. Planktic foraminiferal species turnover across deep-sea Aptian/Albian boundary sections. *Journal of Foraminiferal Research*, 41: 53-95.

Huber, B.T., Leckie, R.M., Norris, R.D., Bralower, T.J. and Cobabe, E. 1999. Foraminiferal assemblage and stable isotopic change across the Cenomanian-Turonian boundary in the subtropical North Atlantic. *Journal of Foraminiferal Research*, 29: 329-417.

Korchagin, O.A. 2003. Mesozoic planktonic foraminifer classification (superfamilies Planomalinacea, Rotaliporacea and Globotruncanacea). *Trudy VNIGRI*, 547, 1-89. [in Russian]

Krasheninnikov, V.A. and Basov, I. A. 1983. Stratigraphy of Cretaceous sediments of the Falkland Plateau based on planktonic foraminifers, Deep Sea Drilling Project, Leg 71. In Ludwig, W.J., Krasheninnikov, V.A. et al. (eds), *Initial Reports of the Deep Sea Drilling Project,* 71: 789–820. United States Government Printing Office, Washington, D. C.

Lapparent, L. de 1918. Étude lithologique des terrains creates de la region d'Hendaye. *Mémoires du Service de la Carte Géologique Détaillée de France*, 1-155.

Leckie, R.M. 1984. Mid-Cretaceous planktic foraminiferal biostratigraphy off central Morocco, Deep Sea Drilling Project Leg 79, Sites 545 and 547. In Hinz, K., Winterer, E.L. et al. (eds), *Initial Reports of the Deep Sea Drilling Project,* 79: 579-620. United States Government Printing Office, Washington, D. C.

Loeblich, A.R.Jr. and Tappan, H. 1961. Cretaceous Planktic foraminifera: Part I – Cenomanian. *Micropaleontology*, 7: 257-304.

Loeblich, A.R.Jr. and Tappan, H. 1964. Sarcodina Chiefly "Thecamoebians" and Foraminifera. In Moore, R.C. (ed.), *Treatise on Invertebrate Paleontology, Part C, Protista*, 1-900. The Geological Society of America and The University of Kansas Press, Lawrence, Kansas.

Loeblich, A.R.Jr. and Tappan, H. 1988. *Foraminiferal Genera and Their Classification*. Van Nostrand Reinhold Company, New York.

Loetterle, G.J. 1937. The Micropaleontology of the Niobrara Formation in Kansas, Nebraska, and South Dakota. *Nebraska Geological Survey Bulletin*, 12: 1-73.

Longoria, J.F. 1974. Stratigraphic, morphologic and taxonomic studies of Aptian planktic foraminifera. *Revista Española de Micropaleontología, Numero extraordinario*: 11-107.

Luterbacher, H. 1975. Early Cretaceous foraminifera from the northwestern Pacific: Leg 32 of the Deep Sea Drilling Project. In Larson, R.L., Moberly, R. et al. (eds), *Initial Reports of the Deep Sea Drilling Project,* 32: 703-718. United States Government Printing Office, Washington, D. C.

Luterbacher, H.P. and Premoli Silva, I. 1962. Note préliminaire sur une revision du profil de Gubbio, Italie. *Rivista Italiana di Paleontologia*, 68: 253-288.

Marianos, A.W. and Zingula, R.P. 1966. Cretaceous planktonic foraminifers from Dry Creek, Tehama County, California. *Journal of Paleontology*, 40: 328-342.

Masella, L. 1960. Le *Schackoina* (Foraminifera, Globigerinacea) del Cretaceo di Patti (Messina). *Rivista Mineraria Siciliana*, 11(61): 16-30.

Masters, B.A. 1976. Planktic foraminifera from the Upper Cretaceous Selma Group, Alabama. *Journal of Paleontology*, 50: 318-330.

Masters, B.A. 1977. Mesozoic Planktic Foraminifera: a world–wide review and analysis. In Ramsay, A.T.S., (ed.), *Oceanic Micropaleontology*, 1: 301-733. Academic Press, London-New York-San Francisco.

McNeil, D.H. 1984. The eastern facies of the Cretaceous System in the Canadian Western Interior. In Stott, D.F. and Glass, D.J. (eds), *The Mesozoic of Middle North America*. Canadian Society of Petroleum Geologists, 9: 145-171.

McNeil, D.H. and Caldwell, W.G.E. 1981. Cretaceous rocks and their foraminifera in the Manitoba Escarpment. *The Geological Association of Canada, Special Paper*, 21: 1-435.

Miles, G.A. and Orr, W.N. 1980. Planktic foraminifers from the Bermuda Rise, Deep Sea Drilling Project Legs 51, 52, and 53. In Donelly, T., Francheteau, J. et al. (eds), *Initial Reports of the Deep Sea Drilling Project,* v. 51-53(2): 791-813. United States Government Printing Office, Washington, D. C.

Montanaro Gallitelli, E. 1954. Marne ed argile a *Schackoina* e *Gümbelina* nella formazione a fucoidi ed elmintoidee di Serramazzoni (Modena). *Atti Memorie della Academia di Scienze e Letteri Modena*, 12: 201-210.

Montanaro Gallitelli, E. 1955. *Schackoina* from the Upper Cretaceous of the northern Apennines, Italy. *Micropaleontology*, 1: 141-146.

Mornod, L. 1950. Les Globorotalidés du Crétacé supérieur du Montsalvens (Préalpes fribourgeoises). *Eclogae Geologicae Helvetiae*, 42: 573-596.

Morozova, V.G. 1957. Foraminiferal superfamily Globigerinidea, superfam. nov., and some of its representatives. *Doklady Akademii Nauk SSSR*, 114: 1109-1112. [in Russian]

Morrow, A.L. 1934. Foraminifera and ostracoda from the Upper Cretaceous of Kansas. *Journal of Paleontology*, 8: 186-205.

Moullade, M. 1966. Étude stratigraphique et micropaléontologique du Crétacé Inférieur de la "Fosse Vocontienne". *Documents des Laboratories de Géologie de la Faculté des Sciences de Lyon*, 15: 1-369.

Moullade, M., Bellier, J.-P. and Tronchetti, G. 2002. Hierarchy of criteria, evolutionary process and taxonomic simplification in the classification of Lower Cretaceous planktic foraminifera. *Cretaceous Research*, 23: 111-148.

Neagu, T. 1965. Albian foraminifera of the Rumanian Plain. *Micropaleontology*, 11: 1-38.

Neagu, T. 1966. *Schackoina* from the Cenomanian of the eastern Carpathians. *Micropaleontology*, 12: 365-369.

Neagu, T. 1970. Micropaleontological and Stratigraphical study of the Upper Cretaceous deposits between the upper valleys of the Buzău and Rîul Negru rivers (eastern Carpathians). *Mémoires Institut Géologique*, 12: 1-109.

Neagu, T. 1979. *Micropaleontologie*. Editura Tehnică, Bucureşti.

Neagu, T. 2005. Albian foraminifera of the Romanian Plain. Planktonic foraminifera. *Acta Paleontologica Romaniae*, 5: 311-332.

Nederbragt, A.J. 1991. Late Cretaceous biostratigraphy and development of Heterohelicidae (planktic foraminifera). *Micropaleontology*, 37: 329-372.

Noth, R. 1951. Foraminiferen aus Unter- und Oberkreide des Österreichischen anteils an Flysch, Helvetikum und Vorlandvorkommen. *Jahrbuch der Geologischen Bundesanstalt Sonderband*, 3: 1-91.

Orbigny, A. d' 1840. Mémoire sur les foraminifères de la craie blanche du basin du Paris. *Mémoires de la Société Géologique de France*, 4: 1-51.

Patterson, R.T., Haggart, J.W. and Dalby, A.P. 2010. A guide to the late Albian-Cenomanian (Cretaceous) foraminifera from the Queen Charlotte Islands, British Columbia, Canada. *Palaeontologia Electronica*, 13: 1-21.

Pessagno, E.A.Jr. 1967. Upper Cretaceous planktonic foraminifera from the western Gulf Coastal Plain. *Palaeontographica Americana*, 5(37): 245-445.

Petrizzo, M.R. 2000. Upper Turonian-lower Campanian from southern mid-high latitudes (Exmouth Plateau, NW Australia): biostratigraphy and taxonomic notes. *Cretaceous Research*, 21: 479-505.

Petrizzo, M.R. and Huber, B.T. 2006. Biostratigraphy and taxonomy of Late Albian planktic foraminifera from ODP Leg 171B (western North Atlantic Ocean). *Journal of Foraminiferal Research*, 36: 166-190.

Pflaumann, U. and Krasheninnikov, V.A. 1977. Early Cretaceous planktic foraminifers from Eastern North Atlantic, DSDP Leg 41. In Lancelot, Y., Seibold, E. et al. (eds), *Initial Reports of the Deep Sea Drilling Project,* 41: 539-564. United States Government Printing Office, Washington, D. C.

Playford, G., Haig, D.W. and Dettermann, M.E. 1975. A mid-Cretaceous microfossil assemblage from the Great Artesian Basin, northwestern Queensland. *Neues Jahrbuch für Geologie und Paläontologische Abhandlungen*, 149: 333-362.

Pokorný, V. 1958. *Grundzüge der Zoologischen Mikropaläontologie, Band 1*. VEB Deutscher Verlag der Wissenschaften, Berlin.

Premoli Silva, I. and Boersma, A. 1977. Cretaceous planktonic foraminifers – DSDP Leg 39 (South Atlantic). In Supko, P.R., Perch-Nielsen, K. et al. (eds), *Initial Reports of the*

Deep Sea Drilling Project, 39: 615-641. United States Government Printing Office, Washington, D. C.

Premoli Silva, I. and Bolli, H.M. 1973. Late Cretaceous to Eocene planktonic foraminifera and stratigraphy of Leg 15 sites in the Caribbean Sea. In Edgar, N.T., Saunders, J.B. et al. (eds), *Initial Reports of the Deep Sea Drilling Project,* 15: 499-547. United States Government Printing Office, Washington, D. C.

Premoli Silva, I. and McNulty, C.L. 1984. Planktonic foraminifers and calpionellids from Gulf of Mexico sites, Deep Sea Drilling Project Leg 77. In Buffler, R.T., Schlager, W. et al. (eds), *Initial Reports of the Deep Sea Drilling Project,* 77: 547-584. United States Government Printing Office, Washington, D. C.

Porthault, B. 1969. Foraminifères planctoniques et biostratigraphie du Cénomanien dans le sud-est de la France. In Brönnimann, P. and Renz, H.H. (eds), *Proceedings of the First International Conference on Planktonic Microfossils, Geneva 1967*, 2: 526-546. E. J. Brill, Leiden.

Quilty, P.G. 1973. Cenomanian-Turonian and Neogene sediments from northeast of Kerguelen Ridge, Indian Ocean. *Journal of the Geological Society of Australia*, 20: 361-371.

Quilty, P.G. 1992. Upper Cretaceous planktonic foraminifers and biostratigraphy, Leg 120, southern Kerguelen Plateau. In Wise, S.W.Jr., Schlich, R. et al. (eds), *Proceedings of the Ocean Drilling Program Scientific Results,* 120: 371-392. College Station, Texas (Ocean Drilling Program).

Reichel, M. 1948. Les Hantkéninidés de la Scaglia et des Couches rouges (Crétacé supérieur). *Eclogae Geologicae Helvetiae*, 40: 390-409.

Renz, H.H., Luterbacher, H. and Schneider, A. 1963. Stratigraphisch-paläontologische Untersuchungen im Albien und Cénomanien des Neuenburger Jura. *Eclogae Geologicae Helvetiae*, 56: 1073-1116.

Reuss, A.E.R. von 1845. *Die Versteinerungen der böhmischen Kreideformation.* E. Schweizerbart, Stuttgart.

Robaszynski, F. and Caron, M. 1995. Foraminifères planctoniques du Crétacé: commentaire de la zonation Europe-Mèditerranée. *Bulletin de la Société Géologique de France*, 6: 681-692.

Robaszynski, F., Caron, M., Gonzalez Donoso, J.M., Wonders, A.A.H. and The European Working Group on Planktonic Foraminifera 1984. Atlas of Late Cretaceous Globotruncanids. *Revue de Micropaléontologie*, 26: 145-305.

Saavedra, J.L. 1965. La evolución de los Globigerináceos. Boletín de la Real Sociedad Española de Historia Natural, *63: 317-349.*

Salaj, J. and Samuel, O. 1966. Foraminifera der Westkarpaten-Kreide. *Geologický Ústav Dionýza* Š*túra, Bratislava.*

Samuel, O. 1962. Mikrobiostratigrafické pomery kreidových sedimentov vnútorného bradlonvého pásma v Okolí Beňatíny. *Geologické Práce*, 24: 153-167.

Schacko, G. 1897. Beitrag über Foraminiferen aus der Cenoman–Kreide von Molzow in Mecklenburg. *Archiv des Vereins der Freunde de Naturgeschichte in Mecklenburg (1896)*, 50: 161–168.

Scheibnerová, V. 1968. Middle and Upper Cretaceous microbiostratigraphy of the Klippen Belt (West Carpathians). *Acta Geologica et Geographica Universitatis Comenianae*, 17: 5-98.

Sigal, J. 1948. Notes sur les genres de foraminifères *Rotalipora* Brotzen 1942 and *Thalmanninella. Revue de l'Institut Français du Pétrole et Annales des Combustibles Liquides*, 3: 95-103.

Sigal, J. 1952. Aperçu stratigraphique sur la micropaléontologie du Crétacé. *19th International Geological Congress, Monographies regionales, 1re ser., Algérie*, 26: 1-52.

Sigal, J. 1958. La classification actuelle des familles de foraminifères planctoniques du Crétacé. *Compte Rendu des Séances, Société Géologique de France*, 1958: 262–265.

Sigal, J. 1966. Contribution à une monographie des Rosalines. 1. Le genre *Ticinella* Reichel, souche des Rotalipores. *Eclogae Geologicae Helvetiae*, 59: 185–217.

Sigal, J. 1979. Chronostratigraphy and ecostratigraphy of Cretaceous formations recovered on DSDP Leg 47B, Site 398. In Laughter, F.H. and Fagerberg, E.M. (eds), *Initial Reports of the Deep Sea Drilling Project,* 47: 287-326. United States Government Printing Office, Washington, D. C.

Sliter, W.V. 1968. Upper Cretaceous foraminifera from souther California and northwestern Baja California, Mexico. *The University of Kansas Paleontological Contributions*, 49: 1-141.

Sliter, W.V. 1977. Cretaceous forsaminifers from the southwestern Atlantic Ocean, Leg 36, Deep Sea Drilling Project. In Barker, P., Dalziel, W.D. et al. (eds), *Initial Reports of the Deep Sea Drilling Project,* 36: 519-573. United States Government Printing Office, Washington, D. C.

Sliter, W.V. 1980. Mesozoic foraminifers and deep-sea benthic environments from Deep Sea Drilling Project sites 415 and 416, Eastern North Atlantic. In Lancelot, Y., Winterer, E.L. et al. (eds), *Initial Reports of the Deep Sea Drilling Project,* 50: 353-428. United States Government Printing Office, Washington, D. C.

Steineck, P.L. and Fleisher, R.L. 1978. Towards the classical evolutionary reclassification of Cenozoic Globigerinacea (Foraminiferida). *Journal of Paleontology*, 52: 618-635.

Subbotina, N.N. 1971. *Fossil foraminifera of the USSR. Globigerinidae, Hantkeninidae and Globorotaliidae*. Collet's (Publishers) Ltd, London and Wellingborough. [original edition published in Russian in 1953]

Takayanagi, Y. 1965. Upper Cretaceous planktonic foraminifera from the Putah Creek subsurface section along the Yolo-Solano County Line, California. *Science Reports of the Tohoku University, Sendai*, 36: 161-237.

Tappan, H. 1940. Foraminifera from the Grayson Formation of northern Texas. *Journal of Paleontology*, 14: 93-126.

Tappan, H. 1943. Foraminifera from the Duck Creek Formation of Oklahoma and Texas. *Journal of Paleontology*, 17: 476-517.

Thalmann, H.E. 1932. Die Foraminiferen-Gattung *Hantkenina* Cushman, 1924, und ihre regional-stratigraphische Verbreitung. *Eclogae Geologicae Helvetiae*, 26: 287–292.

Todd, R. and Low, D. 1964. Cenomanian (Cretaceous) Foraminifera from the Puerto Rico Trench. *Deep-Sea Research*, 11: 395-414.

Verga, D. and Premoli Silva, I. 2002. Early Cretaceous planktonic foraminifera from the Tethys: the genus *Leupoldina. Cretaceous Research*, 23: 189-212.

Verga, D. and Premoli Silva, I. 2005. Early Cretaceous planktonic foraminifera from the Tethys: the Upper Aptian, planispiral morphotypes with elongate chambers. *Cretaceous Research*, 26: 239-259.

Viterbo, I. 1965. Examen micropaléontologique du Crétacé du Maroc meridional (Bassin côtier de Tarfaya). *Mémoires du Bureau de Recherches Géologiques et Minières*, 32: 61-100.

Weidich, K.F. 1990. Die kalkalpine Unterkreide und ihre Foraminiferenfauna. *Zitteliana*, 17: 3-312.

Wonders, A.A.H. 1992. Cretaceous planktic foraminiferal biostratigraphy, Leg 112, Exmouth Plateau, Australia. In von Rad, U., Haq, B.U. et al. (eds), *Proceedings of the Ocean Drilling Program Scientific Results,* 122: 587-599. College Station, Texas (Ocean Drilling Program)

Zanzucchi, G. 1955. Su una placca di argilla medio-cretacica nella "Formazione ofiolitifera" del Monte Prinzera (Parma). *Atti della Societa Italiana di Scienze Naturali*, 94: 368-377.

In: Deep-Sea
Editors: D. R. Bailey and S. E. Howard

ISBN: 978-1-62257-173-4

Chapter 2

A REVIEW OF THE KNOWLEDGE RELATED TO THE NOMENCLATURE, ETYMOLOGY, MORPHOLOGY, DISTRIBUTION, AND BIOLOGICAL CHARACTERISTICS OF THE SKILFISH, *ERILEPIS ZONIFER* (ANOPLOPOMATIDAE), IN THE NORTH PACIFIC OCEAN*

A. M. Orlov*[1]*,
***A. M. Tokranov*[2]**
***and B. A. Megrey*[3]**

[1]Russian Federal Research Institute of Fisheries and Oceanography (VNIRO), Moscow, Russia
[2]Kamchatka Branch of Pacific Institute of Geography, Far East Branch of Russian Academy of Science (KBPIG FEB RAS), Petropavlovsk-Kamchatsky, Russia
[3]National Oceanic and Atmospheric Administration, National Marine Fisheries Service, Alaska Fisheries Science Center, Seattle, Washington, US

ABSTRACT

In this chapter we review the nomenclature, morphology, distribution, and biological characteristics including aquarium observations of the skilfish, *Erilepis zonifer*, in the North Pacific Ocean. Results are based on the analysis of all published sources and new unpublished data from 600 captures taken in the North Pacific between 1956 and 2001. New data sources were used to derive life history parameters, namely relationships between capture depth vs.

* The materials of this paper were presented at the International Symposium on the North Pacific Transition Areas (La Paz, Mexico, 23-25 April 2002) and the 10th Deep-Sea Biology Symposium (Coos Bay, Oregon, USA, 25-29 August 2003.).

been taken in Sagami Bay off the Pacific coast of Honshu. They described this species using the name *Ebisus sagamius*, which referred to *Ebisu* (恵比寿), the Japanese god of wealth and prosperity, and to Sagami Bay (相模) as the site of capture, and placed it in family Serranidae (Jordan and Snyder, 1901). It was later determined (Jordan and Snyder, 1906) that the species they had described was an adult skilfish, *Erilepis zonifer*. Jordan (1917) later identified this species as representing a separate family Erilepidae though *Erilepis* had been placed into the family Anoplopomatidae in an earlier publication (Jordan and Evermann, 1905). According to present thinking (Eschmeyer et al., 1998; Mecklenburg et al., 2002; Mecklenburg, 2003), the skilfish is one of two members of the family Anoplopomatidae, which is endemic to the North Pacific. This family also includes the better-known species, the sablefish (*Anoplopoma fimbria)*.

Despite its rare appearance in fish catches, the skilfish has acquired numerous common or local names. Most of them have to do with its exceedingly fatty meat, and coloration, which, in the adult fish, reminds some of a clergyman's priestly clothes (Vasilieva, 1999). The name "skilfish" is used in the USA and Canada. Some English names used previously were "giant skilfish", "priestfish", "fat priest", "giant sea bass", "marine monk" (Phillips, 1966), "deepwater cod" (Jordan, 1918), "black cod" (Wada et al., 1978, 1979a, 1979b), "mystery fish" and "Mr. E." (Newman, 1959, 1963). In Poland it is called "skil" (Froese and Pauly, 2001); in France it is "morue barioleé" (Reshetnikov et al., 1989), in China "dochzhi in' syue" 多脂銀鱈; in Vietnam it is "cá than béo"; in Russia the name is "морской монах" (morskoi monakh) that means "sea monk" (Lindberg et al., 1964, 1980; Kotlyar, 1984). The Japanese gave it the names of "aburabodzu" アブラボズ (fatty priest), アブライナギ "aburainagi" (fatty perch) (Phillips, 1966). It is also called "aburabou" アブラボウ,"aburabo" アブラボ and "aburame" アブラメ in Hokkaido, and "kuro" クロ or "kurouo" クロウオ off Odawara (a town on the Pacific coast of Honshu Island in Kanagawa Prefecture) (Amaoka et al., 1995). In some papers (Lindberg et al., 1964, 1980; Fedorov, 1973) one could encounter the Japanese word "aburako" アブラコ as a name for skilfish, which is a mistake since this name is applied in Japan to rock greenling *Hexagrammos ladocephalus* (Lisovenko, 1990).

Skilfish in Drawings and Photographs

The first description of skilfish (Lockington, 1880) was not supplemented with a drawing; the first photographic images of this species were made by Jordan and Snyder (1901) from a 137.5 cm long individual found by them in the Tokyo Imperial Museum. One of these two photographs remains the only current dorsal view image of skilfish. The same scientists are also the authors of the first drawing of skilfish made from a 87.5 cm fish taken from the waters of Japan (Jordan and Snyder, 1906).

The next photograph of this species appeared after over forty years in the well-known monograph by Clemens and Wilby (1949). The first recorded capture of skilfish in Russian waters was documented by Andriashev (1955) who also provided a picture of a 54.7 cm individual, which remains the only portrayal of a specimen found in Russian waters. In 1964, an outline drawing of the skilfish taken from Jordan and Snyder (1906) was included in the dictionary of commercial fishes of the Northwest Pacific Ocean (Lindberg et al., 1964). This drawing was later reproduced in the dictionary of names of commercial marine fishes (Lindberg et al., 1980).

The first photograph of a young 44.5 cm skilfish was published by Phillips (1966) caught in Californian waters. Two drawings of young skilfish appeared nearby at the same time in guides to fishes of California and Canada in 1972 (Miller and Lea, 1972) and 1973 (Hart, 1973). The latter drawing was recently used in "Fishes of Alaska" (Mecklenburg et al., 2002). The photo of a 1 m long skilfish shown in the atlas of "Fishes of the Japanese Archipelago" (Masuda et al., 1984) was subsequently reproduced in the book "Sea Fishes of the World" (Masuda and Allen, 1987) and is likely the first color photograph of skilfish. The most recent photographs of young individuals are from the waters of North Japan (Amaoka et al., 1995), Monterey Bay (photograph by D.W. Gotshall in Froese and Pauly, 2001), and Alaska (Mecklenburg et al., 2002).

Distribution and Migrations

Until now, data on the occurrence of skilfish have been mostly limited by incomplete references to the site and depth of capture, and length and weight of specimens. Near the coast of Japan, skilfish inhabits areas characterized by rather even underwater relief, with occasional protruding rocks (Mitani et. al., 1986).

On the Pacific side of the northern Kuril Islands, skilfish from our samples were captured only around underwater plateaus with rather steep rocky slopes. Very little is known about the spatial distribution of skilfish. Only fragmentary data are available on its capture near the Asian and North American coasts.

Figure 1. Skilfish caught by Russian long-liner in 2008 on the Emperor Seamounts (Photo courtesy A.V. Buslov).

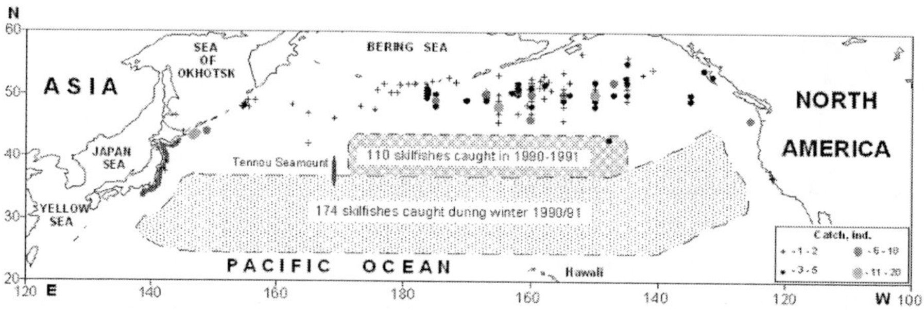

Figure 2. Distribution of skilfish (*Erilepis zonifer*) in the North Pacific Ocean according to published and unpublished data, 1879-2001 (shaded areas show continuous distribution).

Only two papers (Neave, 1959; LeBrasseur, 1967) provide the position of skilfish catches on the high seas. Other publications have not specified capture sites when they discuss the occurrence of the species captured on the high seas (Larkins, 1964; Pearcy et. al., 1996; Broudeur et. al., 1999; Ishida et. al., 1999). Available data show that skilfish are distributed continuously from Central Japan (Sagami Bay, southern part of the Pacific coast of Honshu Island), along the Kuril and Aleutian Islands, down to Monterey Bay, central California (Figure 2). As for the Asian coast, the species is most abundant near the Japanese coast where its harvests have increased from year to year (Mitani et. al., 1986). Off the Kuril Islands and in the southern Sea of Okhotsk, most of the skilfish captures between 1995 and 2001 were near the bottom and were probably related to a northerly extension of the species' range due to the recent climatic restructuring in the North Pacific (Hare and Mantua, 2000). In contrast, adult individuals have rarely been taken by the fishing gear near the North American coast. Adult skilfish are also found around seamounts. Mitani et. al., (1986) mention occurrences around the Tennou seamount (170°E). Japanese, Korean, and Russian fishermen (Figure 1) have recently observed adults and juveniles of this species in their bottom trawl and longline gears on the banks of the Emperor Seamount (Saruwatari et al., 2005; Kwon et al., 2008; our data).

Though our data are fragmentary, we speculate that the major area for the pelagic young of the species is the subtropical and transitional waters, while the subarctic zone is visited by the skilfish only during the warming periods. Data on the capture of skilfish with salmon driftnets in 1959-1969 (Robert Walker, unpubl. data) show that young pelagic skilfish are most numerous in the area between the middle of the Aleutian Archipelago and the central Gulf of Alaska (Figure 2). Throughout the fourteen year period described in this paper there were 227 specimens taken in total, i.e. 16 fish a year on average. Meanwhile, the main habitat of the young, in our view, is found more southerly in the transitional and subtropical waters. For example, McKinnell's data (Stewart McKinnell, pers. comm.) indicate that between 38°N and 46°N, and 170°E and 145°W driftnet catches taken in summer and autumn of 1990 and 1991 contained 110 skilfish (Figure 2). The Japanese driftnet fishery for tuna and other co-occurring species in subtropical waters during the winter of 1990-1991 (Stewart McKinnell, pers. comm.) caught an additional 174 skilfish.

Skilfish migrations have not been studied. Based on length differences in various regions near the Japanese coast, Mitani et al. (1986) suggested that skilfish spawn near the southern coast of Honshu Island and that the larvae are transported with drifting algae towards the northern coast of Hokkaido Island. The fish move south as they grow and change from a pelagic to a demersal way of life. The migration pattern of skilfish, according to our speculations, is somewhat different. Based on our new interpretation of data on captures of various sizes (Appendix 1, Figure 2), the species' **life history** appears to be as follows. Skilfish spawn on the continental slope at 300-500 m in the southern part of its range. Off the Asian coast, spawning regions are near southern and central Honshu Island, while off the North American coast spawning occurs in coastal waters off California. Most likely, skilfish spawn off Japan in early spring since, a short period later in coastal waters, skilfish are seen as juveniles of several centimeters (Patrik Safran, pers. comm.; Michio Omori, pers. comm.) in some catches. This species probably spawns also on seamounts. Eggs and larvae of skilfish have not been observed or described. We suggest that skilfish eggs and larvae rise to the upper layers of the sea, and are gradually transported to the high seas by the currents while they develop and grow. Juveniles become widely distributed in the subtropical and intermediate waters, and are often caught in driftnets (Neave, 1959; Larkins, 1964). Larvae may be carried by currents far from the spawning grounds. If they are transported to regions having unfavorable habitat or feeding conditions they will certainly die. For example, in July 1960 one skilfish larva was found in the eastern Bering Sea at 57°41' N and 174°25' W (Waldron, 1981), however, accurate identification of this larva remains uncertain (Morgan Busby, pers. comm.). Young skilfish are residents of the upper layers on the high seas for several years. Most individuals remain in pelagic waters until they reach about 60 cm. When skilfish reach the length of at least 45 cm and the age of 4-6 years they move shoreward and change to a demersal way of life. In our view, it is only in summer months or during conditions of large-scale warming that both adults and pelagic young penetrate subarctic waters.

Morphology

The coloration of skilfish changes significantly with age. Young individuals are dark grey, or light blue/blackish. On the sides the fish can be found light spots and crosswise stripes characteristic of the species (Figure 3). Adult skilfish are typically colored somewhat differently: light spots on body sides are absent; the upper part of the body is darker than the belly; and the crosswise light stripes, if present, are not as bright as in young fish (Figure 4). It is noteworthy that the spots and stripes may be observed in rather large adult individuals.

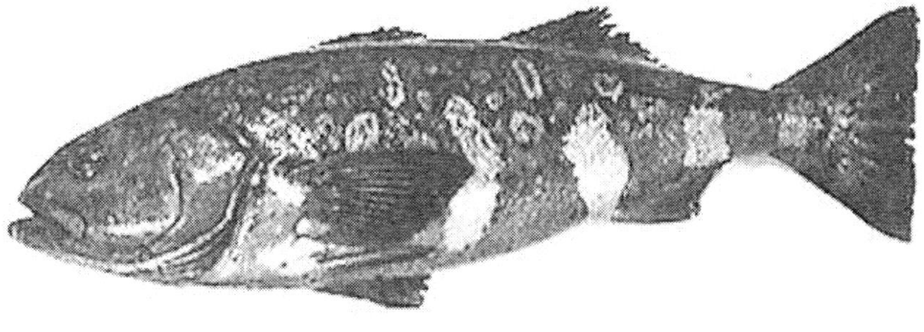

Figure 3. Juvenile skilfish specimen (57 cm FL) from Pacific waters off northern Kuril Islands (photo courtesy I.A. Biryukov).

Table 1. Comparison of some skilfish moprhometrics from Asian and North American coasts (μ – mean value, n – number of specimens examined). t_d are Student-t scores calculated from the data and t_{st} is Student-t score corresponding to $p < 0.05$. Rows in bold are statistically significant. The origination of moprhometric data is the same as for Appendix 2

Meristics	Asia			North America			p<0.05	
	Min-Max	μ	n	Min-max	μ	n	t_d	t_{st}
Standard length (SL), mm	154-1215	801.9	8	244-980	535.3	3	1.21	2.26
In %SL								
Length of head	29.2-37.2	31.8	8	30.0-32.0	31.0	3	1.37	2.26
Depth of head	24.5-36.6	28.5	5	30.0	30.0	1	0.18	2.18
Depth of caudal peduncle	7.8-10.1	8.8	7	8.5-10.0	9.3	3	1.16	2.31
Body depth	28.4-32.5	30.2	6	31.5-33.3	32.4	2	1.91	2.45
Anteanal distance	65.0-74.3	69.2	6	66.1	66.1	1	1.78	2.57
Antedorsal distance	37.0-42.5	39.1	6	37.1	37.1	1	1.63	2.57
Eye diameter, horizontal	4.7-6.2	5.0	8	4.5-5.9	5.2	3	0.29	2.26
Eye diameter, vertical	3.5-4.3	3.8	4	4.1	4.1	1	2.50	3.18
Upper jaw length	10.1-14.1	12.1	6	11.8-13.0	12.3	3	0	2.37
Lower jaw length	10.8-17.2	14.3	4	14.5-16.0	15.2	2	0.23	2.78
Length of snout	8.2-13.3	10.4	8	7.5-11.0	9.4	3	1.18	2.26
Interorbital space	9.7-12.4	11.4	6	10.5-12.7	11.7	3	0.23	2.37
Length of pectoral fin	16.8-24.7	20.7	8	18.0-24.6	21.9	3	0.45	2.26
Depth of ventral fin	12.0-16.9	14.0	8	12.5-15.4	14.2	3	0.16	2.26
Length of anal fin base	13.2-15.7	14.4	5	16.0-16.6	16.4	3	3.77	2.45
Height of anal fin	11.3-15.3	12.6	3	12.0-14.1	13.1	3	1.18	2.45
Length of 1st dorsal fin base	19.5-23.2	21.8	4	21.3-25.8	23.5	2	0.81	2.78
Depth of the 1st dorsal fin	8.9-13.4	10.5	5	8.5-12.8	10.8	3	0	2.45
Length of second dorsal fin base	21.7-27.6	24.2	4	25.0-26.6	25.9	3	1.03	2.57
Depth of the second dorsal fin	9.8-13.8	12.4	4	11.0-14.7	12.9	3	0.17	2.57

For instance, one 83 cm fish taken off Southeast Sakhalin (Kim Sen Tok, 2000) had a coloration typical of the young. White stripes on the body were found in a 97.5 cm fish kept at the Vancouver Aquarium (Hewlett, 1966). White stripes were also noticeable in individuals of 126 cm and 130 cm from off the Kuril Islands (Tokranov and Diakov, 1996).

Figure 4. Adult skilfish specimen (121 cm FL) from the Pacific waters off the northern Kuril Islands (photo A.M. Orlov.).

These age-related differences in coloration of young and adult skilfish resulted in them being identified as different species (Lockington, 1880; Jordan and Snyder, 1901). It was believed that skilfish in the North Pacific were distributed amphi-pacifically, i.e. it inhabited waters along both the Asian and North American coasts (Andriashev, 1939, 1955), but not in the gaps at subarctic northern and tropical southern areas.

This implied the existence of separate populations off North America and Asia, which would suggest that there may be perhaps morphological differences. A comparative morphological analysis was made using previously published data (Lockington, 1880; Jordan and Snyder, 1901, 1906; Thompson, 1917; Andriashev, 1955; Phillips, 1966; Tokranov and Diakov, 1996; Moukhametov and Volodin, 1999; Kim Sen Tok, 2000) from twelve specimens from the coasts of Asia and North America (Appendix 2). No significant differences between the nine Asian and three North American specimens were found (Table 1). Variations in all features of the Asian and North American fish overlap. Some differences were recorded in the body height (Asian fish: 28.4-32.5% SL, average 30.2%, North American fish: 31.5-33.3%, average 32.4%); in the snout length (8.2-13.3%, average 10.4%, and 7.5-11.0%, average 9.4%, respectively); length of the base of the anal fin (13.2-15.7 %, average 14.4 %, and 16.0-16.6%, average 16.4% respectively); length of base of the first dorsal fin (19.5-23.2%, average 21.8%, and 21.3-25.8%, average 23.5% respectively). However, the difference in all the characteristics examined, except the base length of the anal fin, proved statistically insignificant ($p<0.05$, Table 1). The only difference was found in the anal fin base length. However this result is somewhat inconclusive due to the limited amount of data available for comparison. Moreover, the comparison was made using literature data and it is especially difficult to make comparisons of morphometric data across publications.

Composition of Catches That Include Skilfish

Skilfish are caught with various fishing gears. In subtropical waters (30-38° N) it may be caught incidentally with driftnets in tuna and billfish fisheries. In the winter of 1990-1991 their frequency of occurrence in driftnets in this area was around 21% (Stewart McKinnell, pers. comm.). In more northerly transitional waters (38°-46° N) the species occurs in the pelagic driftnet squid fisheries where its occurrence in 1990 and 1991 made up 3.3% and 0.5% respectively (Stewart McKinnell, pers. comm.). As for the subarctic high seas waters of the ocean (north of 35° N), skilfish are taken in driftnets in the Pacific salmon fisheries, but it occurs much less often there than in the subtropical zone. 227 skilfish were found in salmon driftnet catches in the subarctic waters of the North Pacific during the 14 year period 1956-1969. This species was found in 7.8% of net hauls (Robert Walker, pers. comm.).

About 40 fish species, including skilfish, were recorded in 7 m high driftnets (50-130 mm mesh) by the salmon research program in spring and summer of 1955-1961 on the high seas adjacent to the Aleutian Islands and the Gulf of Alaska (Larkins, 1964). The most frequent species, depending on the area, were: steelhead (*Oncorhynchus mykiss*) (17-54%), Pacific pomfret (*Brama japonica*) (11-64%), and albacore (*Thunnus alalunga*) (12-25%). Skilfish occurred in catches with a 4-30% frequency (Table 2).

During summer-autumn of 1956-1958, skilfish were captured in the high sea off British Columbia in association with drifting gill nets along with North Pacific daggertooth (*Anotopterus nikparini*), Pacific bluefin tuna (*Thunnus orientalis*), skipjack (*Katsuwonus pelamis*), smalleye squaretail (*Tetragonurus cuvieri*), and pelagic armorhead (*Pseudopentaceros richardsoni*) (Neave, 1959).

At ocean station P (50° N, 145° W) during 1958-1966 skilfish appeared in the catches of line fishing along with salmon (*Oncorhynchus* spp.), Pacific pomfret, longnose lancetfish (*Alepisaurus ferox*), Pacific jack mackerel (*Trachurus symmetricus*), salmon shark (*Lamna ditropis*), and blue shark (*Prionace glauca*) (LeBrasseur, 1967) In subarctic and transitional waters between the Asian and North American coasts in 1978-1993, the driftnet catches (7 m high, mesh 19-204 mm) contained skilfish, 40 other fish species, and four species of cephalopods (Pearcy et al., 1996). The most frequent fish species, depending on the area, were Pacific saury *Cololabis saira* (30-92%), boreal chubhook squid *Onychoteuthis borealijaponica* (24-91%), and Pacific pomfret (4-99%). The skilfish occurrence in the catches reached 8% (Table 3).

Table 2. The most frequently caught species taken during the salmon research program (driftnets, 7 m high, 50-130 mm mesh) in the high seas adjacent to the Aleutian Islands and the Gulf of Alaska, spring-summer 1955-1961 (based on data of Larkins, 1964)

Taxa	Frequency of occurrence (min-max), %
Albacore *Thunnus alalunga*	12-25
Atka mackerel *Pleurogrammus monopterygius*	0-15
Blue shark *Prionace glauca*	0-44
North Pacific daggertooth *Anotopterus nikparini*	1-12
Pacific jack mackerel *Trachurus symmetricus*	0-30
Pacific pomfret *Brama japonica*	11-64
Sablefish *Anoplopoma fimbria*	0-12
Skilfish *Erilepis zonifer*	4-30
Smalleye squaretail *Tetragonurus cuvieri*	0-11
Spiny dogfish *Squalus acanthias*	5-11
Steelhead *Oncorhynchus mykiss*	17-54

As for the high seas area in the North Pacific (36-46° N 166°E-150°W), the driftnet catches (about 10 m high, mesh 72-120 mm) in June-October 1990-1991 in the flying squid fishery contained over 40 species of fish and cephalopods (Brodeur et al., 1999). Neon flying squid composed the bulk of the catches (occurrence 99% in 1990 and 100% in 1991). Other most frequently caught species were Pacific pomfret (96 and 95% respectively), blue shark (89%), and albacore (86-80% respectively). Skilfish occurrence in catches during the period of observation made up 3% (Table 4).

About 30 species of fish and 4 species of cephalopods were found together with the skilfish in catches of driftnets (mesh 48-157 mm; height unknown) in the open waters of the northwest Pacific limited by 165°E between the southernmost tip of Honshu Island (Japan) and South Kamchatka (Russia) in the summer of 1992-1998 (Ishida et. al., 1999). The most frequent fish in the nets were pink (49.7%), chum (7.2%) and sockeye (5.7%) salmon, Pacific saury (16.3%) and Atka mackerel (5%). The rate of occurrence of other species did not exceed 3%, and in most cases it was much lower. Skilfish were found only in 0.01% of the catch.

Table 3. The most frequent species caught by driftnets (7 m high, 19-204 mm mesh) in the subarctic and transitional waters between the Asian and North American coasts, 1978-1993 (based on data of Pearcy et al., 1996)

Taxa	Frequency of occurrence (min-max), %
Albacore *Thunnus alalunga*	0-72
Amberjack *Seriola lalandi*	0-40
Blue shark *Prionace glauca*	1-81
Boreal chubhook squid *Onychoteuthis borealijaponica*	24-91
Chinook salmon *Oncorhynchus tschawytscha*	0-70
Chum salmon *Oncorhynchus keta*	0-100
Coho salmon *Oncorhynchus kisutch*	0-99
Eight-armed squid *Gonatopsis borealis*	0-88
Longnose lancetfish *Alepisaurus ferox*	7-9
Neon flying squid *Ommastrephes bartrami*	2-96
Ocean sunfish *Mola mola*	0-14
Pacific pomfret *Brama japonica*	4-99
Pacific saury *Cololabis saira*	30-92
Pelagic armorhead *Pseudopentaceros richardsoni*	0-8
Pink salmon *Oncorhynchus gorbusha*	0-96
Salmon shark *Lamna ditropis*	4-54
Skilfish *Erilepis zonifer*	0-8
Skipjack *Katsuwonus pelamis*	0-78
Sockeye salmon *Oncorhynchus nerka*	0-89
Steelhead *Oncorhynchus mykiss*	0-95

Table 4. The most frequent species caught by driftnets (10 m high, 72-120 mm mesh) during the neon flying squid fishery in the North Pacific (36-46° N, 166° E-150° W), June-October 1990-1991 (based on data of Brodeur et al., 1999)

Taxa	Frequency of occurrence, %	
	1990	1991
Albacore *Thunnus alalunga*	86	80
Amberjack *Seriola lalandi*	53	
Black fathead *Cubiceps baxteri*	0	16
Blue shark *Prionace glauca*	89	
Bluefin driftfish *Psenes pellucidus*	5	10
Bluefin tuna *Thunnus thunnus*	9	19
Boreal chubhook squid *Onychoteuthis borealijaponica*	9	12
Butter fish *Hyperoglyphe japonica*	13	
Chum salmon *Oncorhynchus keta*	11	7
Coho salmon *Oncorhynchus kisutch*	14	15
Common dolphinfish *Coryphaena hyppurus*	10	
Longnose lancetfish *Alepisaurus ferox*	57	60
Luvar *Luvarus imperialis*	31	27
Neon flying squid *Ommastrephes bartrami*	99	100
Ocean sunfish *Mola mola*	66	58
Opah *Lampris guttatus*	16	11
Pacific pomfret *Brama japonica*	96	95
Pacific saury *Cololabis saira*	31	39
Pelagic armorhead *Pseudopentaceros richardsoni*	40	39
Pelagic octopus *Ocythoe tuberculata*	9	24
Pilotfish *Naucrates ductor*	10	15
Ragfish *Icosteus aenigmaticus*	11	
Rough pomfret *Taractes asper*	7	17
Salmon shark *Lamna ditropis*	67	58
Shortfin mako *Isurus oxyrinchus*	11	7
Skilfish *Erilepis zonifer*	up to 3	
Skipjack *Katsuwonus pelamis*	24	26
Swordfish *Xiphias gladius*	30	32
Other species	< 10	

Table 5. Species composition of bottom trawl catches containing skilfish in the Pacific waters off the northern Kuril Islands (W – percent of total catch weight, N – percent of total specimen number, FO – percent of frequency of occurrence)

Taxa	W	N	FO
Pacific ocean perch *Sebastes alutus*	43.8	40.3	80.0
Dimdisc snailfish *Elassodiscus tremebundus*	22.2	27.2	70.0
Undescribed snailfish *Careproctus* cf. *cyclocephalus*	7.0	6.2	70.0
Shortraker rochfish *Sebastes borealis*	4.4	1.5	90.0
Red squid *Berryteuthis magister*	3.4	5.5	80.0
Broadbanded thornyhead *Sebastolobus macrochir*	2.4	2.4	80.0
Forktail snailfish *Careproctus furcellus*	2.4	1.2	50.0
Whiteblotched skate *Bathyraja maculata*	2.2	0.3	50.0
Kamchatka flounder *Atheresthes evermanni*	1.6	0.4	70.0
Darkfin sculpin *Malacocottus zonurus*	1.3	4.8	100.0
Matsubara skate *Bathyraja matsubarai*	1.2	0.2	50.0
Gian grenadier *Albatrossia pectoralis*	0.7	0.1	50.0
Skilfish *Erilepis zonifer*	0.3	< 0.1	100.0
Other species combined	5.7	9.5	
Number of hauls	10		

Catches taken from near the surface off the Japanese coast southeast of Hokkaido Island, using 7 m high nets with 2 m openings and 2 mm mesh, showed that young skilfish found among floating algae co-occurred with the young of four species (rate of occurrence, percent): jacopever (*Sebastes schlegeli*) (93%), butter fish (73%), three-spined stickleback (*Gasterosteus aculeatus*) (63%) and spotty-belly greenling (*Hexagrammos decagrammus*) (48%) (Safran and Omori, 1990). In addition, skilfish were caught periodically in the pollock trawl fishery at about 200 m (Mitani et. al., 1986). The same authors point out that skilfish were taken incidentally in the rockfish longline fishery off northeast Honshu Island at the rate of 10%, in the rockfish trawl and longline fishery off central Honshu Island at depths 300 to 900 m (5%) and in the grenadiers longline fishery near the Tennou seamount at 550-1030 m (1.7%). However, maximum skilfish catches were recorded near southeast Honshu Island where it constitutes 100% of catches in the longline fishery (Mitani et. al., 1986). In the Pacific waters of the northern Kuril Islands, catches containing skilfish also had some 60 other species of fish and 4 species of cephalopods. By weight and abundance the bulk of catches were comprised of Pacific Ocean perch *Sebastes alutus*, dimdisc snailfish *Elassodiscus tremebundus*, undescribed snailfish *Careprocus* cf. *cyclocephalus,* and red squid *Berryteuthis magister* (Table 5). The darkfin sculpin *Malacocottus zonurus* was found most frequently in catches (100%).

Vertical and Thermal Distribution

Data on the thermal conditions associated with the occurrence of skilfish are scarce. The near-bottom temperatures off the shores of Japan at recorded skilfish capture sites are 3-4°C (Mitani et al., 1986).

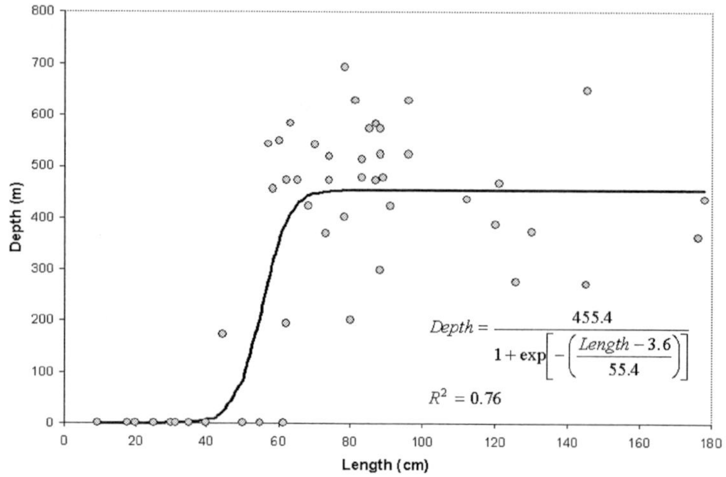

Figure 5. Relationship between skilfish length and capture depth compared to an estimated sigmoid curve.

Our observations of the temperatures at the sites where skilfish were found are sparse (16 observed cases near the surface, and 8 near the bottom). In the surface layers the species was recorded at water temperatures of 8.4-13.2°C (average 10.21°C); in the near-bottom layers skilfish were found at 0.9-3.45°C (average 2.71°C). The known adult skilfish depths of capture are 0-1030 m (Mitani et al., 1986). No information exists on the captures of skilfish eggs or larvae, and early life stages of this species have not been described (Matarese et al., 1989).

The capture of single larva was reported in July 1960 in the eastern Bering Sea at 57°41' N, 174°25' W (Waldron, 1981) but its identification is inconclusive (Morgan Busby, pers. comm.). Young skilfish occur near the coast at the surface among floating algae (Safran, 1990; Safran and Omori, 1990).

Skilfish stay in the upper layers for several years after which they change over to a near-bottom mode of life, where they settle down on the continental slope. The depth of habitation then switches over to living near the bottom as size increases. The body-length/capture-depth relationship is well described with a sigmoid function

$$Y = \frac{\alpha}{1 + \exp\left[-\left(\frac{X - \beta}{\gamma}\right)\right]} \qquad (1)$$

where Y is habitation depth (m), X is total length (cm), α is the asymptotic depth of habitation, γ is the parameter that adjusts the curves inflection points, and β is the length value where the curve is increasing at its maximum rate. Skilfish remain in the pelagic layers until reaching about 60 cm in length, though some individuals descend to the bottom at 45 cm. Figure 5 indicates the sigmoid curve that describes the data well (estimated parameter values given in the Figure 5), with the curve explaining 76% of the variation in depth of habitation (R^2 = 0.76).

Biological Characteristics

Growth

According to Froese and Pauly (2001), the maximum total length of skilfish is 183 cm, though there are reports of capturing 184 cm TL skilfish near south Honshu Island (Mitani et al., 1986). The skilfish length-weight relationship from Japanese waters (Mitani et al., 1986) is described by the equation:

$$W=3.80\ 10^{-5}\ TL^{2.8247} \quad (2)$$

where W is weight (kg) and TL is total length (cm).

According to this relationship, the weight of skilfish is 0. 025 kg at 10 cm, 2.39 kg at 50 cm, 16.95 kg at 100 cm, 53.28 kg at 150 cm, and 89.17 kg at 180 cm (Mitani et al., 1986).

Based on the data available from this study (22 measurements), the skilfish allometric length-weight relationship was fit to the equation

$$W=\alpha\ TL^{\beta} \quad (3)$$

using nonlinear regression, where W is weight (kg) and TL is total length (cm). Results (Figure 6) indicate a very good agreement between the curve and the data (R^2 = 0.99). Parameter estimates (and their standard errors) were α = 8.334E-05 (S.E. = 4.4137E-05) and β = 2.65 (S.E. = 0.1048). The curve fit to the data from this study is very close to the Japanese relationship (Mitani et al., 1986) described above. Both curves are shown along with data from this study in Figure 6. These values are compared with sablefish, the other commercially important member of the family Anoplopomatidae (Table 7).

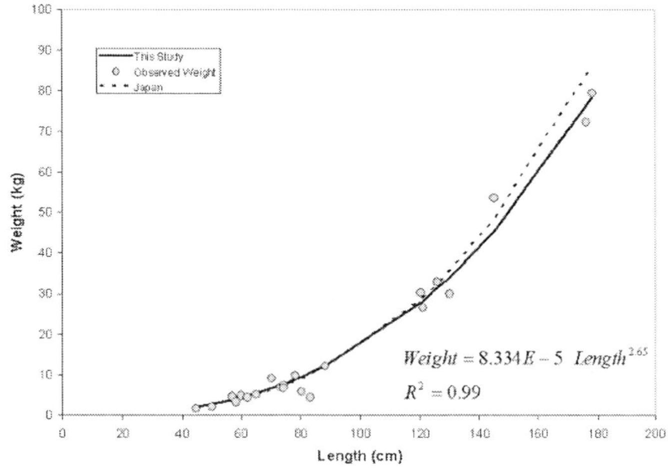

Figure 6. Relationship between skilfish length and body weigth. Compared are data from this study (grey circles), a curve from a Japanese analysis (Mitani et al., 1986) (dashed line) and the estimated curve fit to the data from this study (solid line).

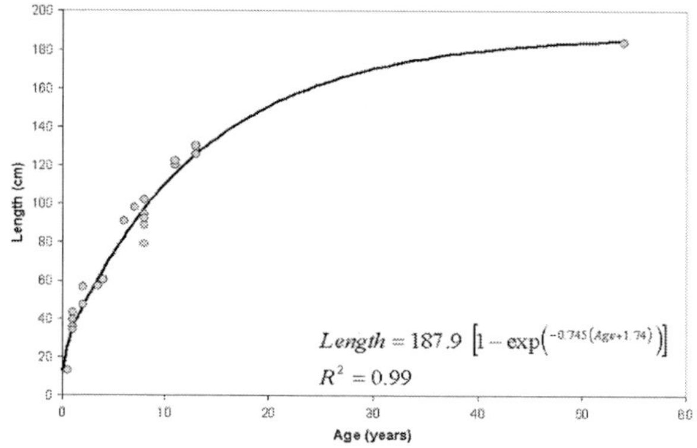

Figure 7. Growth curve of skilfish based on published information (Mitani et al., 1986; Tokranov and Diakov, 1996; Moukhametov and Volodin, 1999) and our own data. Plotted are the observations (grey circles) and the estimated von Bertalanffy curve (solid line).

Age

Age data for this species are scarce. Japanese experts (Mitani et al., 1986) used the count of annual rings in the vertebra and aquarium observations for age determination. They produced the following growth equation on the basis of those data:

$$L_t=186.3\ (1 - \exp^{-0.2351-0.0744t}) \quad (4)$$

where L_t is the total length of fish (cm) at age t (years).

According to this equation at age 5 years skilfish grows to 85 cm; at 10 years they reach 116 cm; at 15 years – 138 cm; and at 20 years – 153 cm. The age of a 57 cm long fish caught by us near the Kuril Islands was 3.5 years, as determined from reading otoliths. Specimens of 126 cm and 130 cm taken in the same area were aged to be 13 years old by scales (Tokranov and Diakov, 1996). Another 120 cm fish from the same area was 11 years old as counted by a cross section of the spine of the first dorsal fin (Moukhametov and Volodin, 1999).

Based on 22 length-age data observations collected in this study, we estimated the parameters of the vonBerlanffy growth equation (vonBertalanffy, 1938)

$$L_t = L_\infty[1-\exp(-K(t-t_0))] \quad (5)$$

using nonlinear regression, where L_t is the total length at age t (cm), t is age in years, L_∞ is the asymptotic length (cm), K is the growth rate (year^{-1}), and t_0 is the theoretical age (typically negative) the fish would have length zero had they always grown as described by Equation 5. Parameter estimates (and standard errors) are $L_\infty = 187.9$ cm (S.E. = 8.98), $K = 0.075$ yr^{-1} (S.E. = 0.00079), and $t_0 = 1.74$ yr (S.E. = 0.396). The ageing results of the Kuril Island fish agree very well ($R^2 = 0.99$) with the above equation for the skilfish from the waters of Japan (Figure 7). In Table 6, these values are compared with sablefish, the other commercially important member of the Anoplopomatidae.

According to the growth formula suggested by Mitani et al. (1986), the asymptotic length of skilfish is 186.3 cm. This is very close to our estimate of 187.9 cm and is very close to the result of 186.4 cm obtained by Froese and Pauly (2001). Data on the observed and calculated age of the largest skilfish vary greatly. Mitani et al. (1986) report that the age of a 184 cm individual determined by counting rings in the vertebra was assumed to be 54 years since some rings were not distinct. Using growth equations they showed that the maximum age for skilfish may reach over 80 years.

The growth rate of skilfish is rapid, especially during the first years of life. One skilfish at Vancouver Aquarium grew by 57 cm (30 to 87.5 cm) during 8 years (1956 to 1963); another individual grew from 30 cm to 97.5 cm (by 67.5 cm) during the same time period between 1959 and 1966 (Newman, 1963; Hewlett, 1966). A ten centimeter juvenile at Muroran Aquarium (Hokkaido Island, Japan) reached 140 cm/55 kg after 28 years (Anon., 2002). The instantaneous growth rate (G) can be estimated from these observations by rearranging the equation

$$W_{t+\Delta t} = W_t \exp(G\Delta t) \quad (6)$$

to

$$G = \frac{\ln\left[\dfrac{W_{t+\Delta t}}{W_t}\right]}{\Delta t} \quad (7)$$

where W_t is weight at age t (kg), Δt is the time that has elapsed between weight measurements (years), G is the instantaneous growth rate (year^{-1}), and t is time (years) and calculating weight from length using equation (3) above. Results indicate growth rates of 0.366 yr^{-1}, 0.403 yr^{-1}, and 0.257 yr^{-1} for the examples given above, respectively. The von Bertalanffy growth equations of sablefish from the Bering Sea during the 1960s and 1970s in terms of body weight are given in paper of Sasaki (1985). These can be compared to sablefish from the Bering Sea and Aleutian Islands (Hanselman et al., 2006a), the Gulf of Alaska (Hanselman et al., 2006b), and results from this study (Table 6). In comparison to sablefish, skilfish appear to have a larger asymptotic size (L_∞) and a slower growth rate (K).

Mortality

Natural mortality can be estimated from the life history parameters estimated above by using the method of Zhang and Megrey (2006). We used estimates of K = 0.0746 and t_0 = 1.74 from equation (5) above, β = 2.7263 from equation (3), the constant relating t_{mb} to t_{max} for demersal species ($C_{demersal}$ = 0.44), and the maximum age observed in our length-age data set (t_{max}) (Figure 7) of 54 years. These give t_{mb} = 0.44x54 = 23.76. Substituting these values into the equation

$$M = \frac{\beta K}{\exp^{K(t_{mb}-t_0)} - 1} \quad (8)$$

gives, as a first approximation, an estimate of M = 0.00358. There are no other estimates of this important parameter for the skilfish. However we can compare it to a sablefish (*Anoplopoma fimbria*) estimate, which was used in a Bayesian stock assessment. The value was set to $M = 0.1$ for the Bering Sea, Aleutian Island, and Gulf of Alaska stock assessment (Hanselman et al., 2006a, 2006b). In the Bayesian analysis, Hanselman et al. (2006a, 2006b) assumed that they knew the approximate value of sablefish natural mortality very precisely (c.v. = 0.001 for prior probability distribution) and that the approximate value was 0.10. At this level of prior precision, it was essentially a fixed parameter. In the Gulf of Alaska, sablefish over 40 years old are regularly recorded (Kimura et al., 1993). The sablefish M estimate from Zhang and Megrey (2006) is based on a maximum age of 94 years old (Kimura et al., 1998; Munk, 2001). The previously reported maximum age was 62 (Siegler et al., 1997). Canadian researchers report age determinations up to 55 years (McFarlane and Beamish, 1983). Maximum age, added to $K = 0.141$, $t_0 = 6.05$, and $\beta = 3.19$ (Lowe et al., 2001) gives an $M = 0.001$ for sablefish, which is not unlike our estimate of $M = 0.00358$ for skilfish.

Reproduction

Data on reproductive biology for this species are very scarce. The Japanese experts (Mitani et al., 1986) assessed the maturity stage of gonads by an index calculated by the formula given by

$$IM = \frac{W}{L^3 10^4} \tag{9}$$

where IM is the index of gonad maturity, W is the weight of gonads (kg), and total body weight is estimated from a form of Equation 3 given L as the total length of fish (cm) assuming that $\beta = 3.0$ and a conversion factor of $1\text{x}10^{-4}$. The maximum maturity index for females (10.8) was recorded in January (Figure 8). For males, the greatest maturity index (4.9) was found in April: it expressed with milt when the belly was pressed. Most of the fish examined had high gonad maturity indices (3-5) from April to June. An exponential relationship between the index of gonad maturity was assumed and the curve

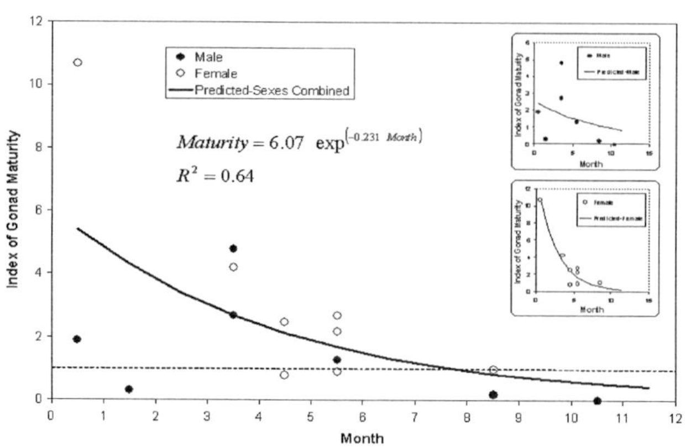

Figure 8. Seasonal dynamics of a skilfish gonad maturity index in Japanese waters (after Mitani et al., 1986). Plotted are exponential relationships between the gonad maturity index and month for males and female skilfish separately and a curve estimated from the combined data.

$$Y = \alpha \exp(\beta X) \tag{10}$$

was fit to the data, where Y is the index of gonadal maturity and X is the month. Testing for sexual dimorphism, one curve was fit to each sex and one for sexes combined (Figure 8). Results show that estimated parameters for the male (α = 2.466, S.E. = 1.452, 95% confidence intervals = -1.266 - 6.199, β = -0.096, S.E. = 0.147, 95% confidence intervals = -0.473 - 0.281, R^2 = 0.57) and female data (α = 12.876, S.E. = 1.280, 95% confidence intervals = 9.745 - 16.01, β = -0.370, S.E. = 0.046, 95% confidence intervals = -0.481 - 0.258, R^2 = 0.97) were not significantly different ($p < 0.01$), thus one curve was estimated for sexes combined. Parameters from the combined curve compared to the individual fits (Figure 8) give (α = 6.07, S.E. = 1.840, 95% confidence intervals = 2.094 - 10.041, β = -0.231, S.E. = 0.106, 95% confidence intervals = -0.460 - 0.002, R^2 = 0.64). Most likely, skilfish spawn off Japan in early spring since, a short period later in coastal waters, skilfish are seen as juveniles of several centimeters (Patrik Safran, pers. comm.; Michio Omori, pers. comm.) in some catches. The floating algae reach their peak in May-June, and in July in coastal waters their number goes down sharply (Safran and Omori, 1990). All the specimens caught in recent years near the Kuril Islands turned out to be mature males. The gonad maturity index of one 126 cm male captured in November (Tokranov and Diakov, 1996) calculated by the above formula was 0.006. Length at maturity data are quite contradictory. Froese and Pauly (2001) report that the skilfish total length at maturity is 91.4 cm. Mitani et al. (1986) report a 123 cm fish had mature gonads. There are data on the capture of a mature 83 cm female (Kim Sen Tok, 2000), whereas all males over 120 cm had mature gonads (Tokranov and Diakov, 1996; Moukhametov and Volodin, 1999).

Diet

There are practically no data on the diet of skilfish. LeBrasseur (1967) noted that stomachs of juvenile skilfish caught in line fishing in the high sea contained zooplankton and squid. The only paper (Moukhametov and Volodin, 1999) that mentions adult skilfish diet reported that the stomach of one 120 cm adult skilfish contained two octopi weighing a total of 0.75 kg. The stomach of a 57 cm specimen (estimated age from otolith reading is 3.5 years) collected by us near the Kuril Islands contained one beak of the red squid, *Berryteuthis magister* (lower rostral length of beak 5.69 mm). Using a formula for the relationship between the sizes of a red squid mantle and beak size (Walker, 1996) we estimated the squid mantle length as 249 mm.

Observations on Skilfish in Captivity

The first data on keeping this species in captivity appeared in the late 1950s when several young 20-30 cm long fish caught with salmon driftnets near British Columbia were delivered to the Vancouver Public Aquarium (Newman, 1959, 1963; Hewlett, 1966).

When in captivity, skilfish typically take food directly from the hands of their keepers and there are reports (Newman, 1959) that they respond in a positive way to being physically stroked while captive in their tanks.

Figure 9. A picture of a skilfish, a symbol of the aquaria in Muroran city Japan (photo is taken from website of Muroran aquarium (http://travel.mimo.com/aquarium/2850muroran/area.html).

Such observations regarding the features of their captive behavior compelled the first mate of "Fort Ross," Dave Stevenson, to oppose delivering the objects of his care to the aquarium after he fed the fish and become familiar with them during transportation (Newman, 1959). The newly delivered juvenile skilfish were kept at the Vancouver Aquarium in a separate tank where they aggregated in a school. When other species were introduced, the skilfish displayed some nervousness and agitation. It was decided therefore to keep the skilfish separate from other species in the future. One skilfish lived in the Vancouver Aquarium for 8 years, and died only because of a severe parasite infestation.

In 1966, three young skilfish, with one fish being 97.5 cm, were kept in the Vancouver Aquarium. The largest grew 67.5 cm during eight years in captivity. There are no additional data regarding maintaining skilfish in the Vancouver Aquarium. In August of 1996, seven young skilfish of about 7 cm were caught near Astoria, Oregon, in a tuna fishing area. The skilfish were swimming among surface floating aggregations. Within a month, they were transported to the Oregon Coast Aquarium (Newport), being fed tuna meat during their journey from point of capture to the Aquarium. Skilfish grew rapidly, doubled their length during the month-long transfer period. For several months they were displayed at Newport, and in autumn of 1997 they were sent to the Seattle Aquarium (James Burke, pers. comm.). At the present time, there are four skilfish over 90 cm at the Seattle aquarium, and they continue to be a significant local attraction (Anon., 1999; Cowan, 2000). Jeff Christiansen (pers. comm.), offered several observations from the Seattle Aquarium. He states that the skilfish can often be seen protruding from among various aquarium structures and often swim together. They do not seem to demonstrate any schooling instincts but rather they typically make up a loosely dispersed group in the mid-water, usually near incoming water openings or under kelp domes. He also remarked that the procedure of feeding skilfish in aquaria was of special interest. Divers feeding sturgeons and Pacific halibut repeatedly observed the snout of one of the four skilfish in the aquarium to appear suddenly and directly in front of the diver's mask. A fish would approach the diver, bring its mouth to some 20 cm from the diver's mask, as if demanding food. If the fish did not get fed, they started to bite the diver's mask. After such encounters the fish did not need to "persuade" the diver's to feed them. Watching the skilfish swimming is quite impressive. As observers state, the eyes of skilfish express its strong nature.

Despite the large number of aquariums in Japan, skilfish are being kept now only at the Muroran Aquarium on Hokkaido Island (Anon., 2002). Skilfish is the symbol of this aquarium (Figure 9). In Japan, it is only here that one can simultaneously watch the young and adult individuals of the species. In addition, it was at this Japanese aquarium that the greatest success was achieved in keeping skilfish in captivity where 10 cm individuals were grown into 140 cm fish weighing 55 kg during 28 years. Those skilfish were kept in 7-8°C water. These successes were not achieved at any other aquarium in the world.

In summary, we reviewed past information on skilfish nomenclature and etymology, morphological features, distribution, and observations made in aquaria in the USA, Canada, and Japan, combined this information with 600 new observations to update past information and describe skilfish biological characteristics, used the new data sources to derive life history parameters, namely depth of habitation vs. length relationship, length-age relationship and weight vs. length relationships, derived estimates of natural mortality and growth from empirical relationships, and compared life history values to previously published information.

ACKNOWLEDGMENTS

This chapter would not be possible without the assistance of our numerous colleagues. Robert Walker (High Seas Salmon Program, School of Aquatic and Fishery Sciences, University of Washington, Seattle, USA), Yoshiyuki Kajiwara (T/S Oshoro-Maru, Faculty of Fisheries, Hokkaido University, Hakodate, Japan), Yukimasa Ishida (Hokkaido National Fisheries Research Institute, Kushiro, Japan), James Orr (Alaska Fisheries Science Center, Seattle, USA), Stewart (Skip) McKinnell (PICES, Sidney, Canada), Michio Omori (Graduate School of Agricultural Science, Tohoku University, Sendai, Japan), James Burke (Oregon Coast Aquarium, Newport, USA), Anatoly Dekshtein (Kamchatka Research Institute of Fisheries and Oceanography, Petropavlovsk-Kamchatsky, Russia), Boris Sheiko (Zoological Institute, Russian Academy of Science, Sankt-Peterburg, Russia) and Vladimir Touponogov (Pacific Fisheries Research Centre, Vladivostok, Russia) provided to us the data on the captures of skilfish. Patrick Safran (Asia Development Bank, Manila, Philippines), Treva Ricou (Vancouver Aquarium Marine Science Centre, Vancouver, Canada), Boris Sheiko, Oleg Bulatov (Russian Federal Research Institute of Fisheries and Oceanography, Moscow, Russia), Michio Omori, Toru Taniuchi (Nihon University, Fujisawa, Japan) and Fumiji Kasai (Kanai Fisheries Co., Ltd., Kushiro, Japan) provided some rare publications on the skilfish. Masaru Yagi (Kanai Fisheries Co., Ltd., Kushiro, Japan) and Alexander Yarzhombek (Russian Federal Research Institute of Fisheries and Oceanography, Moscow, Russia) translated from the Japanese. James Burke, Jeff Christiansen (Seattle Aquarium, Seattle, USA) and Kei-ichi Mito (Hokkaido National Fisheries Research Institute, Kushiro, Japan) provided information on aquarium observations. Alexander Buslov and Igor Birukov (Sakhalin Research Institute of Fisheries and Oceanography, Yuzhno-Sakhalinsk, Russia) gave us photographs. Beth Sinclair (Alaska Fisheries Science Center, Seattle, USA) gave us data on the beak/mantle size ratio in red squid. We are deeply thankful to all these colleagues. We would also like to thank Morgan Busby and Jay Orr of the Alaska Fisheries Science Center for reviewing an early version of the chapter.

Appendix 1. Data on skilfish *Erilepis zonifer* captures in the North Pacific Ocean (in the temperature column, all data represent surface measurements except as otherwise noted)

Date (DD.MM.YYY)	Coordinates	Number of fish	Depth, m	Total length, cm	Weight, kg	Temperature, °C	Source
??.??.????	45° 51' N 154° 53' W	?	?				Fisheries Center, UBC (1996)
??.??.????	49° 59' N 160° 03' W	?	?				Fisheries Center, UBC, (1996)
??.08.1879	36° 45' N 123° 01' W	1	?	29.4			Lockington (1880)
??.??.1915	52° 54' N 131° 33' W	1	?				Clemens and Wilby (1961)
??.??.1916	52° 54' N 131° 33' W	1	?				Clemens and Wilby (1961)
??.??.1917	52° 54' N 131° 33' W	3	364	176.2	72.1		Clemens and Wilby (1949)
??.01.1916	53° 50' N 132° 50' W	1	273	145	53.6		Thompson (1916)
??.10.1916	53° 50' N 132° 50' W	1	437	178	79.4		Thompson (1917); Clemens and Wilby (1961)
??.10.1916	53° 50' N 132° 50' W	1	437	112			Thompson (1917); Clemens and Wilby (1961)
21.08.1951	44° 00' N 148° 48' E	5	0	9.2-17.7			Andriashev (1955)
07.08.1953	48° 23' N 162° 27' E	1	0	54.7			Andriashev (1955)
16.07.1956	51° 00' N 145° 00' W	1	0				Robert Walker, pers. comm.
20.07.1956	50° 00' N 150° 00' W	1	0				Neave (1959)
21.07.1956	48° 00' N 150° 00' W	3	0				Robert Walker, pers. comm.
22.07.1956	49° 00' N 150° 00' W	4	0				Robert Walker, pers. comm.
24.07.1956	51° 00' N 150° 00' W	1	0				Robert Walker, pers. comm.
25.07.1956	50° 00' N 135° 00' W	1	0	31			Newman (1963); Clemens and Wilby (1961)
25.07.1956	50° 00' N 155° 00' W	1	0				Robert Walker, pers. comm.
26.07.1956	50° 00' N 135° 00' W	1	0				Neave (1959)
26.07.1956	49° 00' N 155° 00' W	1	0				Robert Walker, pers. comm.
29.07.1956	45° 51' N 154° 53' W	1	0				Robert Walker, pers. comm.

Date (DD.MM.YYY)	Coordinates	Number of fish	Depth, m	Total length, cm	Weight, kg	Temperature, °C	Source
31.07.1956	45° 59' N 160° 03' W	6	0				Robert Walker, pers. comm.
01.08.1956	47° 00' N 160° 00' W	1	0				Robert Walker, pers. comm.
02.08.1956	48° 00' N 160° 00' W	3	0				Robert Walker, pers. comm.
03.08.1956	56° 00' N 144° 50' W	1	0				Robert Walker, pers. comm.
04.08.1956	54° 57' N 145° 00' W	1	0				Robert Walker, pers. comm.
05.08.1956	51° 00' N 160° 10' W	1	0				Robert Walker, pers. comm.
06.08.1956	51° 57' N 160° 00' W	1	0				Robert Walker, pers. comm.
22.08.1956	51° 00' N 155° 00' W	1	0				Robert Walker, pers. comm.
23.08.1956	50° 00' N 155° 00' W	1	0				Robert Walker, pers. comm.
26.08.1956	49° 00' N 155° 00' W	2	0				Robert Walker, pers. comm.
28.08.1956	47° 00' N 165° 00' W	1	0				Robert Walker, pers. comm.
28.08.1956	48° 00' N 155° 00' W	1	0				Robert Walker, pers. comm.
30.08.1956	49° 00' N 165° 00' W	1	0				Robert Walker, pers. comm.
02.09.1956	51° 00' N 150° 00' W	1	0				Robert Walker, pers. comm.
03.09.1956	52° 00' N 150° 00' W	1	0				Robert Walker, pers. comm.
12.09.1956	49° 00' N 165° 00' W	1	0				Robert Walker, pers. comm.
13.09.1956	52° 00' N 145° 00' W	1	0				Robert Walker, pers. comm.
15.09.1956	50° 00' N 145° 00' W	1	0				Robert Walker, pers. comm.
17.09.1956	49° 00' N 150° 00' W	1	0				Robert Walker, pers. comm.
19.07.1957	50° 00' N 160° 00' W	8	0				Robert Walker, pers. comm.
21.07.1957	50° 00' N 150° 00' W	17	0				Robert Walker, pers. comm.
08.08.1957	50° 00' N 175° 00' W	2	0				Robert Walker, pers. comm.
09.08.1957	50° 00' N 165° 00' W	1	0				Robert Walker, pers. comm.
09.08.1957	50° 00' N 175° 00' W	1	0				Robert Walker, pers. comm.
18.08.1957	53° 00' N 165° 00' W	1	0				Robert Walker, pers. comm.
20.08.1957	50° 00' N 165° 00' W	1	0				Robert Walker, pers. comm.
20.08.1957	50° 00' N 175° 00' W	1	0				Robert Walker, pers. comm.

Appendix 1. (Continued)

Date (DD.MM.YYY)	Coordinates	Number of fish	Depth, m	Total length, cm	Weight, kg	Temperature, °C	Source
20.08.1957	50° 00' N 175° 01' W	1	0				Robert Walker, pers. comm.
24.05.1958	50° 00' N 145° 00' W	1	0				Neave (1959)
23.06-25.06.1958	50° 00' N 155° 00' W	8	0				Neave (1959)
01.08.1958	50° 00' N 155° 00' W	1	0				Neave (1959)
07.08.1958	49° 00' N 175° 00' W	1	0				Robert Walker, pers. comm.
09.08.1958	49° 00' N 170° 00' W	4	0				Robert Walker, pers. comm.
17.08.1958	52° 03' N 157° 15' W	1	0			10.3	Robert Walker, pers. comm.
18.08.1958	55° 00' N 145° 00' W	2	0				Neave (1959)
19.08.1958	50° 00' N 165° 00' W	1	0				Robert Walker, pers. comm.
25.08.1958	52° 35' N 145° 03' W	4	0				Robert Walker, pers. comm.
??.??.1958-??.??.1966	50° 00' N 145° 00' W	33	0		< 0.9		LeBrasseur (1967)
05.07.1959	50° 00' N 135° 00' W	2	0	20-25			Newman (1959)
17.07.1959	48° 01' N 165° 05' W	10	0				Robert Walker, pers. comm.
19.07.1959	47° 01' N 160° 02' W	1	0				Robert Walker, pers. comm.
20.07.1959	48° 02' N 160° 00' W	4	0				Robert Walker, pers. comm.
21.07.1959	49° 02' N 160° 00' W	1	0				Robert Walker, pers. comm.
07.08.1959	49° 00' N 175° 00' W	5	0				Robert Walker, pers. comm.
09.08.1959	48° 00' N 175° 00' W	4	0				Robert Walker, pers. comm.
12.08.1959	49° 05' N 170° 10' W	2	0				Robert Walker, pers. comm.
01.08.1960	50° 20' N 178° 00' E	1	0				Robert Walker, pers. comm.
02.08.1961	53° 09' N 142° 29' W	1	0	35		13.2	Robert Walker, pers. comm.
22.08.1961	53° 30' N 155° 00' W	1	0				Robert Walker, pers. comm.
23.08.1961	51° 01' N 149° 56' W	1	0				Robert Walker, pers. comm.
25.08.1961	50° 00' N 147° 00' W	2	0				Robert Walker, pers. comm.
27.08.1961	52° 00' N 147° 00' W	7	0				Robert Walker, pers. comm.
21.09.1961	54° 00' N 141° 00' W	1	0				Robert Walker, pers. comm.

Date (DD.MM.YYY)	Coordinates	Number of fish	Depth, m	Total length, cm	Weight, kg	Temperature, °C	Source
08.08.1962	50° 19' N 178° 59' E	1	0				Robert Walker, pers. comm.
10.08.1962	50° 41' N 175° 54' E	1	0				Robert Walker, pers. comm.
25.08.1962	50° 00' N 153° 58' W	2	0				Robert Walker, pers. comm.
26 08.1962	50° 43' N 156° 52' W	1	0				Robert Walker, pers. comm.
12.09.1962	51° 00' N 160° 00' W	3	0				Robert Walker, pers. comm.
14.09.1962	50° 16' N 162° 57' W	3	0				Robert Walker, pers. comm.
29.11.1962	49° 03' N 135° 00' W	2	0				Robert Walker, pers. comm.
11.07.1963	48° 20' N 160° 35' W	1	0				Robert Walker, pers. comm.
25.07.1963	51° 00' N 162° 00' W	2	0				Robert Walker, pers. comm.
26.07.1963	50° 00' N 162° 00' W	4	0				Robert Walker, pers. comm.
28.07.1963	50° 30' N 162° 00' W	5	0				Robert Walker, pers. comm.
02.08.1963	50° 00' N 176° 26' W	4	0				Robert Walker, pers. comm.
03.08.1963	50° 28' N 176° 20' W	3	0				Robert Walker, pers. comm.
04.08.1963	51° 04' N 176° 20' W	1	0				Robert Walker, pers. comm.
13.08.1963	50° 34' N 176° 19' W	1	0				Robert Walker, pers. comm.
14.08.1963	50° 00' N 176° 19' W	3	0				Robert Walker, pers. comm.
15.08.1963	50° 47' N 176° 18' W	1	0				Robert Walker, pers. comm.
19.08.1963	51° 40' N 162° 00' W	2	0				Robert Walker, pers. comm.
20.08.1963	51° 00' N 162° 00' W	2	0				Robert Walker, pers. comm.
21.08.1963	50° 20' N 162° 00' W	1	0				Robert Walker, pers. comm.
25.08.1963	50° 40' N 162° 00' W	2	0				Robert Walker, pers. comm.
09.09.1963	51° 30' N 162° 00' W	1	0				Robert Walker, pers. comm.
04.12.1963	36° 56' N 122° 03' W	1	174	44.5	1.63		Phillips (1966)
20.04.1965	42° 50' N 147° 40' W	2	0	40-50		10.3	Phillips (1966)
18.07.1966	49° 00' N 167° 01' W	1	0				Robert Walker, pers. comm.
18.07.1966	49° 30' N 158° 00' W	1	0				Robert Walker, pers. comm.
26.07.1966	49° 57' N 167° 00' W	7	0				Robert Walker, pers. comm.

Appendix 1. (Continued)

Date (DD.MM.YYY)	Coordinates	Number of fish	Depth, m	Total length, cm	Weight, kg	Temperature, °C	Source
27.07.1966	49° 30' N 167° 01' W	1	0				Robert Walker, pers. comm.
29.07.1966	50° 15' N 176° 22' W	1	0				Robert Walker, pers. comm.
09.08.1966	50° 30' N 167° 00' W	1	0				Robert Walker, pers. comm.
09.08.1966	52° 10' N 158° 36' W	1	0				Robert Walker, pers. comm.
10.08.1966	49° 33' N 167° 02' W	2	0				Robert Walker, pers. comm.
10.08.1966	51° 10' N 158° 45' W	1	0				Robert Walker, pers. comm.
11.08.1966	49° 00' N 166° 59' W	1	0				Robert Walker, pers. comm.
12.08.1966	50° 01' N 167° 07' W	9	0				Robert Walker, pers. comm.
15.08.1966	51° 30' N 157° 50' W	3	0				Robert Walker, pers. comm.
19.08.1966	49° 30' N 176° 22' W	2	0				Robert Walker, pers. comm.
22.08.1966	52° 00' N 150° 00' W	2	0				Robert Walker, pers. comm.
24.08.1966	52° 57' N 158° 00' W	1	0				Robert Walker, pers. comm.
26.08.1966	51° 05' N 158° 00' W	1	0				Robert Walker, pers. comm.
01.09.1966	49° 30' N 167° 01' W	1	0				Robert Walker, pers. comm.
01.09.1966	51° 40' N 158° 00' W	1	0				Robert Walker, pers. comm.
02.09.1966	49° 02' N 167° 04' W	3	0				Robert Walker, pers. comm.
21.11.1966	50° 00' N 145° 00' W	3	0	25-30			Hewlett (1966)
09.02.1967	36° 48' N 122° 50' W	1	403	78			Phillips (1967)
11.07.1967	50° 30' N 176° 22' W	1	0				Robert Walker, pers. comm.
24.07.1967	49° 45' N 176° 22' W	4	0				Robert Walker, pers. comm.
04.08.1967	49° 30' N 176° 22' W	4	0				Robert Walker, pers. comm.
05.08.1967	50° 45' N 176° 22' W	1	0				Robert Walker, pers. comm.
13.08.1967	51° 18' N 176° 26' W	1	0	61		10.1	Robert Walker, pers. comm.
16.08.1967	51° 00' N 176° 22' W	1	0				Robert Walker, pers. comm.
18.08.1967	51° 15 N 176° 22' W	1	0				Robert Walker, pers. comm.
06.09.1969	50° 20' N 162° 30' W	1	0				Robert Walker, pers. comm.
07.09.1969	51° 00' N 160° 00' W	1	0				Robert Walker, pers. comm.

Date (DD.MM.YYY)	Coordinates	Number of fish	Depth, m	Total length, cm	Weight, kg	Temperature, °C	Source
08.09.1969	51° 30' N 160° 00' W	1	0				Robert Walker, pers. comm.
10.09.1969	52° 30' N 160° 00' W	1	0				Robert Walker, pers. comm.
16.07.1976	36° 48' N 122° 08' W	1	?				Anonymous (1993)
31.01.1981	37° 33' N 142° 13' E	1	650	145.2		3.3*	B.A. Sheiko, unpubl. data
19.07.1990	56° 29' N 152° 03' W	1	180				James Orr, pers. comm.
30.07.1992	52° 00' N 154° 00' W	1	0			9.8	Yoshiyuki Kajiwara, pers. comm.
17.04.1993	52° 36' N 172° 57' W	1	350				James Orr, pers. comm.
25.07.1993	51° 44' N 178° 25' W	1	292				James Orr, pers. comm.
06.07.1995	51° 30' N 145° 00' W	1	0			9.9	Yoshiyuki Kajiwara, pers. comm.
13.07.1995	47° 00' N 165° 00' E	1	0	50.0	2.2	8.4	Yukimasa Ishida, pers. comm.
30.07.1995	47° 30' N 175° 30' E	1	0			10.4	Yoshiyuki Kajiwara, pers. comm.
03.11.1995	52° 08' N 175° 39' W	1	270				James Orr, pers. comm.
11.11.1995	48° 04' N 154° 46' E	1	277	125.8	33		Tokranov and Diakov (1996)
27.11.1995	48° 09' N 154° 50' E	1	375	130	30	0.8-1.0*	Tokranov and Diakov (1996)
06.07.1996	51° 38' N 179° 18' W	1	270				James Orr, pers. comm.
??.08.1996	46° 00' N 125° 30' W	7	0	7			James Burke, pers. comm.
10.08.1996	51° 39' N 176° 57' E	1	300				James Orr, pers. comm.
25.05.1997	48° 32' N 154° 48' E	1	202	80	5.7	3.0*	This study
08.07.1997	52° 00' N 145° 00' W	2	0			12.2	Yoshiyuki Kajiwara, pers. comm.
18.07.1997	46° 08' N 169° 05' E	1	0			8.8	Yoshiyuki Kajiwara, pers. comm.
19.06.1998	48° 28' N 155° 04' E	1	388	120.1	30.3	3.8*	Moukhametov and Volodin (1999)

Appendix 1. (Continued)

Date (DD.MM.YYY)	Coordinates	Number of fish	Depth, m	Total length, cm	Weight, kg	Temperature, °C	Source
10.07.1998	49° 00' N 145° 00' W	1	0			11.0	Yoshiyuki Kajiwara, pers. comm.
16.07.1998	42° 02' N 165° 03' E	1	0	39.4		11.7	Yukimasa Ishida, pers. comm.
01.08.1998	48° 00' N 173° 16' E	1	0			9.8	Yoshiyuki Kajiwara, pers. comm.
02.08.1998	47° 30' N 175° 30' E	1	0			10.6	Yoshiyuki Kajiwara, pers. comm.
11.08.1998	48° 58' N 156° 36' E	1	0	50			Anatoly Dekshtein, unpubl. data
12.08.1998	47° 58' N 155° 36' E	1	0	50			Anatoly Dekshtein, unpubl. data
22.10.1998	46° 35' N 143° 54' E	1	515	83	4.5		Kim Sen Tok (2000)
27.10.1998	48° 05' N 154° 25' E	1	195	62	4.5	2.1-2.4*	This study
29.12.1998	49° 06' N 155° 27' E	1	300	88	12.3	1.6*	This study
25.05.1999	51° 55' N 171° 49' W	1	350				James Orr, pers. comm.
27.07.1999	45° 30' N 165° 00' W	1	0			8.4	Yoshiyuki Kajiwara, pers. comm.
28.07.1999	48° 04' N 154° 59' E	1	469	121	26.5	3.3-3.5*	This study
15.08.2000	51° 35' N 179° 24' E	1	335				James Orr, pers. comm.
22.09.2000	47° 57' N 154° 37' E	1	457	58	3.34	8.5	This study
05.11.2000	47° 55' N 154° 31' E	1	371	73	7	3.4-3.5*	This study
05.10.2001	43° 33' N 147° 14' E	1	520	74	7.4	-	Vladimir Touponogov, unpubl. data
06.10.2001	43° 35' N 147° 19' E	1	695	78	9.7	-	Vladimir Touponogov, unpubl. data
08.10.2001	43° 28' N 147° 00' E	13	425	68-91	123.5**	-	Vladimir Touponogov, unpubl. data
08.10.2001	43° 28' N 147° 01' E	15	585	63-87	100.5**	-	Vladimir Touponogov, unpubl. data

Date (DD.MM.YYY)	Coordinates	Number of fish	Depth, m	Total length, cm	Weight, kg	Temperature, °C	Source
08.10.2001	43° 32' N 147° 09' E	1	475	65	5.2	-	Vladimir Touponogov, unpubl. data
09.10.2001	43° 33' N 147° 13' E	2	575	85-88	20.5**	-	Vladimir Touponogov, unpubl. data
09.10.2001	43° 24' N 146° 47' E	1	550	60	4.9	-	Vladimir Touponogov, unpubl. data
10.10.2001	43° 23' N 146° 45' E	1	545	57	4.7	-	Vladimir Touponogov, unpubl. data
15.10.2001	43° 28' N 147° 00' E	10	475	62-87	67**	-	Vladimir Touponogov, unpubl. data
16.10.2001	43° 27' N 146° 57' E	1	475	74	6.7	-	Vladimir Touponogov, unpubl. data
21.10.2001	43° 23' N 146° 45' E	1	544	70	9.2	-	Vladimir Touponogov, unpubl. data
26.10.2001	43° 22' N 146° 43' E	6	480	83-89	66.7**	-	Vladimir Touponogov, unpubl. data
27.10.2001	43° 21 N 146° 41' E	2	630	81-96	23.2**	-	Vladimir Touponogov, unpubl. data
31.10.2001	43° 22' N 146° 43' E	10	525	88-96	100**	-	Vladimir Touponogov, unpubl. data

** - total weight of all specimens caught.

* - bottom temperature.

Appendix 2. Some morphometrics and meristics of skilfish *Erilepis zonifer* from various parts of the North Pacific Ocean (morphometrics are given in mm)

№	Characters	Information Source											
		1	2	3	4	5	6	7	8	9	10	11	12
1	Fork length (FL), (mm)	-		-	-	573	830	-	-	1201	1204	1250	1372
2	Total length (TL),(mm)	177	293.7	453	547	-	-	875	1120	-	1258	1300	1400
3	Standard length (SL), (mm)	142*	244*	382	466*	511	714	753*	980	1102	1073	1130	1215
4	Weight of body, (g)	-		1.63	-	3.34	4.5	-	-	30.3	33.0	30.0	-
5	Number of rays in I D	-	12	13	-	11	12	12	13	13	-	-	13
6	Number of rays in II D	-	16	18	-	18	16	16	19	17	-	-	-
7	Number of rays in P	-	18	18	-	17	16	-	19	17	-	-	17
8	Number of rays in V	-	6	6	-	6	-	-	-	7	-	-	-
9	Number of rays in A	-	13	14	-	13	12	15	15	12	-	-	13
10	Number of rays in C	-		-	-	13	15	-	-	-	-	-	-
11	Gill rays	-	7	-	-	-	7	-	7	-	-	-	-
12	Gill rakers on the 1st gill arch	-		22	-	20	21	-	-	-	-	-	-
13	incl.upper	-		6	-	6	-	-	-	-	-	-	-
14	incl.lower	-		16	-	14	-	-	-	-	-	-	-
15	Pores in lateral line	-	128134	133	-	-	-	122	126	-	-	-	122
16	Rows of scales over lateral line	-		-	-	-	-	-	30	-	-	-	-
17	Rows of scales under lateral	-		-	-	-	-	-	51	-	-	-	-
18	Head length	45.0	75.5	114.5	149.3	154	-	250	313.6	328	335	420	392

Appendix 2. (Continued)

№	Characters	Information Source											
		1	2	3	4	5	6	7	8	9	10	11	12
19	Head depth	-		-	-	125	-	276	294.0	-	270	280	382
20	Head width	-		-	-	-	-	-	-	-	198		-
21	Depth of forehead	-		-	-	-	-	-	-	-	130		-
22	Depth of caudal peduncle	15.0	24.5	36.0	43.2	40	-	76	83.3	92	_85	95	-
23	Length of caudal peduncle	25.0		-	69.5	87	-	-	-	-	100	-	-
24	Maximum body depth	50.0	81.2	120.5	140.0	145	-	-	-	337	328	340	-
25	Antepectorial distance	-	75.0	-	-	167	-	-	-	-	307	-	-
26	Anteventral distance	-	83.0	-	-	179	-	-	.	396	-	430	-
27	Anteanal distance	103.0	161.2	-	310.1	361	-	-	-	741	765	840	-
28	Antedorsal distance	57.0	90.5	-	177.8	191	-	-	-	435	440	480	-
29	Pectoventral distance	-		-	.	61	-	-	-	-	-	-	-
30	Ventroanal distance	-		-	-	177	-	-	-	-	-	-	-
31	Body thickness	-	43.0	71.0	-	102	-	-	-	-	-	-	-
32	Body girth	-		-	-	385	-	-	-	-	-	-	-
33	Eye diameter, horizontal	9.6	14.5	20.0	23.5	25	-	37	44.1	53	50	55	60
34	Eye diameter, vertical			15.5	-	22	-	-	-	42	40	40	-
35	Upper jaw length	15.6	28.7	46.1	56.9	63	-	89	127.4	155	134	-	-
36	Lower jaw length	21.2	35.5	-	68.9	79	-	-	156.8	-	185	-	131
37	Length of snout	12.6	18.2	37.0	49.8	48	-	83	107.8	115	113	150	118

№	Characters	Inform	№	Charact	Informa	№	Cha	Inform	№	Charact	Informati	№	Charact
		1	2	3	4	5	6	7	8	9	10	11	12
38	Interorbital space	15.0	25.7	48.5	59.1	64	-	89	117.6	114	123	-	-
39	Post eye length of the head	-		-	-	88	-	-	-	-	172	215	-
40	Length of base P	-	22.5	31.5	-	40	-	-	73.5	-	-	-	-
41	Length of P	38.0	60.0	88.5	112.1	111	-	167	176.4	201	200	190	245
42	Length of base V	-		-	-	19.5	-	-	-	-	-	-	-
43	Height of V	26.0	37.5	56.0	72.7	75	-	117	122.5	144	129	140	147
44	Exterior inter-base distance	-		-	-	43	-	-	-	-	-	-	-
45	Interior inter-base distance	-		-	-	13	-	-	-	-	-	-	-
46	Length of base A	-	40.5	63.2	-	80	-	-	156.8	146	150	160	-
47	Height A	-	34.5	51.0	-	78	-	111	117.6	125	121	-	-
48	Length of base DI	-	52.0	98.5	-	115	-	-	-	256	235	220	-
49	Height DI	-	31.2	42.0	-	47	-	87	83.3	104	96	-	163
50	Length of base DII	-	65.0	100.0	-	141	-	-	245.0	247	272	245	-
51	Height DII	-	36.0	49.5	-	68	-	104	107.8	108	139	-	-
52	Distance between I and II D	-		10.0	-	14	-	30	-	15	25	-	-
53	Length of C	-		-	-	62	-	156	-	-	185	170	-
54	Height C	-		-	-	165	-	-	-	-	-	-	314

1. South Kurils (Andriashev, 1955); 2. Monterrey Bay (Lockington, 1880); 3. Monterrey Bay (Phillips, 1966); 4. North Kurils (Andriashev, 1955); 5. North Kurils (this study); 6. South Sea of Okhotsk (Kim, 2000); 7. Misaki Prefecture, Japan (Jordan and Snyder, 1906); 8. Queen Charlotte Islands, B.C., Canada (Thompson, 1917); 9. North Kurils (Moukhametov and Volodin, 1999); 10, 11. North Kurils (Tokranov and Diakov, 1996); 12. Sagami Bay, Japan (Jordan and, Snyder, 1901).

Note*: - standard length is calculated from the fork length by the formula: SL=0.8764xTL-13.44 (correlation coefficient 0.999).

REFERENCES

Amaoka, K., Nakaya, K., and Yabe, M. (1995). *The fishes of northern Japan*. Sapporo: Kita Nihon Kaijo Center Co. Ltd.

Andriashev, A.P. (1939). On amphi-pacific (Japanese-Oregonian) distribution of marine fauna in the North Pacific Ocean. *Zoologichesky Zhurnal*. Vol. 2. P. 181-195 (In Russian).

Andriashev, A.P. (1955). New for fauna of the USSR fish – skilfish [*Erilepis zonifer* (Lock.) Pisces, Anoplopomatidae] from Kamchatka waters off the Pacific Ocean. *Voprosy Ikhtiologii*. Vol. 4. P. 3-9 (In Russian).

Anonymous. (1993). *Computerized catalog of the fish collection*. San Francisco, CA, USA: California Academy of Sciences.

Anonymous. (1999). The Oregon Coast Aquarium one year after Keiko. *World Wide Web Electronic Publication*. www.aquarium.org/press/newsarchive99.htm, 08 September, 1999.

Anonymous. 2002. Aburabozu. *World Wide Web Page*. http://travel.mimo. com/aquarium /2850muroran/area.html (In Japanese).

Brodeur, R., McKinnell, S., Nagasawa, K., Pearcy, W., Radchenko, V., and Takagi, S. (1999). Epipelagic nekton of the North Pacific Subarctic and transition zones. *Progress in Oceanography*. Vol. 43. P. 365-397.

Clemens, W.A., and Wilby G.V. (1949). Fishes of the Pacific coast of Canada. *Bulletin of the Fisheries Research Board of Canada*. No. 68. P. 1-368.

Clemens, W. A., and Wilby G.V. (1961). Fishes of the Pacific coast of Canada. *Bulletin of the Fisheries Research Board of Canada*. No. 68. P. 1-443.

Cowan, R. (2000). Passages of the deep will show visitors the Pacific Ocean. *Statesman Journal. World Wide Web Electronic Publication*. www.aquarium.org/thedeep/news/ Openingnews/StatesmanJournal_2.htm. 21 May, 2000.

Egerton, Sir P. de M. Grey. (1864). On some Ichtyolites from New South Wales, forwarded by the Rev. W.B. Clarke. *Journal of Geological Society*. Vol. 12. P. 1-5.

Eschmeyer, W.N., Herald E.S., and Hamman, H. (1983). *A field guide to Pacific coast fishes*. Boston: Houghton Mifflin Company.

Eschmeyer, W.N., Ferraris, C.J. Jr., Hoang, M.D., and Long, D.J. (1998). *Catalog of fishes*. Vol. 1. Species of fishes. San Francisco: California Academy of Sciences.

Fedorov, V.V. (1973). List of fishes of the Bering Sea. *Izvestiya TINRO*. Vol. 87. P. 42-71 (In Russian).

Fisheries Centre, UBC. (1996). *University of British Columbia Fish Collection*. Vancouver, Canada: Fisheries Centre, University of British Columbia.

Froese, R., and Pauly, D., eds. (2001). FishBase. *World Wide Web Electronic Publication*. www.fishbase.org. Version March 2001.

Gill, T. (1894). (*Myriolepis zonifer* Lockington). *Science*. No. 23(573). P. 54.

Hanselman, D.H., Lunsford, C.R., Fujioka, J.T. and Rodgveller, C.J. (2006). Alaska Sablefish Assessment for 2006. In: *Stock Assessment and Fishery Evaluation Report for the Groundfish Resources of the Gulf of Alaska* (pp. 341-428). Anchorage, AK, USA: North Pacific Fishery Management Council.

Hanselman, D.H., Lunsford, C.R., Fujioka, J.T. and Rodgveller, C.J. (2007). Alaska Sablefish Assessment for 2007. In: *Stock Assessment and Fishery Evaluation Report for the Groundfish Resources of the Bering Sea/Aleutian Islands Regions* (pp. 329-446). Anchorage, AK, USA: North Pacific Fishery Management Council.

Hare, S.R. and Mantua, N.J. (2000). Empirical evidence for North Pacific regime shifts in 1977 and 1989. *Progress in Oceanography.* Vol. 47. P. 103-145.

Hart, J.L. (1973). Pacific fishes of Canada. *Bulletin of the Fisheries Research Board of Canada No.* 180. P. 1-740.

Hewlett, G. (1966). New skilfish. *Aquarium Newsletter, Vancouver Aquarium.* Vol. 10. P. 4.

Ishida, Y., Azumaya, T., and Fukuwaka, M. (1999). Summer distribution of fishes and squids caught by surface gillnets in the western North Pacific Ocean. *Bulletin of Hokkaido National Fisheries Research Institute.* No. 63. P. 1-18.

Jordan, D.S. (1917). Changes in names of American fishes. *Copeia.* No. 49. P. 85-89.

Jordan, D.S. (1918). The "fat-priest fish" in California. *Copeia.* No. 54. P. 29-30.

Jordan, D.S., and Snyder, J.O. (1901). Descriptions of nine new species of fishes contained in museums of Japan. *Journal of College of Science, Imperial University of Tokyo.* Vol. 15. P. 301-311.

Jordan, D.S., and Evermann, B.W. (1905). *American food and game fishes.* New York: Doubleday, Page and Company.

Jordan, D.S., and Snyder, J.O. (1906). *The giant bass of Japan. Proceedings of the U.S. National Museum.* No. 30(1476). P. 841-845.

Kim Sen Tok. 2000. Capture of skilfish *Erilepis zonifer* (Anoplopomatidae) in waters off the southeastern Sakhalin. *Voprosy Ikhtiologii.* Vol. 40. P. 277 (In Russian).

Kimura, D.K., Shimada A.M., and Lowe, S.A. (1993). Estimating von Bertalanffy growth parameters of sablefish, *Anoplopoma fimbria*, and Pacific cod, *Gadus macrocephalus*, using tag-recapture data. *Fishery Bulletin.* Vol. 91. P. 271-280.

Kimura, D.K., Shimada, A.M., and Shaw, F.R. (1998). Stock structure and movement of tagged sablefish, *Anoplopoma fimbria*, in offshore northeast Pacific waters and the effects of El Niño- Southern Oscillation on migration and growth. *Fishery Bulletin.* Vol. 96. P. 462-481.

Kinumaki, T., Arai, K., Sugii, K., and Iseki, S. (1977). Nutritive value of fish containing a large amount of alkoxydiglyceride or wax ester in meat. *Bulletin of Tokai Regional Fisheries Research Laboratory.* Vol. 91. P. 73-91.

Kotlyar, A.N. 1984. *Dictionary of names of marine fishes in six languages.* Moscow: Russkii Yazyk (In Russian).

Kwon, Y-J., Hwang S-D., Kim Y-S., and Moon, D-Y. (2008). Recent stock status of fishes on Emperor seamounts in the Pacific Ocean. *Poster presented at the 17th PICES Annual Meeting.* Oct. 24 – Nov. 2, 2008. Dalian, People's Republic of China.

Larkins, H.A. (1964). Some epipelagic fishes of the North Pacific Ocean, Bering Sea, and Gulf of Alaska. *Transactions of American Fisheries Society.* Vol. 93. P. 286-290.

LeBrasseur, R.J. (1967). Line fishing at ocean station P, 50°00'N, 145°00'W. *Journal of Fisheries Research Board of Canada.* Vol. 24. P. 2201-2203.

Lindberg, G.U., Tran Kong Tam, Chzhu Shu-pin, Kim Min Su and Dashidorzhi, A., eds. (1964). *Illustrated dictionary of commercial fish names of the western Pacific Ocean in Latin, Russian, Chinese, Korean, Vietnamese, Mongolian, and Japanese languages.* Beijing: Commission on Fisheries Research in the Western Pacific Ocean.

Lindberg, G.U., and Krasyukova, Z.V. (1987). *Fishes of the Sea of Japan and neighboring parts of Okhotsk and Yellow Seas*. Part 5. Leningrad: Nauka (In Russian).

Lindberg, G.U., Gerdt, A.S., and Rass, T.S. (1980). *Dictionary of names of marine commercial fishes*. Leningrad: Nauka (In Russian).

Lisovenko, L.A. (1990). *Japanese-Russian and Russian-Japanese fisheries dictionary*. Moscow: Russkii Yazyk.

Lockington, W.N. (1880). Description of a new chiroid fish, *Myriolepis zonifer*, from Monterey Bay, California. *Proceedings of the U.S. National Museum*. Vol. 4. P. 248-251.

Lowe, S.A., Fritz, L.W., and Zenger, H. (2001). Atka mackerel. In: *Stock Assessment and Fishery Evaluation Report for the Groundfish Resources of the Bering Sea/Aleutian Islands Regions* (pp. 1-51). Anchorage, AK, U.S.A.: North Pacific Fisheries Management Council.

Masuda, H., Amaoka, K., Araga, C., Uveno, T.,amd Yoshino, T., eds. (1984). *The fishes of the Japanese Archipelago*, 3 Vols. Tokyo: Tokai University Press.

Masuda, H., and Allen, G.R. (1987). *Sea fishes of the World (Indo-Pacific Region)*. Tokyo: Yama-Kei Publishers Co., Ltd.

Matarese, A.C., Kendall, A.W., Jr., Blood, D.M., Vinter, B.M. (1989). *Laboratory guide to early life history stages of Northeast Pacific fishes*. U.S. Department of Commerce, NOAA Technical Report NMFS 80. P. 1-652.

McFarlane, G.A. and Beamish, R.J. (1983). Preliminary observations on the juvenile biology of sablefish (*Anoplopoma fimbria*) in waters off the west coast of Canada. In: B. Meltef (Ed.), Proceedings of International Sablefish Symposium: *Alaska Sea Grant Report 83-8* (pp. 119-136). Anchorage: University of Alaska Fairbanks.

Mecklenburg, C.W. (2003). Family Anoplopomatidae Jordan and Gilbert 1883 – sablefishes. *California Academy of Science Annotated Checklist of Fishes*. No. 2. P. 1-3.

Mecklenburg, C.W., Mecklenburg, T.A., and Thorsteinson, L.K. (2002). *Fishes of Alaska*. Bethesda, MD, USA: American Fisheries Society.

Miller, D.J., and Lea, R.N. (1972). Guide to the coastal marine fishes of California. *California Department Fish and Game Fishery Bulletin*. Vol. 157. P. 1-235.

Mitani, I., Kamei, M., and Shimizu, T. (1986). Some aspects of biology and fisheries of skilfish (*Erilepis zonifer* (Lockington)) off Japan. *Kamisuikenhou*. Vol. 7. P. 23-27 (In Japanese).

Monakhtina, S. (2010). Skilfish (*Erilepis zonifer*): traits of biology from a fishery near the Emperor Seamounts in the north-west Pacific Ocean. In: *Abstracts of PICES-2010 annual meeting*. October 22-31, 2010. Portland, OR, U.S.A. (p. 22.). Sidney, B.C., Canada: PICES Headquarters.

Moukhametov, I.N., and Volodin, A.V. (1999). Capture of two rare and one new to fauna of the northern Kuril Islands fish species. *Voprosy Ikhtiologii*. Vol. 39. P. 426-427 (In Russian).

Munk, K.M. (2001). Maximum ages of groundfishes in waters off Alaska and British Columbia and considerations of age determination. *Alaska Fisheries Research Bulletin*. Vol. 8. P. 12-21.

Neave, F. (1959). Records of fishes from waters off the British Columbia coast. *Journal of Fisheries Research Board of Canada*. Vol. 16. P. 383-384.

Newman, M.A. (1959). New mystery fish. *Aquarium Newsletter, Vancouver Aquarium*. Vol. 3. P. 2.

Newman, M.A. (1963). Passing of the "mystery fish". *Aquarium Newsletter, Vancouver Aquarium*. Vol. 7. P. 3.

Orlov, A., Moukhametov, I., and Volodin, A. (1998). New records of heat-loving fishes in the Pacific waters of the northern Kuril Islands and southeastern Kamchatka. In: *Actual Problems of Fish Taxonomy: Abstracts of International Conference*. Sankt-Peterburg, November 7-19, 1998 (pp. 45-47). Sankt-Peterburg: Zoological Institute (In Russian).

Pearcy, W.G., Fisher, J.P., Anma, G., and Meguro, T. (1996). Species associations of epipelagic nekton of the North Pacific Ocean, 1978-1993. *Fisheries Oceanography*. Vol. 5. P. 1-20.

Phillips, J.B. (1966). Skilfish, *Erilepis zonifer* (Lockington), in Californian and Pacific Northeast waters. *California Fish and Game*. Vol. 52. P. 151-156.

Phillips, J.B. (1967). A longfin sanddab, *Citharichthys xanthostigma*, and a skilfish, *Erilepis zonifer*, taken in Monterey Bay. *California Fish and Game*. Vol. 53. P. 297-298.

Reshetnikov, Yu.S., Kotlyar, A.N., Rass, T.S., and M.I. Shatunovsky. (1989). *Dictionary of animal names in five languages*. Fishes. Moscow: Russkii Yazyk.

Safran, P. (1990). Drifting seaweed and associated ichthyofauna: floating nursery in Tohoku waters. *La Mer*. Vol. 28. P. 225-239.

Safran, P., and Omori, M. (1990). Some ecological observations on fishes associated with drifting seaweed off Tohoku coast, Japan. *Marine Biology*. Vol. 105. P. 395-402.

Saruwatari, T., Hoshino, K., and Yanagimoto, T. (2005). A preliminary study on the bottom fishes of the Emperor seamount chain. *Poster presented at the 7th Indo-Pacific Fish Conference*. May 16-20, 2005. Howard International House, Taipei, Taiwan.

Sasaki, T. (1985). Studies on the sablefish resources of the North Pacific Ocean. *Bulletin of Far Seas Fisheries Research Laboratory*. Vol. 22. P. 1-107.

Sigler, M.F., S.A. Lowe, and Kastelle, C. 1997. Area and depth differences in the age-length relationship of sablefish *Anoplopoma fimbria* in the Gulf of Alaska. In: M. Saunders, and M. Wilkins (Eds.), *Proceedings of the International Symposium on the Biology and Management of Sablefish* (pp. 55-63). U.S. Department of Commerce, NOAA Technical Report NMFS 130.

Smith, H.M., and Pope, Th.E.B. (1907). List of fishes collected in Japan in 1903, with escriptions of new genera and species. *Proceedings of the U.S. National Museum*. Vol. 31. P. 459-499.

Tanaka, Sh. (1931). On the distribution of fishes in Japanese waters. *Journal of the Faculty of Science, Imperial University of Tokyo (Zoology)*. Vol. 3. P. 1-90.

Thompson, W.F. (1916). A second occurrence of *Erilepis* in American waters. *Copeia*. Vol. 30. P. 29-30.

Thompson, W.F. (1917). Further notes on *Erilepis*, the giant bass-like fish of the North Pacific. *Copeia*. Vol. 40. P. 9-13.

Tokranov, A.M. 2000. On occurrence of skilfish *Erilepis zonifer* (Lock.) in the Pacific waters off the northern Kuril Islands. In: R.S. Moiseev (ed.), *Materials of Regional Scientific Conference*. Petropavlovsk-Kamchatsky, April 11-12, 2000 (pp. 139-141). Petro pav lovsk-Kamchatsky: Izdatel'stvo Goskomkamchatecologii (In Russian).

Tokranov, A.M., and Diakov, Yu.P. (1996). New record of *Erilepis zonifer* (Anoplo pomatidae) in Russian waters. *Voprosy Ikhtiologii*. Vol. 36. P. 708-709 (In Russian).

Ueno, T. (1971). List of marine fishes from the waters of Hokkaido and its adjacent regions. *Scientific Reports of Hokkaido Fisheries Experimental Station.* Vol. 13. P. 61-102.

Vasilieva, E.D. (1999). *Nature of Russia: animal life. Fishes.* Moscow: AST Publishing, Ltd. (In Russian).

Vinnikov, A.V., and Terentiev, D.A. (2000). Data on drift net catches of blue shark *Prionace glauca* in the Pacific part of the northern Kuril Islands in the August 1998. In: *Abstracts of Interntional Pelagic Shark Workshop.* Asilomar Conference Center, Pacific Grove, CA, USA, 14-17 February 2000 (p. 28).

von Bertalanffy, L. (1938). A quantitative theory of organic growth. *Human Biology.* Vol. 10. P. 181-243.

Wada, S., Koizumi, C., Takiguchi, A., and Nonaka, J. (1978). Recognition of triolein in black cod lipid. *Bulletin of Japanese Society of Fisheries Science.* Vol. 44. P. 1167.

Wada, S., Koizumi, C., Takiguchi, A., and Nonaka, J. (1979a). Triglyceride composition of black cod lipid – I. Triglyceride composition based on the acyl carbon number and degree of unsaturation. *Bulletin of Japanese Society of Fisheries Science.* Vol. 45. P. 611-614.

Wada, S., Koizumi, C., Takiguchi, A., and J. Nonaka. (1979b). Triglyceride composition of black cod lipid – II. Possible fatty acid combinations in triglycerides and their contents. *Bulletin of Japanese Society of Fisheries Science.* Vol. 45. P. 615-622.

Waldron, K.D. (1981). Ichthyoplankton. In: *The eastern Bering Sea shelf: oceanography and resources.* Vol. 1 (pp. 471-493). Washington D.C.: U.S. Government Printing Office, U.S. Department of Commerce, NOAA, Office of Marine Pollution Assessment.

Walker, W.A. (1996). Summer feeding habits of Dall's porpoise, *Phocoenoides dalli*, in the southern Sea of Okhotsk. *Marine Mammal Science.* Vol. 12. P. 167-181.

Zhang, C.-I., and Megrey B.A. (2006). A revised Alverson and Carney model for estimating the instantaneous rate of natural mortality. *Transactions of American Fisheries Society.* Vol. 135. P. 620-633.

In: Deep-Sea
Editors: D. R. Bailey and S. E. Howard

ISBN: 978-1-62257-173-4

Chapter 3

EVOLUTIONARY CLASSIFICATION OF THE LATE CRETACEOUS (CONIACIAN-MAASTRICHTIAN) PLANKTIC FORAMINIFERA *ARCHAEOGLOBIGERINA* PESSAGNO, 1967 AND *GANDOLFIA* NEW GENUS/DIRECTIONAL LINEAGE

M. Dan Georgescu*
Department of Geosciences, University of Calgary,
Calgary, Alberta, Canada

ABSTRACT

The genus *Archaeoglobigerina* Pessagno, 1967 is reviewed and redefined in evolutionary classification as branched lineage, which evolved in late Coniacian and became extinct in the late Maastrichtian. A new directional lineage, *Gandolfia*, is defined for the double keeled planktics of late Campanian-Maastrichtian age with plano-convex test and ornamentation that concentrated in chamber central portion on the spiral side. This attempt to develop an evolutionary classification framework for a group of globotruncanids planktic foraminifera is equally based on the test gross architecture features as well as on high resolution observations on the wall ultrastructure, ornamentation, and pore size and distribution patterns.

Keywords: Planktic foraminifera, Late Cretaceous, taxonomy, evolutionary classification, new lineage

---

* E-mail: dgeorge@ucalgary.ca.

INTRODUCTION

Cretaceous planktic foraminiferal classification was based on the typological approach for more than one hundred years. According to this method the units of the lower rank are grouped on the basis of the degree of morphological resemblance into units of higher rank. The necessity to recognize natural monophyletic units is acknowledged in the typological classification (Robaszynski et al., 1984), although such units were rarely erected. Caron (1981) and Ion (2001) defined genera in which the component species can be arranged in a lineage in the case of *Falsotruncana* and *Globotruncanitella* respectively, but these are rather isolated cases. The major problem in defining monophyletic units using the typological approach is that there is a high probability to group together species that developed a certain character or group of characters independently in different lineages. This is due to the iterative evolution, a common pattern among the planktic foraminifera (Frerichs, 1971; Steineck and Fleisher, 1978).

Evolutionary classification is based on the species grouping function of a combination of morphological resemblances and differences; the former are the result of the common ancestry, and the latter are the effects of the evolution process. Georgescu (2009a) grouped species into *lineages*, which are units immediately above the species level and with significance in evolutionary classification; a direct ancestor-descendant relationship is necessary to group the species in one lineage. A step forward in developing an evolutionary classification was made by Georgescu (2010a), who recognized two lineage types: directional and branched, with the mention that additional kinds of lineages can be defined as discovered. The existence of two kinds of units at the lineage level represents a major difference when compared to the Linnaean classification, where there is only one kind of unit at each taxonomic level; the unit type uniqueness was never demonstrated and is accepted *a priori* in typological classification.

The development of the evolutionary classification framework in the case of Cretaceous planktic foraminifera is generated by the increasing number of studies that involve the test high resolution morphological features, such as wall ultrastructure, ornamentation and porosity characteristics through the extensive use of the scanning electron microscope (SEM). The evolutionary classification studies in the case of Cretaceous planktics focused on those the heterohelicid group (Georgescu, 2009b, 2010a, 2011a; Georgescu et al., 2009; Georgescu et al., 2011) and trochospiral taxa (Georgescu, 2009a; 2010b, 2010c).

This study represents the beginning of development of an evolutionary classification framework for the globotruncanids planktic foraminifers. Two lineages are considered: *Archaeoglobigerina* Pessagno, 1967 and a new one herein named *Gandolfia*.

MATERIAL STUDIED

Most of the material used in this study was collected from sites and holes drilled under the auspices of the Deep Sea Drilling Project (DSDP)/Ocean Drilling Program (ODP): DSDP Site 152 (Nicaragua Rise, Caribbean region), ODP Leg 174AX (New Jersey coastal plain, USA), DSDP Site 305 (Shatsky Rise, Central Pacific Ocean), DSDP Site 453 (Mid-Pacific Mountains, Central Pacific Ocean), ODP Hole 761B (Wombat Plateau, East Indian Ocean),

ODP Hole 762C (Exmouth Plateau, East Indian Ocean) and ODP Hole 1050C (Blake Plateau, western North Atlantic Ocean). Additional material was collected from the Maastrichtian Kemp Clay drilled by the wells Mullinax I and Mullinax II (Texas, USA). Very well-preserved specimens from the Ripley Formation and Prairie Bluff Chalk of Alabama (USA) were studied in the McGugan Collection (University of Calgary, Canada) (Figure 1).

The type specimens of two species were examined. The holotype and paratypes of *Archaeoglobigerina blowi* Pessagno, 1967 were examined in the Cushman Collection at the National Museum of Natural History, Smithsonian Institution (NMNH), Washington, D.C. The paralectotypes of *Globigerina cretacea* d'Orbigny, 1840 (original designation is followed) were studied in the Alcide d'Orbigny Collection at the Muséum National d'Histoire Naturelle, Paris. The holotype photographs of *Archaeoglobigerina subcircumnodifer* (Gandolfi, 1955), *Gandolfia subpennyi* (Gandolfi, 1955) and *G. wiedenmayeri* (Gandolfi, 1955) were examined on the Mesozoic Planktonic Foraminifera Taxonomic Dictionary (www. chronos.org).

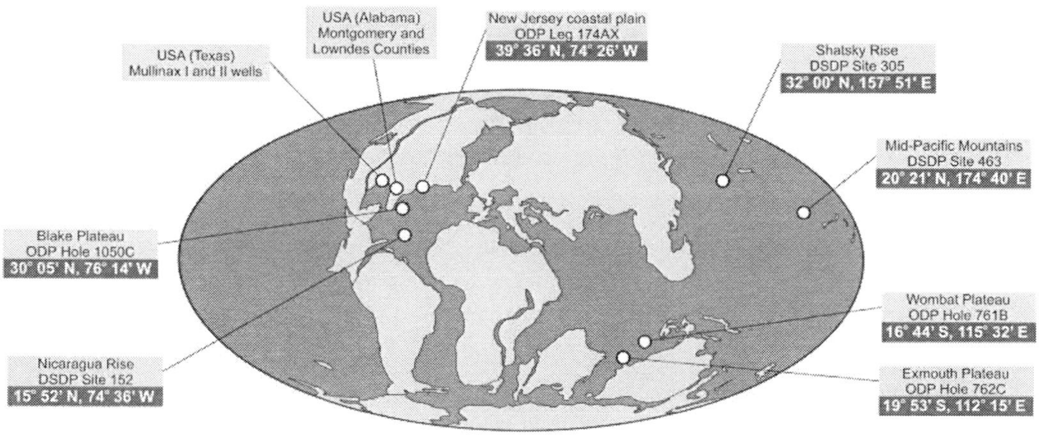

Figure 1. Paleogeographic reconstruction after Hay et al. (1999) showing the material provenance locations.

Systematic Paleontology

Evolutionary classification units are after Georgescu (2009a, 2010a, 2012). The concept of composite paleontological species concept (Georgescu and Huber 2007, 2009) is followed throughout.

Order Foraminiferida Eichwald, 1830

Genus/Branched Lineage *Archaeoglobigerina* Pessagno, 1967 – emended
Type species. –*Archaeoglobigerina blowi* Pessagno, 1967.

1967 *Archaeoglobigerina* Pessagno, p. 315.
1977 *Archaeoglobigerina* Pessagno; Neagu, p. 314.
1979 *Archaeoglobigerina* Pessagno; Robaszynski and Caron, p. 163, 167.
1984 *Archaeoglobigerina* Pessagno; Robaszynski et al., p. 275.
1985 *Archaeoglobigerina* Pessagno; Caron, p. 22.
1988 *Archaeoglobigerina* Pessagno; Loeblich and Tappan, p. 471.
2003 *Archaeoglobigerina* Pessagno; Korchagin, p. 36.

Species included. Initiating species (IS): *Archaeoglobigerina blowi* Pessagno, 1967, first descendant species (FDS): *A.*? *globotruncanoides* (Brotzen, 1936) and second descendant species (SDS): *A. subcircumnodifer* (Gandolfi, 1955).

Emended diagnosis. Late Coniacian-Maastrichtian branched lineage consisting of low trochospiral tests with pustulose ornamentation; peripheral structures consist of an imperforate band bordered by isolated pustules or incipient keels.

Emended description. Test is low trochospiral; chambers in *A. blowi* and *A. cretacea* are globular throughout, and petaloid on the spiral side and subrectangular on the umbilical one in the *A. subcircumnodifer*. Sutures depressed throughout in the *A. blowi* and *A. cretacea*, and raised towards the periphery in the *A. subcircumnodifer*. Test shape in convex-concave to biconvex. Peripheral structures consist of an imperforate band bordered by isolated pustules over the earlier chambers of the final whorl in the *A. blowi*, throughout the final whorl in the *A. cretacea* and an elevated peripheral band bordered by two equally developed keels in incipient stage of development in the *A. subcircumnodifer*. Aperture is a medium high arch, is situated in umbilical position and is bordered by an imperforate or perforate porticus in the *A. blowi*, an imperforate or perforate tegillum in the *A. cretacea* and an imperforate porticus in the *A. subcircumnodifer*. Chamber surface is ornamented with scattered pustules, which can fuse occasionally to form rugosities. Test wall is calcitic, hyaline, simple and perforate; pores are circular, with a diameter of 1.6-3.0 μm.

Remarks. *Archaeoglobigerina* is emended to accommodate a branched lineage, which includes three species ornamented with scattered pustules on both test sides. The main evolutionary trend in this lineage is the development of more complex peripheral structures, namely those lining the peripheral imperforate band. Two rows scattered pustules line the imperforate peripheral band in *A. blowi* and *A. cretacea*; two keels in incipient stage of development (with visible forming pustules) evolved in *A. subcircumnodifer*. *Archaeoglobigerina* differs from *Costellagerina* Petters, El-Nakhal and Cifelli, 1983 mainly by lacking the costellate ornamentation with meridional ornamentation.

Stratigraphical range. Upper Coniacian-Maastrichtian (from the upper part of the *Dicarinella concavata* Biozone to the *Abathomphalus mayaroensis* Biozone).

Geographic distribution. Cosmopolitan.

IS: *Archaeoglobigerina blowi* Pessagno, 1967
Figure 2: 1–9

1927 *Globigerina cretacea* d'Orbigny; Plummer, p. 43, pl. 6, figure 8.
1953 *Globigerina cretacea* d'Orbigny; Shaw, pl. 2, figures 10-12.
1956 ? *Globotruncana pura* Hofker, p. 327, figure 16.
1960 *Globigerina cretacea* d'Orbigny; Takayanagi, p. 137, pl. 10, figure 8.

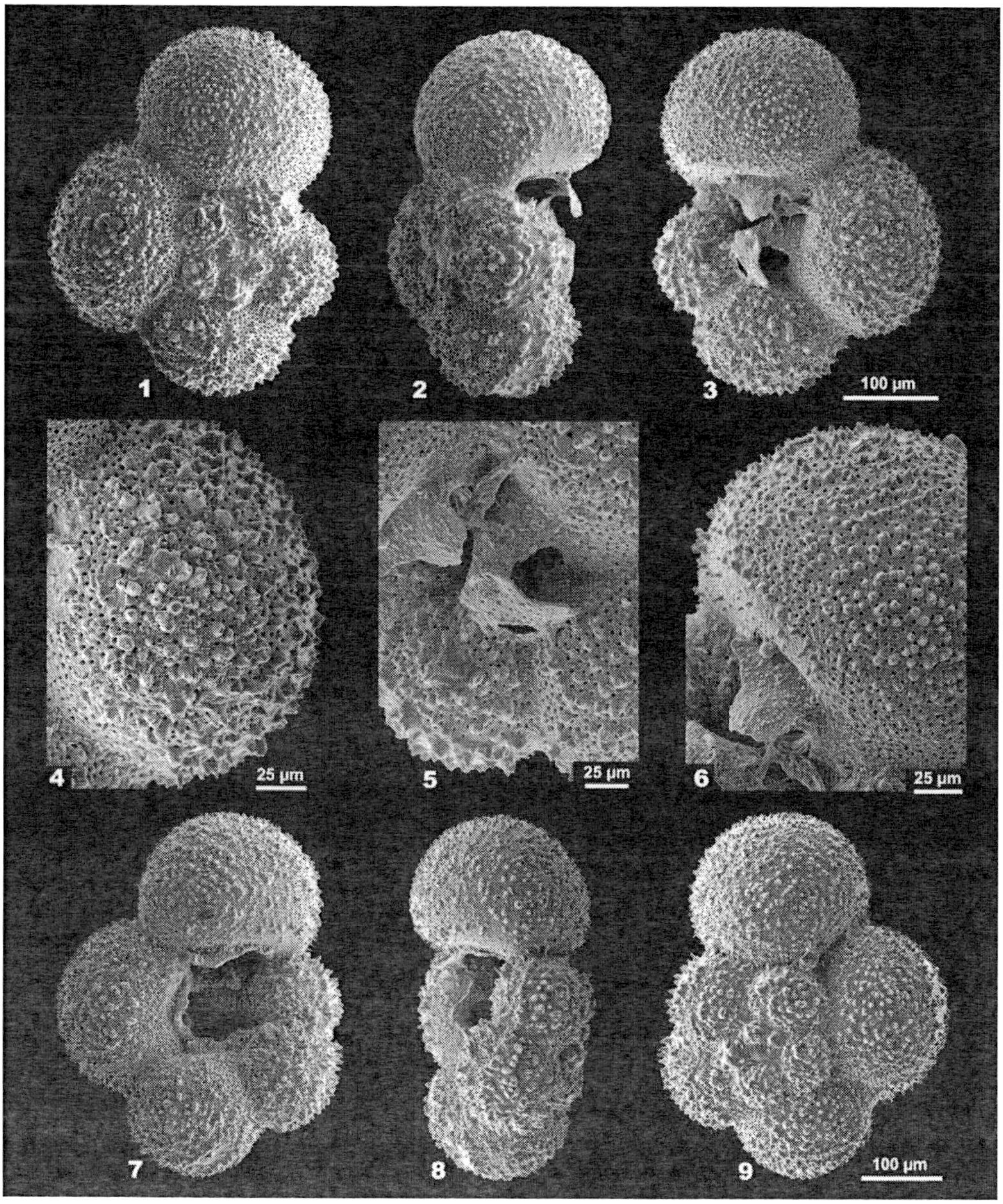

Figure 2. Hypotypes of *Archaeoglobigerina blowi* from the upper Santonian (*Globotruncana arca* Biozone) Merchantville Formation of the New Jersey Coastal Plain, ODP Leg 174AX at Bass River Site. 1-6 Specimen from sample 505.35-505.38 mbsf. 4 Detail micrograph showing the pustulose ornamentation and the sharp separation between the test wall and ornamentation element. 5 Detail micrograph showing the umbilical system composed of imperforate portici. 6 Detail micrograph showing the periumbilical shoulder developed on the umbilical side of the last formed chamber. 7-9 Specimen from sample 505.63-505.66 mbsf. Planktic foraminiferal zonation after Georgescu (2006).

1967 *Archaeoglobigerina blowi* Pessagno, p. 316, pl. 59, figures 1-10, pl. 94, figures 2-3.

1968 *Globotruncana cretacea* (d'Orbigny); Sliter, p. 102, pl. 16, figure 1.

1971 *Rugoglobigerina* (*Archaeoglobigerina*) *blowi* Pessagno; El-Naggar, pl. 4, figures a-c.

1979 *Archaeoglobigerina blowi* Pessagno; Robaszynski and Caron, p. 169, 172, pl. 79, figures 1-2.
1980 *Archaeoglobigerina blowi* Pessagno; Peryt, p. 81, pl. 21, figures 1-3.
1981 *Archaeoglobigerina blowi* Pessagno; Frerichs and Dring, p. 68, pl. 3, figures 10-12.
1982 ? *Archaeoglobigerina* cf. *blowi* Pessagno; Bergstresser and Frerichs, p. 360, pl. 2, figures 1-3.
1984 *Archaeoglobigerina blowi* Pessagno; Robaszynski et al., p. 276, pl. 47, figure 1.
1984 *Archaeoglobigerina blowi* Pessagno; Weidich, p. 105, pl. 21, figures 8-10.
1986 non *Archaeoglobigerina blowi* Pessagno; Abdel-Kireem, p. 35, pl. 2, figures 4-7.
1987 *Archaeoglobigerina blowi* Pessagno; Neagu, p. 304, pl. 12, figures 20-32.
1992 *Archaeoglobigerina blowi* Pessagno; Kaiho, p. 321, figure 3: 4.
1992 *Archaeoglobigerina blowi* Pessagno; Norris, pl. 3, figure e.
1992 *Archaeoglobigerina blowi* Pessagno; Olsson and Usmani, p. 313, figure 7: 3.
1996 *Archaeoglobigerina blowi* Pessagno; Georgescu, p. 324, pl. 10, figure 2.
1998 non *Archaeoglobigerina blowi* Pessagno; Zapeda, p. 133, figure 9: 1-2.
2003 *Archaeoglobigerina blowi* Pessagno; Nishi et al., figure 11: 13.
2005 *Archaeoglobigerina blowi* Pessagno; Georgescu, figure 2: 1-3.
2006 *Archaeoglobigerina blowi* Pessagno; Georgescu, figure 8: 8-10.
2006 *Archaeoglobigerina bosquensis* Pessagno; Georgescu, figure 8: 11-12.

Material. Circa 200 Specimens

Diagnosis. *Archaeoglobigerina* with peripheral structures absent on the last-formed one to three chambers.

Description. Test is low trochospiral, consisting of 11-14 chambers arranged in 2 ½-3 whorls; there are 4-5 (commonly 4 ½) chambers in the final whorl. Chambers are subglobular or globular, and with moderate to high size increase. Sutures are distinct, depressed, straight and radial on both test sides. Test shape is convex-concave, slightly asymmetrical. Periphery is broadly rounded. One imperforate peripheral band occurs on the first one to four chambers of the final whorl; the imperforate band is bordered by two lines of isolated pustules, which can occasionally fuse to form rugosities without an apparent ornamentation. Umbilicus is deep, with a diameter of circa one fourth of the maximum test diameter; umbilical shoulder occurs on the last-formed one to three chambers. Aperture has the shape of a medium high arch, is in umbilical position, and is protected by one narrow imperforate porticus; perforate portici occasionally occur. Chamber surface is ornamented with scattered pustules, which can be dome-like or conical in shape, and with a diameter at the base of 2.9-10.4 μm. Test wall is calcitic, hyaline, simple and perforate; pores are circular in shape, with a diameter of 1.7-2.5 μm.

Remarks. *Archaeoglobigerina blowi* differs from *Costellagerina pilula* (Belford, 1960) mainly by the pustulose rather than costellate ornamentation, and having peripheral structures (imperforate band lined by two rows of scattered pustules) on the earlier chambers rather than simple periphery.

Stratigraphical range. Upper Coniacian-Campanian (from the upper part of the *D. concavata* Biozone to the *Radotruncana calcarata* Biozone).

Geographic distribution. Cosmopolitan.

FDS: *Archaeoglobigerina* ? *globigerinoides* (Brotzen, 1936)
Figure 3:1–8

1931 *Globigerina cretacea* dOrbigny; Cushman, p. 44, pl. 6, figure 6.
1936 *Globotruncana globigerinoides*; Brotzen, p. 138, pl. 12, figure 3.
1946 *Rosalinella globigerinoides* Brotzen; Schijfsma, p. 96, pl. 7, figure 9.
1953 non *Globotruncana globigerinoides* Brotzen; Hagn, p. 94, pl. 8, figure 9.
1954 *Globotruncana globigerinoides* Brotzen; Nakkady and Osman, p. 81, pl. 19, figure 8.
1955 *Globotruncana globigerinoides* Brotzen; Liszka, p. 185, pl. 12, figure 13.
1956 *Globotruncana saratogaensis* (Applin). Brönnimann and Brown, p. 544, pl. 21, figures 1-3.
1957 *Globotruncana* (*Globotruncana*) *globigerinoides* Brotzen; Edgell, p. 112, pl. 2, figures 13-15.
1957 *Globigerina cretacea* dOrbigny; McGugan, p. 343, pl. 34, figure 11.
1957 *Globotruncana globigerinoides* Brotzen; Ziegler, p. 77, pl. 1, figure 10.
1959 *Globotruncana globigerinoides* Brotzen; Eternod Olvera, p. 110, pl. 5, figure 12, pl. 6, figures 1-2.
1961 *Globotruncana cretacea* (dOrbigny); Barr, p. 567, pl. 69, Figure 9, pl. 72, figure 6.
1961 *Globotruncana* (*Marginotruncana*) *globigerinoides* Brotzen; Hofker, figures 65-66.
1962 *Globotruncana globigerinoides* Brotzen; Herm, p. 80, pl. 5, figure 6.
1963 *Globotruncana globigerinoides* Brotzen; Majzon, pl. 6, figure 9.
1963 *Globotruncana* (*Globotruncana*) *cretacea* (dOrbigny); van Hinte, p. 85, pl. 6, figure 3.
1964 *Globotruncana globigerinoide*s Brotzen; Küpper, p. 622, pl. 3, figure 7.
1964 *Globotruncana mariai* Gandolfi. Martin, p. 82, pl. 9, figure 7.
1965 *Globotruncana cretacea* (dOrbigny); van Hinte, p. 21, pl. 3, figure 1.
1966 *Globotruncana globigerinoide*s Brotzen; Banerji, p. 61, pl. 5, figures 3-4, 7.
1969 *Globotruncana* (?) *cretacea* (dOrbigny); Moorkens, p. 451, pl. 3, figure 8.
1969 ? *Globotruncana cretacea* (dOrbigny); Douglas and Rankin, p. 200, figure 12.
1970 ? *Archaeoglobigerina cretacea* (dOrbigny); Porthault in Donze et al., p. 36, pl. 9, figure 17.
1971 *Rugoglobigerina* (*Archaeoglobigerina*) *cretacea* (dOrbigny); El-Naggar, pl. 4, figures d-l.
1973 *Globotruncana cretacea* (dOrbigny); Sliter, pl. 2, figures 14-16.
1974 *Globotruncana cretacea* (dOrbigny); Herb, pl. 6, figure 16.
1975 *Archaeoglobigerina cretacea* (dOrbigny); Frerichs et al., p. 309, pl. 3, figures 13-15.
1976 *Archaeoglobigerina cretacea* (dOrbigny); Evetts, pl. 8, figures 1-4.
1977 *Globotruncana cretacea* (dOrbigny); Masters, p. 551, pl. 41, figures 3-4, pl. 42, Figure 1.
1979 *Archaeoglobigerina cretacea* (dOrbigny); Robaszynski and Caron, p. 173, 176, pl. 80, figure 1.
1980 *Archaeoglobigerina cretacea* (dOrbigny); Peryt, p. 82, pl. 21, figures 5-7.
1981 *Globotruncana cretacea* (dOrbigny); Butt, pl. 17, figures H1-H2.

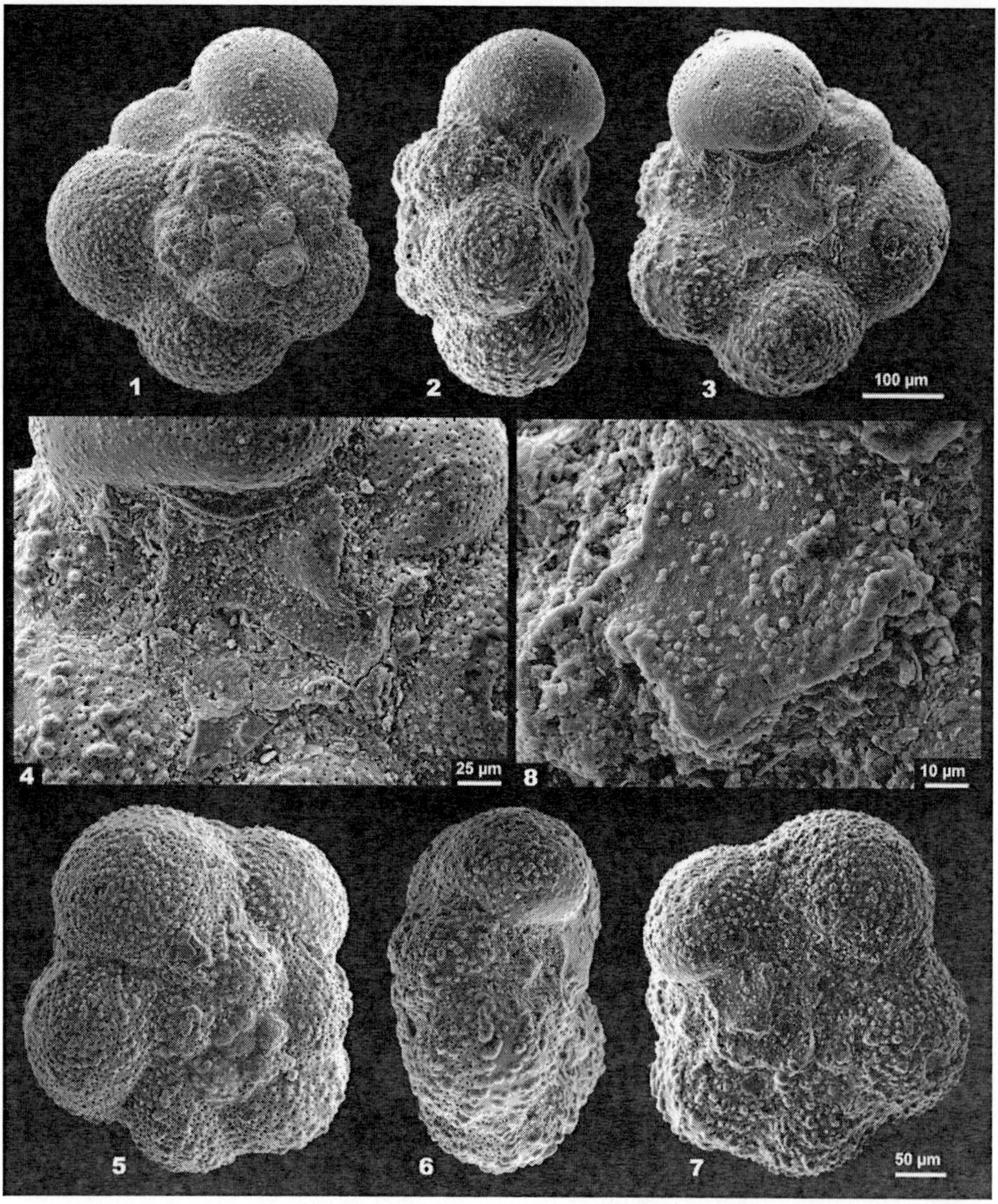

Figure 3. Hypotypes of *Archaeoglobigerina globigerinoides* from the upper Santonian (*Globotruncana arca* Biozone) Merchantville Formation of the New Jersey Coastal Plain, ODP Leg 174AX at Bass River Site. 1-4 Specimen from sample 505.35-505.38 mbsf. 4 Detail micrograph showing the wide tegillum-like perforate periapertural structure. 5-8 Specimen from sample 505.35-505.38 mbsf. 8 Detail micrograph showing the wide tegillum-like imperforate periapertural structure.

1981 *Archaeoglobigerina cretacea* (dOrbigny); Frerichs and Dring, p. 68, pl. 3, figures 13-15.

1981 *Archaeoglobigerina cretacea* (dOrbigny); McNeil and Caldwell, p. 261, p. 21, figure 7.

1982 ? *Archaeoglobigerina cretacea* (dOrbigny); Bergstresser and Frerichs, p. 360, pl. 2, figures 4-6.
1983 *Globotruncana cretacea* (dOrbigny); Beford, p. 13, pl. 2, figures 1-8.
1983 *Archaeoglobigerina cretacea* (dOrbigny); Ion, p. 126, pl. 49, figure 4.
1984 *Archaeoglobigerina cretacea* (dOrbigny); McNeil, figure 16: 3.
1984 *Archaeoglobigerina cretacea* (dOrbigny); Robaszynski et al., p. 278, pl. 47, figures 3-6.
1984 *Archaeoglobigerina cretacea* ((dOrbigny); Weidich, p. 106, pl. 21, figures 5-7.
1989 *Archaeoglobigerina cretacea* (dOrbigny); Hart et al., p. 322, figure 7.4: 4-5.
1992 *Archaeoglobigerina cretacea* (dOrbigny); Gawor Biedowa, p. 90, pl. 16, figures 1-3.
1992 *Archaeoglobigerina cretacea* ((dOrbigny); Kaiho, p. 321, figure 3: 6-8.
1992 *Archaeoglobigerina cretacea* (dOrbigny); Norris, pl. 3, figures f-g.
1996 *Archaeoglobigerina cretacea* (dOrbigny); Georgescu, p. 324, pl. 10, figure 1.
1998 non *Archaeoglobigerina cretacea* (dOrbigny); Robaszynski et al., pl. 15, figure 1.
2001 *Archaeoglobigerina cretacea* (d'ORBIGNY). Ion and Odin, figures 172-173.
2004 *Archaeoglobigerina cretacea* (dOrbigny); Campbell et al., figures 9: G-L.
2005 *Archaeoglobigerina cretacea* (dOrbigny); Georgescu, figures 2: 4-6.
2006 *Archaeoglobigerina cretacea* (dOrbigny); Georgescu, figures 8: 14-16.
2010 *Archaeoglobigerina cretacea* (dOrbigny); Walaszczyk et al., figure 14: 6.

Material. Circa 150 Specimens

Diagnosis. *Archaeoglobigerina* with peripheral structures absent on the last-formed one or two chambers.

Description. Test is low to medium high trochospiral consisting of 14-17 chambers arranged in 3-3 ½ whorls; there are 5 ½-6 (commonly 6) chambers in the final whorl. Chambers are subglobular or globular, and with moderate to low size increase. Sutures are distinct, depressed, straight and radial on both test sides. Test shape is convex-concave, distinctly asymmetrical. Periphery is broadly rounded. An imperforate peripheral band is developed on all the chambers of the final whorl; it is bordered on each side by a row of isolated pustules, which can be occasionally fused forming rugosities that lack an apparent orientation. Umbilicus is wide and shallow, with a diameter of approximately one third of the maximum test diameter. Aperture has the shape of a medium high arch, is in umbilical position, and is protected by a wide tegillum that can be either perforate or imperforate. Chamber surface is ornamented with dome-like or irregular pustules (4.0-9.0 μm in diameter or maximum dimension), which can fuse to form rugosities (10.3-15.8 μm in length) or irregular clusters (13.0-14.7 μm in maximum dimensions). Test wall is calcitic, hyaline, simple and perforate; pores are circular, with a diameter of 1.7-2.8 μm).

Remarks. *Archaeoglobigerina* ? *globigerinoides* differs from *A. blowi* by having more chambers in the final whorl (5 ½-6 rather than 4-5), imperforate peripheral band developed on all the chambers of the final whorl rather than only on the earlier chambers, and aperture protected by one wide tegillum-like plate rather than one narrow imperforate porticus. Following the taxonomical revision by Banner and Blow (1960), this species was frequently confused for *Globigerina cretacea*, a species described from the Campanian sediments of St. Germain by d'Orbigny (1840). Type material of this species deposited in the Alcide d'Orbigny Collection in the Muséum National d'Histoire Naturelle, Paris is re-evaluated (Figure 4). The lectotype selected in 1959 and illustrated by Banner and Blow (1960) is missing (Figure 4-A2). The remaining four specimens (Figure 4, B1-B4) are mounted on glue, and therefore they cannot be reoriented to study the morphological features in edge view. None of these specimens present in-relief peripheral structures, an observation consistent with the description by d'Orbigny (1840, p. 34). The selection of a lectotype with peripheral structures by Banner and Blow (1960) changed completely the meaning of d'Orbigny's species, and led to a forced synonymy between *G. cretacea* and *A. globigerinoides*; therefore it cannot be accepted.

Stratigraphical range. Santonian-Campanian (from the uppermost part of the *D. concavata* biozone to the lower part of the *Gansserina gansseri* Biozone).

Geographic distribution. Cosmopolitan.

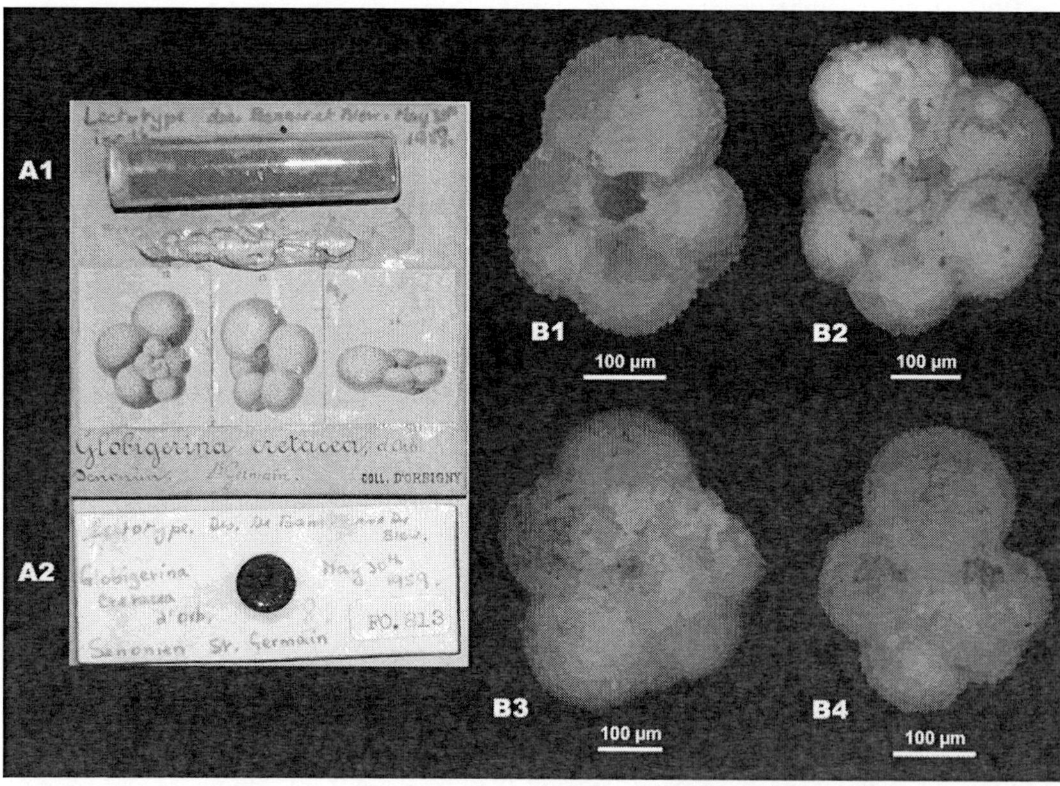

Figure 4. Type specimens of *Globigerina cretacea* d'Orbigny, 1840 (original designation followed) from the Alcide d'Orbigny at the Muséum National d'Histoire Naturelle, Paris. A1 Type specimens protected in the glass bicker. A2 The slide where the lectotype selected by Banner and Blow (1960) was located. B1-B4 The four paralectotypes that still exist after the lectotype loss.

SDS: *Archaeoglobigerina subcircumnodifer* (Gandolfi, 1955)
Figure 5: 1–9

1955 *Globotruncana* (*Rugoglobigerina*) *circumnodifer subcircumnodifer*; Gandolfi; p. 44, pl. 2, figure 8.

1958 *Rugotruncana* sp.; Seiglie, pl. 6, figure 6.
1962 *Globotruncana (Rugotruncana) subcircumnodifer* (Gandolfi); Berggren, p. 67, pl. 10, figure 4.
1964 *Globotruncana subcircumnodifer* Gandolfi; Olsson, p. 170, pl. 6, figures 1-2, 4-6.
1965 *Globotruncana subcircumnodifer* (Gandolfi); Takayanagi, p. 224, pl. 28, figures 1-4.
1967 *Rugotruncana subcircumnodifer* (Gandolfi); Pessagno, p. 369, pl. 62, figures 14-16, pl. 74, Figures 1-3.
1968 *Globotruncana subcircumnodifer* Gandolfi; Ansary and Tewfik, p. 57, pl. 8, figure 8.
1968 *Globotruncana subcircumnodifer* Gandolfi; Barr, p. 318, pl. 38, figures 5-6, pl. 40, figure 4.
1968 *Globotruncana subcircumnodifer* (Gandolfi); Rasheed and Govindan, p. 83, text-figures 5-7.
1968 *Globotruncana subcircumnodifer* (Gandolfi); Rao et al., p. 28, pl. 3, figures 7-9.
1972 *Globotruncana subcircumnodifer* Gandolfi; Barr, p. 25, pl. 10, figure 2.
1972 *Globotruncana subcircumnodifer* Gandolfi; Hanzliková, p. 111, pl. 30, figures 8-10.
1972 *Globotruncana subcircumnodifer* Gandolfi; Govindan, p. 178, pl. 5, figures 16-19.
1975 *Rugotruncana subcircumnodifer* (Gandolfi); Frerichs et al., p. 310, pl. 4, figures 10-12.
1977 *Globotruncana subcircumnodifer* Gandolfi; Masters, p. 609.
1979 non *Rugotruncana subcircumnodifer* (Gandolfi); Abdel-Kireem and Abdou, p. 193, pl. 2, figures 4-5, 7, pl. 8, figures 3-6.
1980 *Globotruncana pozaryskae*; Peryt, p. 79, pl. 20, figures 10-12.
1980 *Archaeoglobigerina blowi* Pessagno; Peryt, p. 81, pl. 20, figures 1-3.
1981 *Rugotruncana subcircumnodifer* (Gandolfi); Frerichs and Dring, p. 69, pl. 4, figures 7-9.
1983 *Globotruncana subcircumnodifer* (Gandolfi); Salaj, pl. 4, figures 5-6.
1983 *Rugotruncana subcircumnodifer* (Gandolfi); Solakius, p. 249, pl. 1, figures 5-7, pl. 2, Figures 1-3.
1984 *Archaeoglobigerina blowi* Pessagno; Robaszynski et al., pl. 47, figure 2.
1985 *Rugotruncana subcircumnodifer* (Gandolfi); Caron, p. 76, figure 34:11-12.
1990 *Globotruncana subcircumnodifer* (Gandolfi); Huber, p. 506, pl. 5, figures 7-8, 12.
1992 ? *Archaeoglobigerina cretacea* **(d'Orbigny); Norris, pl. 3,** figures h-j.
1996 *Globotruncana gagnebini* Brotzen; Ayyad et al., figure 14: b-c.
1996 *Rugotruncana subcircumnodifer* (Gandolfi); Mancini et al., figure 4: 6-7.
1996 *Rugotruncana subcircumnodifer* (Gandolfi); Georgescu, p. 322, pl. 3, figure 2.
1998 *Rugotruncana subcircumnodifer* (Gandolfi); Zapeda, p. 136, figure 9: 5.
2003 *Rugotruncana subcircumnodifer* (Gandolfi); Howe et al., pl. 8, figures 28-30.
2005 *Rugotruncana subcircumnodifer* (Gandolfi); Georgescu, figure 3: 7-9.
2006 *Rugotruncana subcircumnodifer* (Gandolfi); Georgescu, figure 7: 9-12.
2007 *Globotruncana (Rugoglobigerina) circumnodifer subcircumnodifer* Gandolfi; Georgescu and Huber, pl. 2, figure 8.

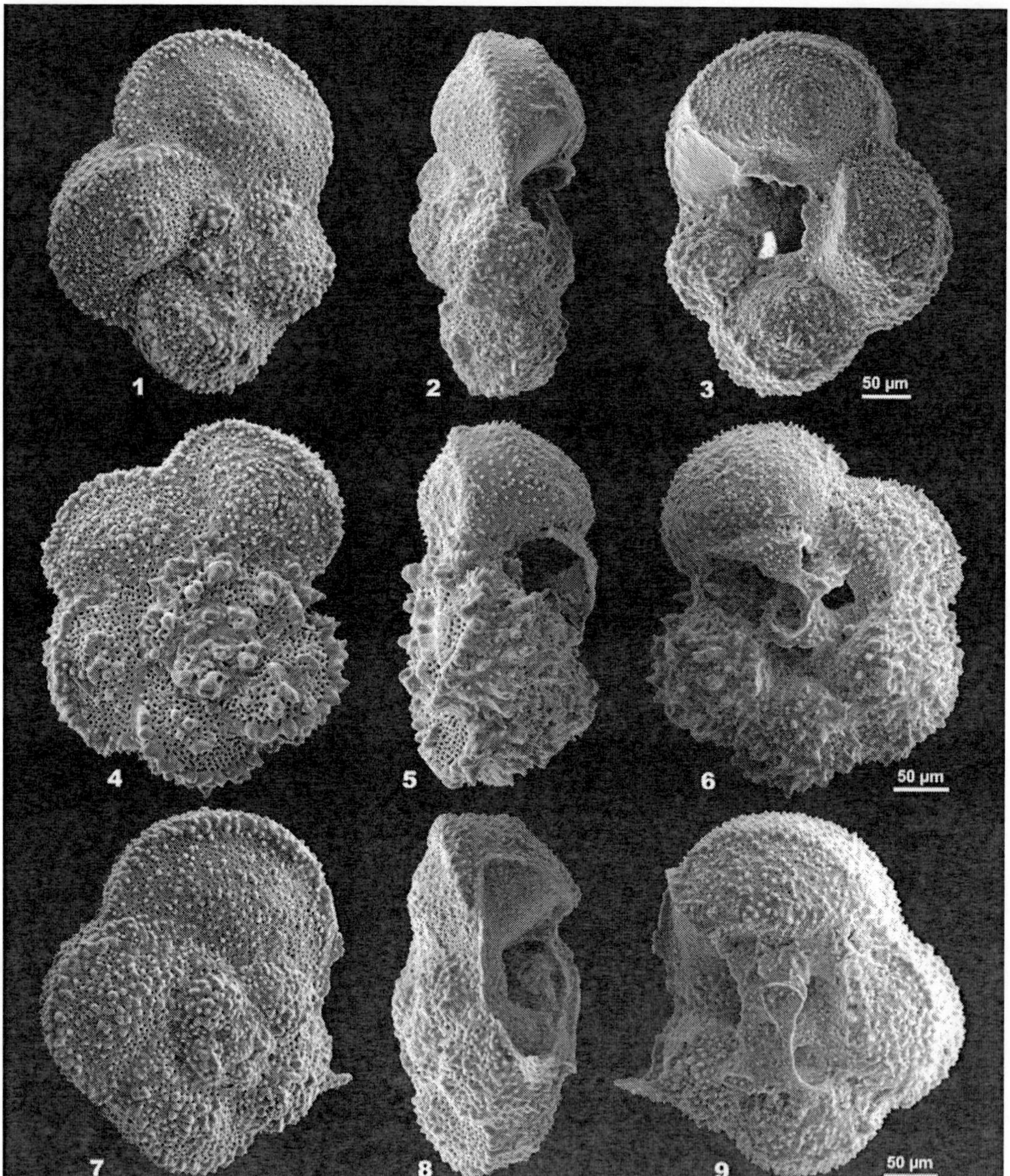

Figure 5. Hypotypes of *Archaeoglobigerina subcircumnodifer* from the uppermost Campanian (*G. gansseri* Biozone) of Central and Eastern Alabama. 1-3 Specimen from the upper part of the Demopolis Chalk, Stop 7/sample 2 (Dublin, Montgomery County). 4-6 Specimen from the upper part of the Ripley Formation, Stop 5/sample 4 (Braggs, Lowndes County). 7-9 Specimen from the upper part of the Ripley Formation, Stop 5/sample 4 (Braggs, Lowndes County).

Material. 80 Specimens

Diagnosis. *Archaeoglobigerina* with raised imperforate band bordered by two keels in incipient state of development.

Description. Test is low to very low trochospiral, consisting of 11-13 chambers arranged in 2 ½-3 whorls; there are 4-5 ½ (commonly 4-4 ½) chambers in the final whorl. Earlier chambers are globular to subglobular, those of the last whorl petaloid in shape on the spiral side and subrectangular on the umbilical one; chambers increase moderately in size. Sutures are curved, raised between the earlier chambers and depressed between the last-formed ones on the spiral side; sutures on the umbilical side are distinct, depressed, straight and radial. Test shape is biconvex, symmetrical or slightly asymmetrical, with the umbilical side slightly more inflated than the spiral one. Periphery is truncated. One elevated imperforate peripheral band is developed on all the chambers; it is bordered by two distinct, equally developed keels. The two keels consist of small-sized and dome-like pustules, which are often fused over the earlier chambers. Umbilicus is deep and narrow, with a diameter of circa one fourth to, more rarely, one third of the maximum test diameter; periumbilical shoulders are developed on all the chambers. Aperture is in the shape of a medium high arch in umbilical position, and is protected by a narrow and imperforate porticus, which is attached to the test wall across the umbilicus or of the previous portici. Chamber surface is ornamented with scattered dome-like pustules, with a diameter of 2.8-8.6 μm, which are concentrated occasionally in the chamber central portion on the spiral side; fused pustules occur over the earlier chambers and form rugosities (maximum 17.7 μm in length), which lack apparent orientation. Test wall is calcitic, hyaline, simple and perforate; pores are circular in shape and with a diameter of 1.6-3.0 μm.

Remarks. *Archaeoglobigerina subcircumnodifer* differs from *A. blowi* mainly by having one elevated imperforate peripheral bordered by two incipient keels on all the chambers, rather than one imperforate peripheral band bordered by isolated dome-like pustules on earlier chambers. The similarities in the general test appearance, wall ultrastructure, ornamentation and porosity indicate that *A. subcircumnodifer* evolved from *A. blowi*. It differs from *A.* ? *globigerinoides* mainly by the stronger peripheral structures and narrower porticus protecting the aperture.

Stratigraphical range. Upper Campanian-Maastrichtian (from the *R. calcarata* to the *A. mayaroensis* Biozone).

Geographic distribution. Cosmopolitan.

Genus/Directional Lineage *Gandolfia* – new

Type species.– Gandolfia subpennyi (Gandolfi, 1955).

Species included. IS: *Gandolfia subpennyi* (Gandolfi, 1955), FDS: *G. gagnebini* (Tilev, 1951) and SDS: *G. wiedenmayeri* (Gandolfi, 1955).

Derivation. Genus and lineage named after Dr Rolando Gandolfi for his outstanding contributions in the study of Cretaceous planktic foraminifera.

Diagnosis. Late Campanian-Maastrichtian lineage of low trochospiral planktic foraminifera with plano-convex tests and peripheral imperforate band bordered by two keels.

Description. Test is very low trochospiral; earlier chambers are globular to subglobular, those of the final whorl petaloid on the spiral side and strongly inflated, with subtriangular to subrectangular-rounded outline. Sutures on the spiral side are depressed throughout in *G. subpennyi* and depressed towards the central test portion and raised towards the periphery in *G. gagnebini* and *G. wiedenmayeri*. Sutures on the umbilical side are distinct, depressed, straight and radial. Test in plano-convex; spiral side is flat, and occasionally depressed in the central portion in *G. subpennyi*. Peripheral structures consist of one imperforate peripheral bordered by two equally developed keels on all the chambers of the final whorl; the two keels are in incipient stage of development in *G. subpennyi*, with rope appearance in *G. gagnebini* and solid appearance in *G. wiedenmayeri*. Umbilicus is deep and medium-sized, with a diameter of one fourth to nearly one half of the maximum test diameter; periumbilical shoulders occur on all the chambers around the umbilicus. Main aperture has the shape of a medium high arch in umbilical position, and is bordered by one perforate porticus, which can be attached to the earlier chambers across the umbilicus, or previous chamber portici. Chambers are ornamented with pustules and rugosities, which can fuse to form short costellae in *G. subpennyi*, and elongate and irregular agglomerations of pustules in *G. gagnebini* and *G. wiedenmayeri*. Test wall is calcitic, hyaline, simple and reticulately-ridged on the earlier chambers in *G. wiedenmayeri*, and perforate; pores are circular, with a diameter of 1.5-4.3 μm.

Remarks. *Gandolfia* differs from *Archaeoglobigerina*, its ancestor, mainly by the test shape, which is plano-convex rather than biconvex or slightly asymmetrical. The species in the *Gandolfia* directional lineage present coarser ornamentation (short costellae in *G. subpennyi* and elongate or irregular structures on the spiral side in *G. gagnebini* and *G. wiedenmayeri*); its ornamentation sharply differs from the pustulose one of *Archaeoglobigerina*.

Stratigraphical range. Upper Campanian-Maastrichtian (from the *R. calcarata* Biozone to the *A. mayaroensis* Biozone).

Geographic distribution. Cosmopolitan.

IS *Gandolfia subpennyi* (Gandolfi, 1955) – emended
Figure 6: 1–9

1955 *Globotruncana* (*Rugoglobigerina*) *pennyi subpennyi*; Gandolfi, p. 73, pl. 7, figure 7.

1960 *Globotruncana* (*Rugotruncana*) *gansseri dicarinata*; Pessagno, p. 103, pl. 2, figures 9-11, pl. 3, figures 1-3, pl. 5, figure 2.

1962 *Globotruncana* (*Rugotruncana*) *gansseri dicarinata* Pessagno; Pessagno, pl. 4, figures 9-10, 12, 15-16.

1967 *Rugotruncana subpennyi* (Gandolfi); Pessagno, p. 370, pl. 76, Figures 12-14, pl. 91, figures 8-15.

2007 *Globotruncana* (*Rugoglobigerina*) *pennyi subpennyi* Gandolfi; Georgescu and Huber, pl. 2, figure 4.

Material. Circa 80 Specimens

Diagnosis. *Gandolfia* with plano-convex test, two keels in early stage of development, and ornamentation consisting of pustules and rugosities.

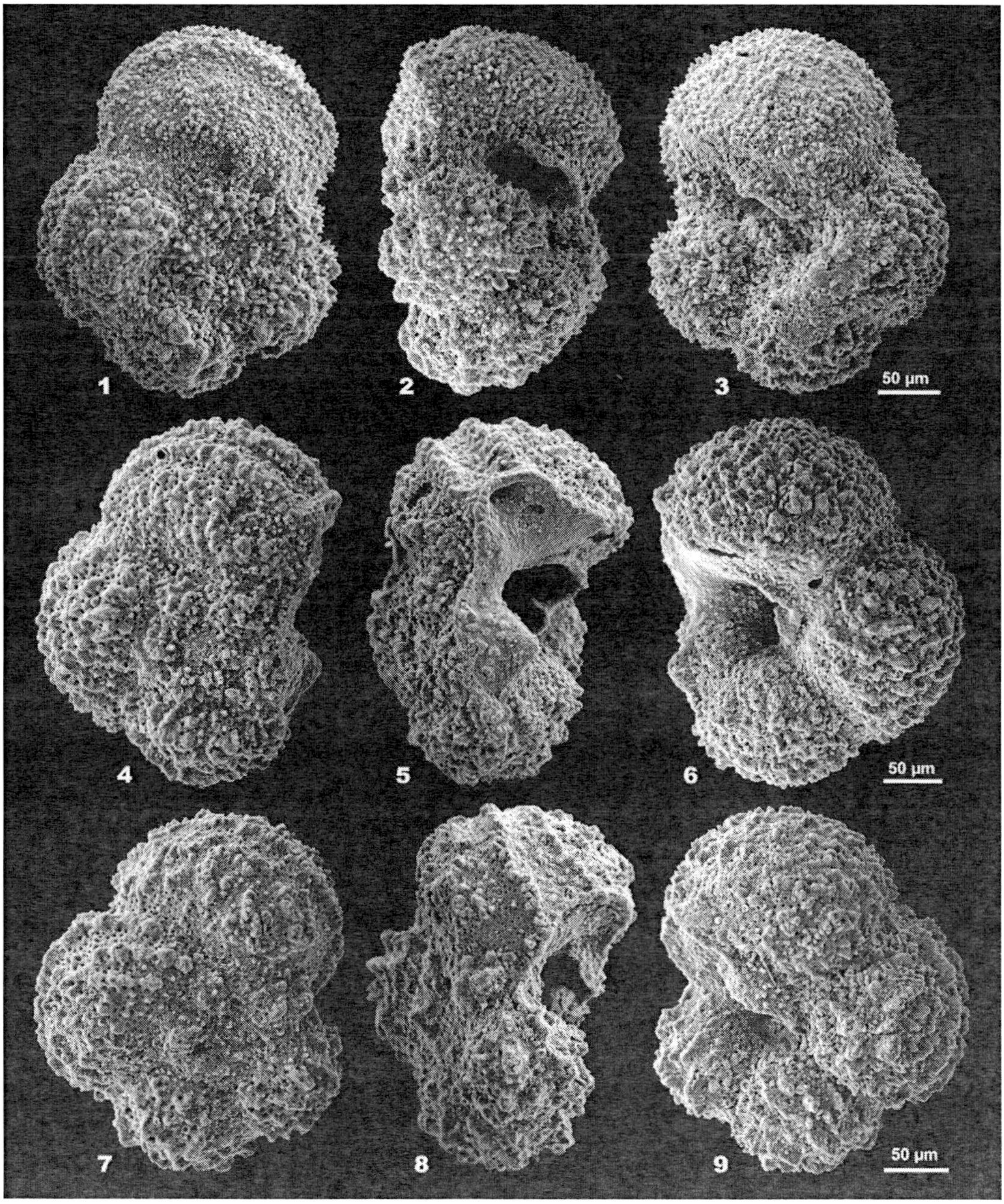

Figure 6. Hypotypes of *Gandolfia subpennyi* from the uppermost Campanian sediments of the Mid-Pacific Mountains (Central Pacific Ocean). 1-3 Specimen from the *R. calcarata* Biozone, sample 62-463-22-4, 74-76 cm. 4-6 Specimen from the *R. calcarata* Biozone, sample 62-463-22-2, 74-76 cm. 7-9 Specimen from the *Globotruncanella subpetaloidea* Biozone, sample 62-463-21-1, 52-54 cm.

Description. Test is very low trochospiral, consisting of 11-15 chambers arranged in 2 ½-3 whorls; there are 4-5 (commonly 4 ½) chambers in the final whorl. Earlier chambers are globular or subglobular, those of the final whorl petaloid on the spiral side and strongly inflated, with subtriangular outline on the umbilical one; chambers increase moderately in size. Sutures on the spiral side are straight or slightly curved, distinct and depressed; those on the umbilical are straight, depressed and radial. Test is asymmetrical in edge view, plano-convex; the spiral side can be occasionally depressed in the central part. Peripheral structures consist of an imperforate band bordered by two equally developed keels; the keels are in incipient stage of development, consisting of fused and, mostly isolated dome-like pustules. Umbilicus is deep and narrow, with a diameter of one fourth to, rarely one third of the maximum test diameter; periumbilical shoulders are developed on the last-formed one to three chambers. Aperture has the shape of a medium high arch, is in umbilical position, and is protected by one narrow and imperforate thin porticus, which is rarely and partly preserved. Chambers are ornamented with dome-like pustules (3.7-13.8 μm in diameter), which can fuse to from rugosities and occasionally short costellae (30.6-42.5 μm in length); there is no apparent ornamentation distribution pattern. Test wall is calcitic, hyaline, simple and perforate; pores are circular, and with a diameter of 1.5-4.0 μm.

Remarks. *Gandolfia subpennyi* differs from *A. subcircumnodifer* mainly by having plano-convex rather than biconvex tests in edge view. The ornamentation presents significant variability: the specimens from the lower part of the species stratigraphical range (late Campanian) are mostly ornamented with pustules, those from the upper part (Maastrichtian) frequently present short costellae. Pessagno (1967) considered that the ornamentation of this species presents a distinct meridional pattern. The holotype examination, together with the well preserved new material from the central Pacific Ocean (Mid-Pacific Mountains) shows that there is no apparent pattern in *G. subpennyi* ornamentation.

Stratigraphical range. Upper Campanian-middle Maastrichtian (from the *R. calcarata* Biozone to the *Gansserina gansseri* Biozone).

Geographic distribution. South America (Colombia), Caribbean region (Puerto Rico), USA (Texas, Arkansas), Mexico, central Pacific Ocean (Mid-Pacific Mountains).

FDS *Gandolfia gagnebini* (Tilev, 1951)
Figure 7: 1–9

1951 *Globotruncana gagnebini*; Tilev, p. 50, pl. 3, figures 2-5, text-figures 14-17.

1955 *Globotruncana wiedenmayeri magdalenaensis*; Gandolfi, p. 72, pl. 7, figure 3.

1957 ? *Globotruncana repanda*; Bolli, p. 56, pl. 13, figure 2.

1962 non *Globotruncana gagnebini* Tilev; Herm, p. 79, pl. 7, figure 5.

1963 *Globotruncana gagnebini* Tilev; Majzon, pl. 3, figure 10.

1964 non *Globotruncana gagnebini* Tilev; Olsson, p. 165, pl. 4, figure 4.

1966 non *Globotruncana gagnebini* Tilev; El-Naggar, p. 111, pl. 2, figures 1-4, pl. 3, figures 1, 3, 6, pl. 4, figure 2.

1971 *Globotruncana gagnebini* Tilev; Postuma, p. 40, thr seven figures on page 41.

1972 *Globotruncana gagnebini* Tilev; Barr, p. 21, pl. 4, figure 8, pl. 8, figure 5.

1972 non *Globotruncana gagnebini* Tilev; Govindan, p. 179, pl. 4, figures 7-9, pl. 5, figures 1-6.

1973 *Globotruncana gagnebini* Tilev; Premoli Silva and Bolli, pl. 6, figures 2-6.

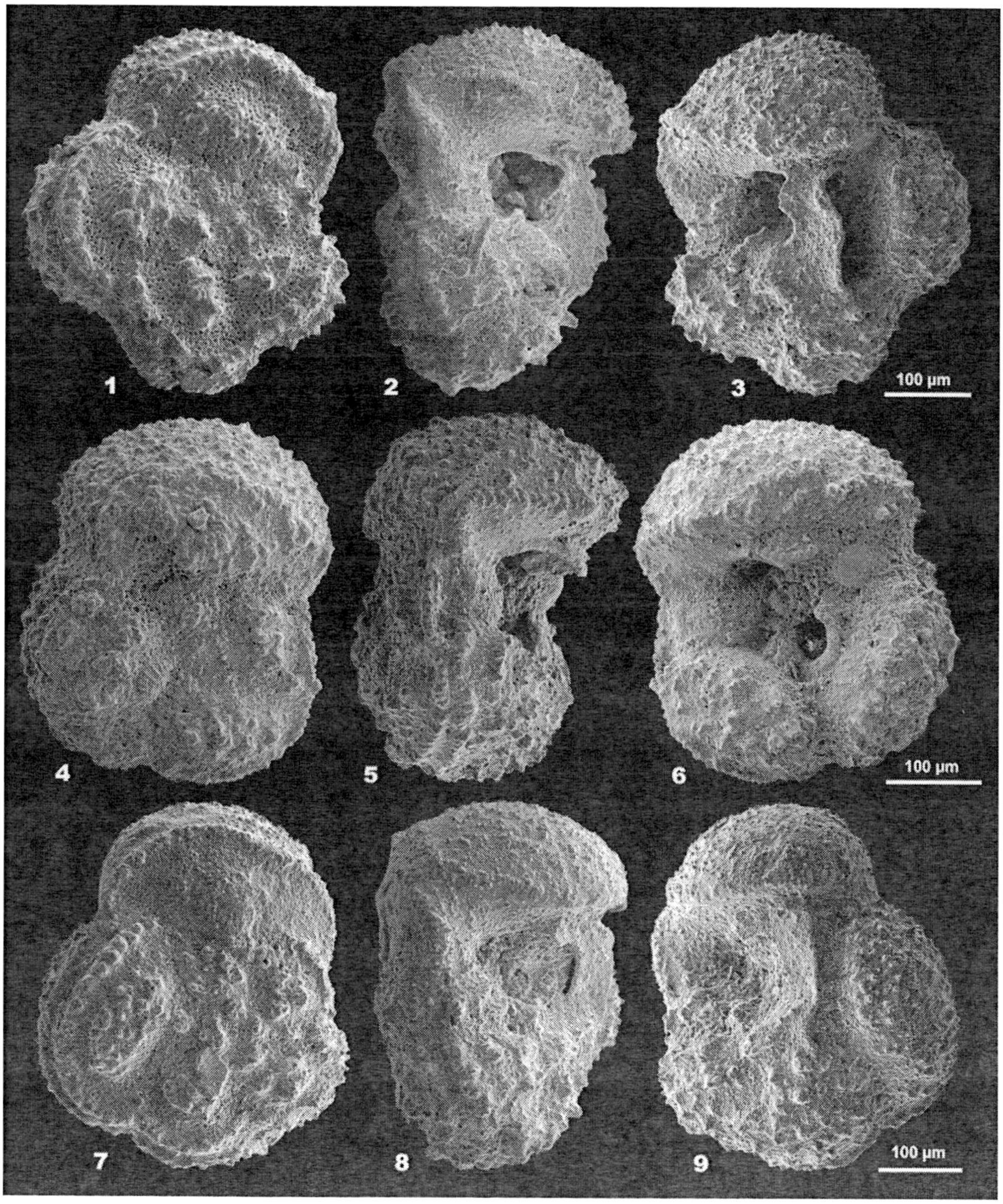

Figure 7. Hypotypes of *Gandolfia gagnebini* from the middle Maastrichtian (*Contusotruncana contusa* Biozone) of the Nicaragua Rise (Caribbean region); all specimens are from sample 15-152-15-2, 24-38 cm.

1979 *Globotruncana aegyptiaca* Nakkady; Abdel-Kireem and Abdou, p. 185, pl. 3, figures 1-4.

1979 non *Globotruncana gagnebini* Tilev; Abdel-Kireem and Abdou, p. 188, pl. 3, Figures 6-8, pl. 7, figures 3-4.

1979 *Globotruncana gagnebini* Tilev; Yassini, p. 25, pl. 6, figures 4-6.

1981 ? *Globotruncana gagnebini* Tilev; Butt, pl. 18, figures M1-M3.

1983 *Globotruncana gagnebini* Tilev; Salaj, pl. 8, figures 1-3.
1984 *Globotruncana aegyptiaca* Nakkady; Robaszynski et al., pl. 2, figure 3.
1986 non *Globotruncana gagnebini* Tilev; Abdel-Kireem, p. 36, pl. 5, figures 8-9.
1986 *Rosita* sp.; Almogi-Labin et al., p. 869, pl. 9, figures 3-5.
1996 non *Globotruncana gagnebini* Brotzen; Ayyad et al., figure 14: b-c.
1996 *Globotruncana aegyptiaca* Nakkady; Mancini et al., figure 5: 6-8.

Material. Circa 120 Specimens

Diagnosis. *Gandolfia* with 4-4 ½ chambers in the final whorl, plano-convex test, two well developed peripheral keels, and ornamentation on the spiral side concentrated in the chamber central portion.

Description. Test is very low trochospiral, consisting of 10-13 chambers arranged in 2 ½-3 whorls; there are 4-4 ½ (commonly 4 ½) chambers in the final whorl. Earlier chambers are globular to subglobular, the last-formed ones petaloid on the spiral side and strongly inflated, with subrectangular outline on the umbilical side; chambers increase in size at moderate rate. Sutures on the spiral side are often indistinct in the test central portion, oblique to the previous whorl, slightly curved, depressed towards the centre and raised towards the periphery; sutures on the umbilical side are distinct, depressed, straight and radial. Test is asymmetrical in edge view, plano-convex; the spiral side is flat, and without depressed central portion. Peripheral structures consist of two well-developed keels on all the chambers on the final whorl; the keels have rope appearance. Umbilicus is wide, with a diameter of one third to one half of the maximum test diameter; periumbilical shoulders are developed on the last-formed one to three chambers of the last whorl. Aperture is a medium high arch in umbilical position, and is bordered by one perforate porticus, which extends over the umbilicus; portici are attached to the periumbilical chamber portion. Chamber surface on the umbilical side is ornamented with dome-like or irregular in shape pustules (10.3-19.6 μm in diameter or maximum dimension respectively), which are denser on the earlier chambers; pustules on the spiral side concentrated in chamber central portion, and are fused and form elongate structures with a maximum dimension of 110 μm. Test wall is calcitic, hyaline, simple and perforate; pores are circular, and with a diameter of 2.2-3.7 μm).

Remarks. *Gandolfia gagnebini* differs from *G. subpennyi* by having two well-developed rather than in incipient stage of development on all the chambers of the final whorl, larger umbilicus, and ornamentation on the spiral side with agglomerated pustules that for elongate or irregular structures. It differs from all the species of *Archaeoglobigerina* by the plano-convex test and well-developed keels throughout the final whorl.

Stratigraphical range. Upper Campanian-middle Maastrichtian (from the *G. havanensis* Biozone to the *Racemiguembelina fructicosa* Biozone).

Geographic distribution. Europe (Germany, Hungary), Middle East (Egypt, Israel, Jordan, Turkey), northern Africa (Libya, Tunisia), southern Asia (India), USA (New Jersey), South America (Colombia), Caribbean region (Nicaragua Rise).

SDS *Gandolfia wiedenmayeri* (Gandolfi, 1955)
Figures 8: 1–8 and 9: 1–8

1955 *Globotruncana wiedenmayeri wiedenmayeri*; Gandolfi, p. 71, pl. 7, figure 4.
1986 *Gansserina wiedenmayeri* (Gandolfi); Almogi-Labin et al., p. 869, pl. 10, figures 18-20.
1996 non *Gansserina wiedenmayeri* (Gandolfi); Georgescu, p. 318, pl. 12, figure 1.
1996 *Gansserina wiedenmayeri* (Gandolfi); Mancini et al., figure 6: 1-3.

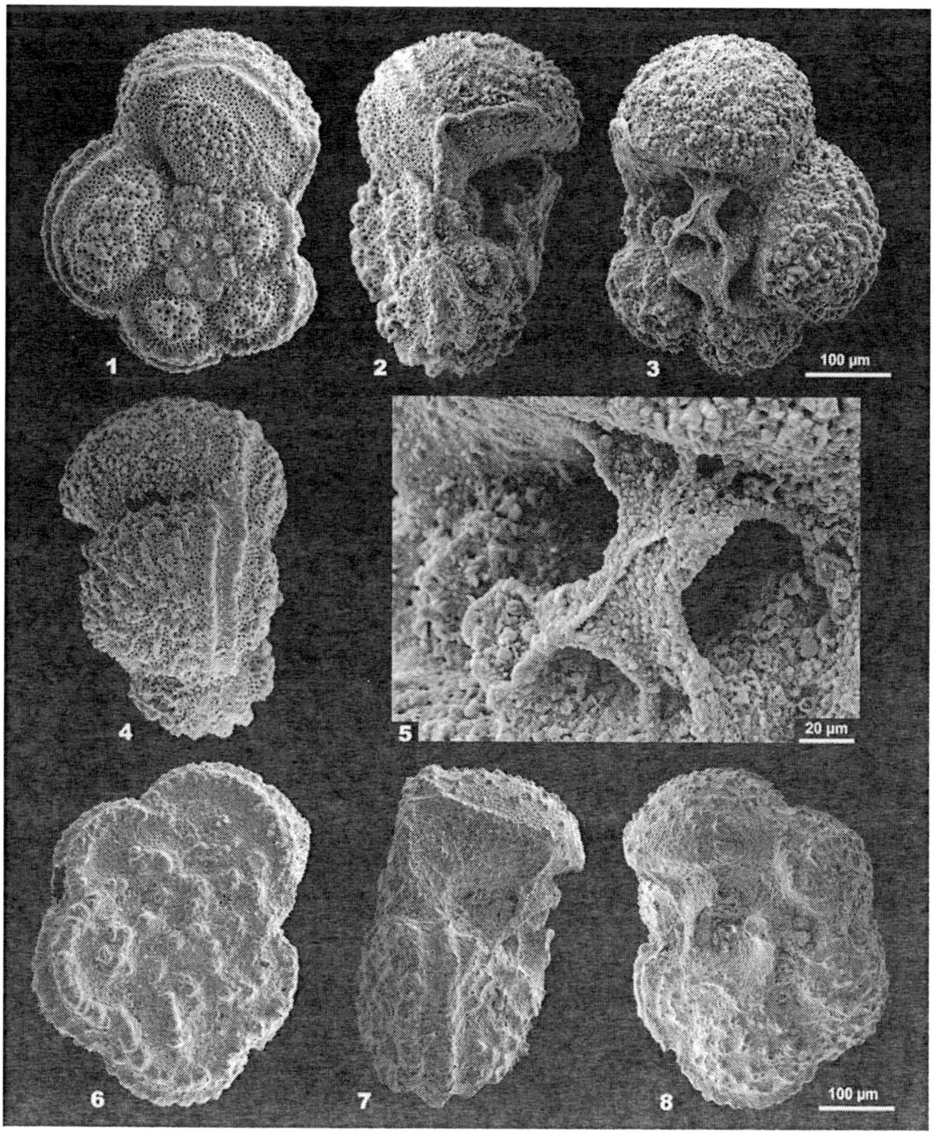

Figure 8. Hypotypes of *Gandolfia wiedenmayeri* from the Mid-Pacific Mountains (Central Pacific Ocean) (1-5) and Nicaragua Rise (Caribbean region) (6-8). 1-5 Specimen from the upper Campanian-lower Maastrichtian sediments (*G. gansseri* Biozone), sample 62-462-11-3, 34-36 cm. 5 Detail micrograph showing the finely perforate porticus. 6-8 Specimen from the middle Maastrichtian (*C. contusa* Biozone), sample 15-152-15-2, 24-38 cm.

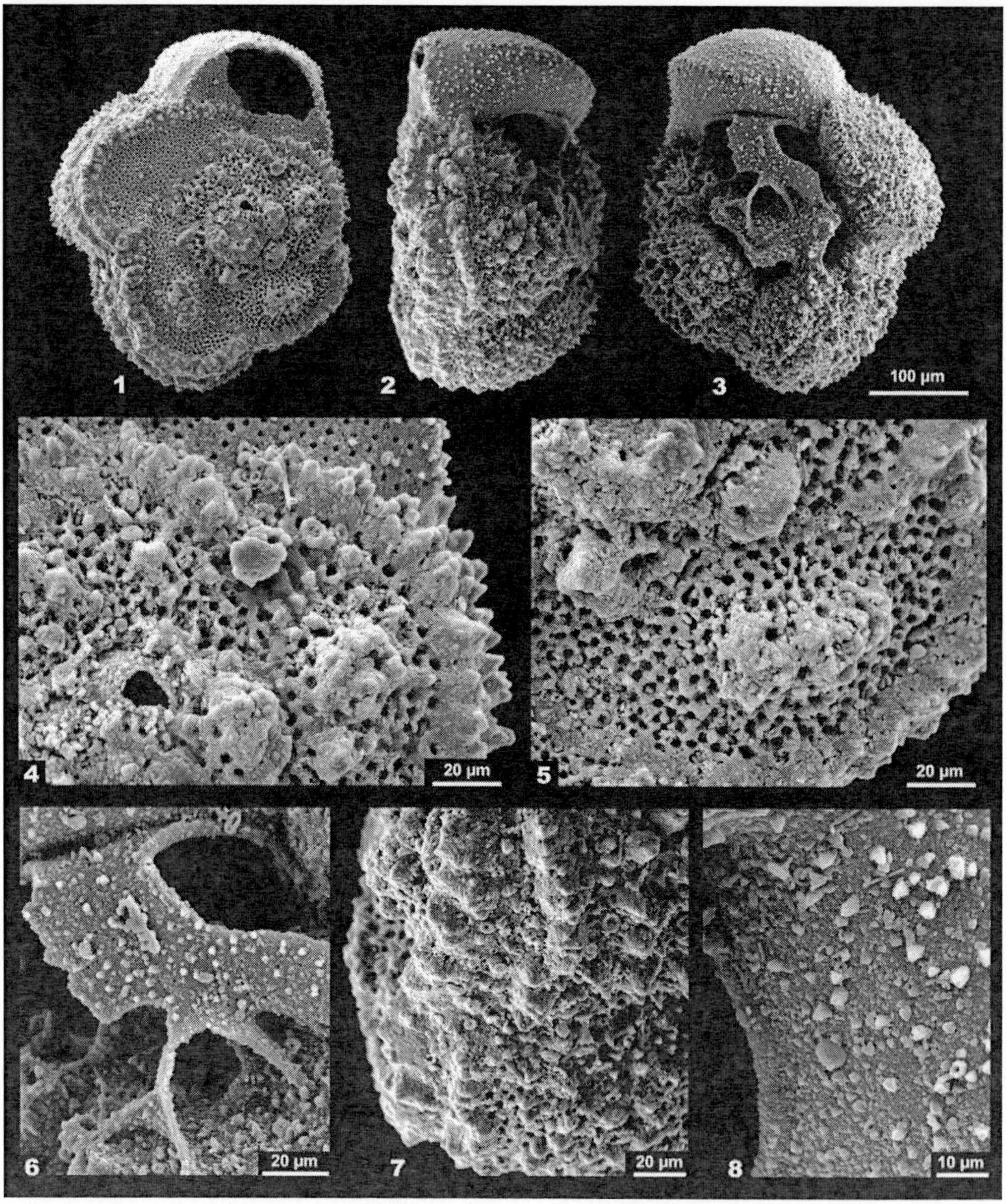

Figure 9. Hypotype of *Gandolfia wiedenmayeri* from the lower Maastrichtian (*Globotruncanita stuarti* Biozone) sediments of Shatsky Rise (Central Pacific Ocean) (1-8), sample 32-305-17-6, 100-102 cm. 4 Detail micrograph showing the different test wall on the spiral side, simple on the last-formed chamber and reticulately-ridged on the earlier chambers. 5 Detail micrograph showing the reticulately-ridged wall on the early chambers of the final whorl. 6 Detail micrograph showing the perforate porticus. 7 Detail micrograph showing the peripheral structures. 8 Detail micrograph showing the imperforate test wall around the aperture.

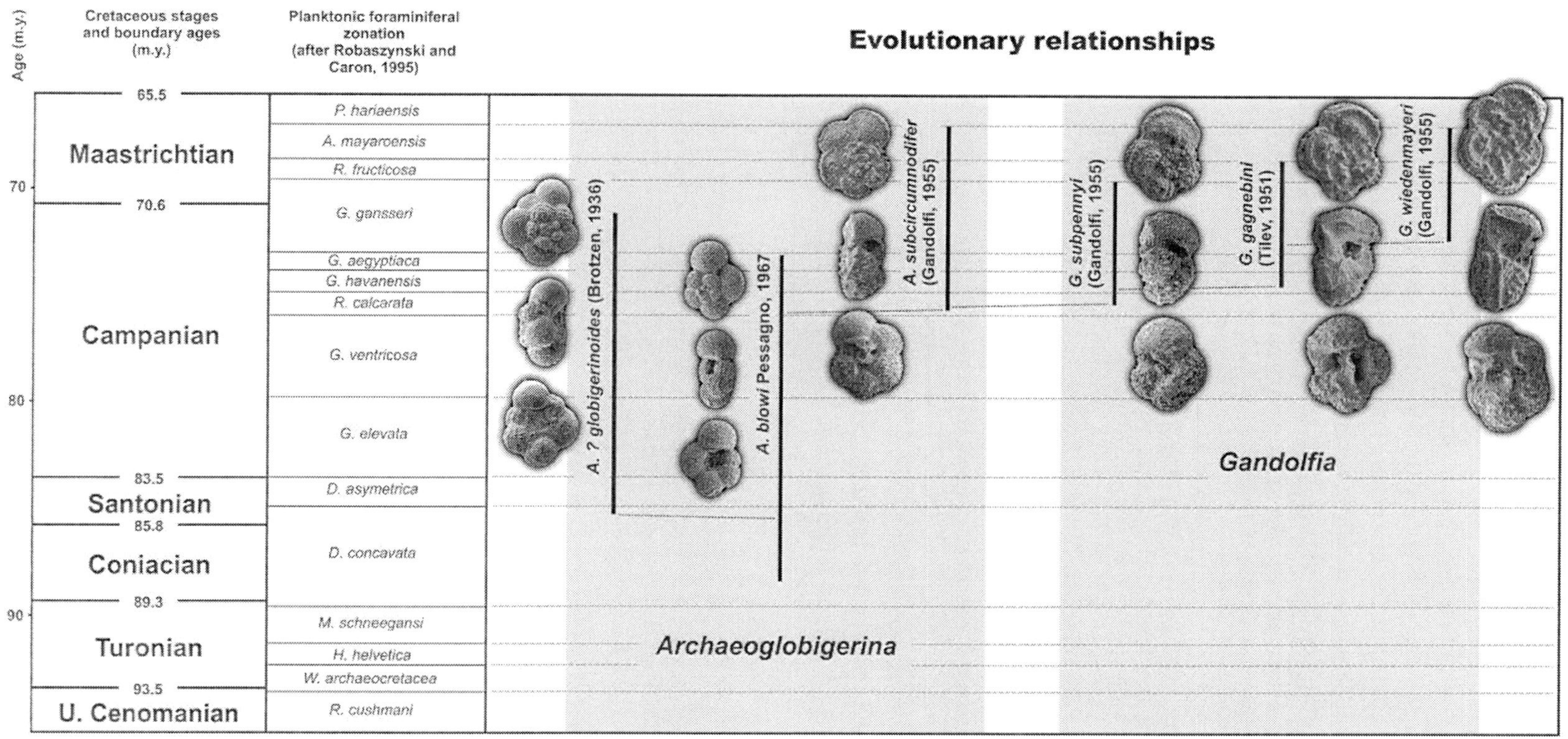

Figure 10. Evolutionary relationships between the species of the branched lineage *Archaeoglobigerina* Pessagno, 1967 and new directional lineage *Gandolfia*. Ages are after Gradstein et al. (2004).

Material. 45 Specimens

Diagnosis. *Gandolfia* with 5-6 ½ chambers in the final whorl, plano-convex test, two well developed peripheral keels, ornamentation on the spiral side concentrated in the chamber central portion, and reticulately-ridged test wall of the earlier chambers.

Description. Test is very low trochospiral, consisting of 15-19 chambers arranged in 2 ½-3 ½ whorls; there are 5-6 ½ (commonly 5 ½-6) chambers in the final whorl. Earlier chambers are globular to subglobular, those of the last whorl petaloid on the spiral side and strongly inflated, and with subrectangular-rounded outline; chambers size increase is at moderate to low rate. Sutures on the spiral side are curved, often indistinct, depressed towards the test central portion and raised towards the peripheral region; sutures on the umbilical side are distinct, depressed, straight, and radial. Test is strongly asymmetrical in edge view; spiral side in flat, and without a depression zone in the central portion. Peripheral structures consist of a narrow imperforate band bordered by two well-developed keels with solid and continuous appearance; the umbilical keel can be weaker on the last-formed one, more rarely two chambers. Umbilicus is wide, with a diameter of circa one third to one half of the maximum test diameter; periumbilical ridges occur on all the chambers of the final whorl. Main aperture has the shape of a medium high arch, is in umbilical position, and is bordered by a perforate porticus, which attached to the previous similar structures. Chamber surface on the umbilical side is ornamented with scattered dome-like, conical, and irregular pustules (10.2-21.0 μm in diameter or maximum dimension), which can fuse to form rugosities and short costellae (51.8 μm in length) that lack an apparent orientation pattern. Test wall is calcitic, hyaline, simple in the last-formed chambers and reticulately-ridged in the earlier ones, and perforate; pores are circular or elliptical in shape, 1.5-4.3 μm in diameter or maximum dimension respectively.

Remarks. This species presents close resemblances to *G. gagnebini* of which it differs mainly by having more chambers in the final whorl (5-6 ½ rather than 4-4 ½), main aperture bordered by a perforate rather than imperforate porticus, and weaker umbilical keel occasionally developed over the last-formed one or two chambers. It differs from *G. subpennyi* mainly by having two well-developed keels and larger umbilici.

Stratigraphical range. Uppermost Campanian-Maastrichtian (from the *G. gansseri* Biozone to the *A. mayaroensis* Biozone).

Geographic distribution. South America (Colombia), USA (Alabama), Middle East (Israel), and Central Pacific Ocean (Shatsky Rise, Mid-Pacific Mountains).

CONCLUSION

Two Cretaceous (Coniacian-Maastrichtian) lineages of keeled planktic foraminifera are reviewed based on SEM-based high resolution morphological observations on the test wall ultrastructure, ornamentation, porosity, as well as discrete features, such as peripheral structures, periapertural structures. The new taxonomical framework documents the evolution from the globular-chambered species into species with keeled periphery, plano-convex shape in edge view and ornamentation concentrated in the central portion of the chambers on the spiral side. This study shows the necessity to re-evaluate the periapertural structure morphology and terminology, and confirms the perforate portici and imperforate tegilla in the Cretaceous planktic foraminifera (Weidich, 1997).

Archaeoglobigerina is reviewed as branched lineage in evolutionary classification, and it consists of three species: *A. blowi*, *A. ? globigerinoides* and *A. subcircumnodifer*. The species of *Archaeoglobigerina* present a peripheral imperforate band bordered by two rows of isolated pustules in *A. blowi* and *A. ? globigerinoides* and incipient keels in *A. subcircumnodifer*. The species of *Archaeoglobigerina* are ornamented with scattered pustules on both spiral and umbilical sides (Figure 10).

Gandolfia is a new directional lineage, which includes *G. subpennyi*, *G. gagnebini* and *G. wiedenmayeri*. The three species present plano-convex tests and periapertural structures consisting of one perforate porticus. The two peripheral keels are weaker in the incipient species (*G. subpennyi*) and stronger in the first and second descendant species, *G. gagnebini* and *G. wiedenmayeri* respectively. Another distinct trend in this directional lineage is the ornamentation concentration in the central portion of the chambers on the spiral side (Figure 10).

Acknowledgments

The author thanks the DSDP/ODP headquarters for the samples from the wells drilled worldwide under their auspices. Dr A. Bartolini is thanked for the permission and help in studying the type material of *Globigerina cretacea* in the Alcide d'Orbigny Collection from the Muséum National d'Histoire Naturelle, Paris. The author thanks Dr L.V. Hills (University of Calgary) for the constructive criticism and suggestions to improve the manuscript. Dr Michael Schoel (Microscope and Imaging Facility, University of Calgary) is thanked for the help during the SEM operations.

References

Abdel-Kireem, M.R. and Abdou, H.F. 1979. Upper Cretaceous-Lower Tertiary planktonic Foraminifera from southern Galala Plateau, Eastern Desert, Egypt. *Revista Española de Micropaleontología*, 11: 175–222.

Almogi-Labin, A., Reiss, Z. and Caron, M. 1986. Senonian Globotruncanidae from Israel. *Eclogae Geologicae Helvetiae*, 79: 849–895.

Ansary, S.F. and Tewfik, N.M. 1968. Planktonic Foraminifera and some new benthonic species from the subsurface Upper Cretaceous of Ezz El Orban area, Gulf of Suez. *Journal of Geology of the United Arab Republic*, 10: 37–76.

Ayyad, S.N., Abed, M.M. and Abu Zied, R.H. 1996. Biostratigraphy and correlation of Cretaceous rocks in Gebel Arif El-Naga, northeastern Sinai, Egypt, based on planktonic Foraminifera. *Cretaceous Research*, 17: 263–291.

Banerji, R.K. 1966. The genus *Globotruncana* and biostratigraphy of the Lower Ariyalur Stage (Upper Cretaceous) of Vridhachalam, South India. *Journal of the geological Society of India*, 7: 51–69.

Banner, F.T. and Blow, W.H. 1960. Some primary types of species belonging to the superfamily Globigerinaceae. *Contributions from the Cushman Foundation for Foraminiferal Research*, 11: 1–41.

Barr, F.T. 1961. Upper Cretaceous planktonic Foraminifera from the Isle of Wight, England. *Palaeontology*, 4: 552–580. [published in 1962]

Barr, F.T. 1968. Late Cretaceous planktonic Foraminifera from the coastal area east of Susa (Apollonia), northeastern Libya. *Journal of Paleontology*, 42: 308–321.

Barr, F.T. 1972. Cretaceous biostratigraphy and planktonic Foraminifera of Libya. *Micropaleontology*, 18: 1–46.

Belford, D.J. 1983. A probable Coniacian (Late Cretaceous) foraminiferal fauna, Carnarvon Basin, western Australia. *Bureau of Mineral Resources, Geology and Geophysics Bulletin*, 217: 11–27.

Berggren, W.A. 1962. Some planktonic Foraminifera from the Maestrichtian and type Danian of southern Scandinavia. *Stockholm Contributions in Geology*, 9: 1–106.

Bolli, H.M. 1945. Zur Stratigraphie der Oberen Kreide in den höheren helvetischen Decken. *Eclogae Geologicae Helvetiae*, 37: 217–329.

Bolli, H.M., 1951. The genus *Globotruncana* in Trinidad, B.W.I. *Journal of Paleontology*, 25: 187–199.

Bolli, H.M. 1957. The genera *Praeglobotruncana*, *Rotalipora*, *Globotruncana* and *Abathomphalus* in the Upper Cretaceous of Trinidad, B.W.I. In Loeblich, A.R. Jr. (ed.), *Studies in Foraminifera*. United States National Museum Bulletin, 215: 51-60.

Brönnimann, P. and Brown, N.J. Jr. 1956. Taxonomy of the Globotruncanidae. *Eclogae Geologicae Helvetiae*, 48: 503–561.

Brotzen, F. 1936. Foraminiferen aus dem Schwedischen, Untersten Senon von Eriksdal in Schonen. *Sveriges Geologiska Undersökning Årsbok*, 30(3): 1–206.

Butt, A. 1981. Depositional environments of the Upper Cretaceous rocks in the northern part of the eastern Alps. *Cushman Foundation for Foraminiferal Research, Special Publication* 20: 5–121.

Campbell, R.J., Howe, R.W. and Rexilius, J.P. 2004. Middle Campanian-lowermost Maastrichtian nannofossil and foraminiferal biostratigraphy of the northwestern Australian margin. *Cretaceous Research*, 25: 827–864.

Carman, K. 1929. Some Foraminifera from the Noibrara and Benton formations of Wyoming. *Journal of Paleontology*, 3: 309–315.

Caron, M. 1981. Un nouveau genre de foraminifère planctonique du Crétacé: *Falsotruncana* nov. gen. *Eclogae Geologicae Helvetiae*, 74: 65–73.

Caron, M. 1985. Cretaceous planktic foraminifera. In Bolli, H.M., Saunders, J.B., and Perch-Nielsen, K. (eds), *Plankton Stratigraphy*, 17-86. Cambridge University Press, Cambridge.

Cushman, J.A. 1931. Cretaceous Foraminifera from Antigua, B.W.I. *Contributions from the Cushman Foundation for Foraminiferal Research*, 7: 33–46.

Donze, P., Porthault, B., Tomel, G. and Volloutreys, O. de 1970. Le Sénonien inférieur de Puget-Theniers (Alpes-Maritimes) et sa microfaune. *Geobios*, 3: 41–106.

Douglas, R.G. and Rankin, C. 1969. Cretaceous planktonic Foraminifera from Bornholm and their zoogeographic significance. *Lethaia*, 2: 185–217.

Edgell, H.S. 1957. The genus *Globotruncana* in Northwest Australia. *Micropaleontology*, 3: 101–126.

Eichwald, C.E. von 1830. *Zoologia specialis II*. D.E. Eichwaldus, Vilnae.

El-Naggar, Z.R. 1966. Stratigraphy and planktonic Foraminifera of the Upper Cretaceous-Lower Tertiary succession in the Esna-Idfu region, Nile valley, Egypt, U.A.R. *Bulletin of the British Museum (Natural History), Geology Supplement*, 2: 1–291.

El-Naggar, Z.R., 1971. On the classification, evolution and stratigraphical distribution of the Globigerinacea. In Farinacci, A. (ed.), *Proceedings of the II Planktonic Conference, Roma 1970*, 1: 421-476. Edizioni Tecnoscienza, Roma.

Eternod Olvera, Y. 1959. Foraminiferos del Cretacico superior de la Cuenca de Tampico-Tuxpan, Mexico. *Boletín de la Asociación Mexicana de Geólogos Petroleros,* 11: 63–134.

Evetts, M.J. 1976. Microfossil biostratigraphy of the Sage Breaks Shale (Upper Cretaceous) in northeastern Wyoming. *The Mountain Geologist*, 13: 115–134.

Frerichs, W.E. 1971. Evolution of planktonic foraminifera and paleotemperatures. *Journal of Paleontology*, 45: 963–968.

Frerichs, W.E., Atherton, C.C. and Shive, P.N. 1975. Planktonic Foraminifera from the chalky members of the Niobrara Formation, Centennial Valley, Wyoming, *Journal of Foraminiferal Research*, 5: 294–312.

Frerichs, W.E. and Dring, N.B. 1981. Planktonic Foraminifera from the Smoky Hill Shale of the west central Texas. *Journal of Foraminiferal Research*, 11: 47–69.

Gandolfi, R. 1955. The genus *Globotruncana* in northeastern Colombia. *Bulletins of American Paleontology*, 36(155): 1–118.

Gawor Biedowa, E. 1992. Campanian and Maastrichtian Foraminifera from the Lublin Upland, eastern Poland. *Palaeontologia Polonica*, 52: 1–187.

Georgescu, M.D. 1996. Santonian-Maastrichtian planktonic foraminifers (Globigerinelloididae, Hedbergellidae, Globotruncanidae and Rugoglobigerinidae). *Micropaleontology*, 42: 305–333.

Georgescu, M.D. 2005. On the systematics of rugoglobigerinids (planktonic foraminifera, Late Cretaceous). In Tyszka, J., Oliwkiewicz-Miklasińska, M., Gedl, P., Kaminski, M.A. (eds), *Methods and Applications in Micropaleontology*. Studia Geologica Polonica, 124: 87-97.

Georgescu, M.D. 2006. Santonian-Campanian planktonic foraminifera in the New Jersey coastal plain and their distribution related to the relative sea-level changes. *Canadian Journal of Earth Sciences*, 43: 101–120.

Georgescu, M.D. 2009a. Upper Albian-lower Turonian non-schackoinid planktic foraminifera with elongate chambers: morphology reevaluation, taxonomy and evolutionary classification. *Revista Española de Micropaleontología*, 41: 255–293.

Georgescu, M.D. 2009b. Taxonomic revision and evolutionary classification of the biserial Cretaceous planktic foraminiferal genus *Laeviheterohelix* Nederbragt, 1991. *Revista Mexicana de Ciencias Geológicas*, 26: 315–334.

Georgescu, M.D. 2010a. Origin, taxonomic revision and evolutionary classification of the late Coniacian-early Campanian planktic foraminifera with multichamber growth in the adult stage. *Revista Española de Micropaleontología*, 42: 59–118.

Georgescu, M.D. 2010b. Transition from the typological to evolutionary classification of the Cretaceous planktic foraminifera: case study of *Anaticinella* Eicher, 1973. *Micropaleontology*, 55: 589–616.

Georgescu, M.D. 2010c. Evolutionary classification of the Upper Cretaceous (Turonian-lower Cretaceous) planktic foraminifera with incipient meridional ornamentation. *Journal of Micropaleontology*, 29: 149–161.

Georgescu, M.D. 2011a. New data on the evolutionary classification of the Late Cretaceous (late Coniacian-Santonian) planktic foraminifera with elongate chambers. *Revista Española de Micropaleontología*, 43: 39–54.

Georgescu, M.D. 2011b. Iterative evolution, taxonomic revision and evolutionary classification of the praeglobotruncanid planktic foraminifera, Cretaceous (late Albian-Santonian). *Revista Española de Micropaleontología*, 43: xxx-xxx. [published March 2012]

Georgescu, M.D. and Huber, B.T. 2007. Taxonomic revision of the Late Campanian-Maastrichtian (Late Cretaceous) planktonic foraminiferal genus *Rugotruncana* Brönnimann and Brown, 1956, and a new paleontological species concept for planktonic foraminifera. *Journal of Foraminiferal Research*, 37: 150–159.

Georgescu, M.D. and Huber, B.T. 2009. Early evolution of the Cretaceous serial planktic foraminifera (late Albian-Cenomanian). *Journal of Foraminiferal Research*, 39: 335–360.

Georgescu, M.D., Saupe, E.E. and Huber, B.T. 2008. Morphometric and stratophenetic basis for phylogeny and taxonomy in Late Cretaceous gublerinid planktonic foraminifera. *Micropaleontology*, 54: 397–424. [published in 2009]

Georgescu, M.D., Quinney, A.E. and Anderson, K.D. 2011. New data on the taxonomy, evolution and biostratigraphical significance of the Turonian-Coniacian (Late Cretaceous) planktic foraminifer *Huberella* Georgescu, 2007. *Micropaleontology*, 57: 247–254.

Gradstein, F.M., Ogg, J.G. and Smith, A.G. (eds.) 2004. *A Geologic Time Scale 2004*. Cambridge University Press, Cambridge.

Hagn, H. 1953. Die Foraminiferen der Pinswanger Schichten (Unteres Obercampan). Ein Beitrag zur Mikropaläontologische der Helvetischen Oberkreide Südbayerns. *Palaeontolographica Abteilung A*, 104: 1–119.

Hanzliková, E. 1972. Carpathian Upper Cretaceous Foraminiferida of Moravia (Turonian-Maastrichtian). *Rozpravy Ústředního Ústavu Geologického*, 39: 1–160.

Hart, M.B., Bailey, H.W., Crittenden, S., Fletcher, B.N., Price, R.J. and Swiecicki, A., 1989. Cretaceous. In Jenkins, D.G. and Murray, J.W. (eds), *Stratigraphical Atlas of Fossil Foraminifera*, 273-371. Ellis Horwood Limited, Chichester.

Hay, W.H., DeConto, R.M., Wold, C.N., Wilson, K.M., Voigt, S., Schulz, M., Wold, A.R., Dullo, W.C., Ronov, A.B., Balukhovsky, A.N. and Söding, E. 1999.

Alternative global Cretaceous paleogeography. In Barrera, E. and Johnson, C.C. (eds), *Evolution of the Cretaceous Ocean-Climate System*. Geological Society of America, Special Publications, 332: 1-47.

Herb, R. 1974. Cretaceous planktic foraminifera from the eastern Indian Ocean. In Davies, T.A., Luydendyk, B.P. et al. (eds), *Initial Reports of the Deep Sea Drilling Project*, 24: 745-769. United States Government Publishing Office, Washington, D.C.

Herm, D. 1962. Stratigraphische und Mikropaläontologische Untersuchungen der Oberkreide um Lattengebirge und im Nierental (Gosaubecken von Reichenhall und Salzburg). *Bayerische Akademie der Wissenschaften, Mathematisch Naturwissenschaftliche Klasse, Abhandlungen*, 104: 1–119.

Hinte, J.E. van 1963. Zur Stratigraphie und Mikropaläontologie der Oberkreide und des Eozäns des Krappfeldes (Kärnten). *Jahrbuch der Geologischen Bundesanstalt, Sonderband*, 8: 1–147.

Hinte, J.E. van 1965. The type Campanian and its planktonic Foraminifera. *Koninklijke Nederlandse Akademie van Wetenschappen*, 68: 8–28.

Hofker, J. 1956. Die Globotruncanen von Nordwest-Deutschland und Holland. *Neues Jahrbuch für Geologie und Paläontologie, Abhandlungen*, 103: 312–340.

Hofker, J. 1961. Les Foraminifères du Tuffeau arénacé de Folx-les-Caves. *Annales de la Société Géologique de Belgique*, 84: 551–579.

Howe, R.W., Campbell, R.J. and Rexilius, J.P. 2003. Integrated uppermost Campanian-Maastrichtian calcareous nannofossil and foraminiferal biostratigraphic zonation of the northwestern margin of Australia. *Journal of Micropaleontology*, 22: 29–62.

Huber, B.T. 1990. Maestrichtian planktonic foraminifer biostratigraphy of the Maud Rise (Weddell Sea, Antarctica): ODP Leg 113 Holes 689B and 690C. In Barker, P.F., Kennett, J.P. et al. (eds), *Proceedings of the Ocean Drilling Program, Scientific Results*, 113: 489-513. Ocean Drilling Program, College Station.

Ion, J. 1983. Étude micropaléontologique (foraminifères planctoniques) du Crétacé supérieur de Ţara Bârsei (Carpates Orientales). *Institut de Géeologie et de Géophysique, Mémoires*, 31: 5–176.

Ion, J. and Odin, G.S. 2001. Planktonic Foraminifera from the Campanian-Maastrichtian at Tercis les Bains (Landes, France). In Odin, G.S. (ed.), *The Campanian-Maastrichtian Boundary*, 349-378. Elsevier B.V.

Kaiho, K. 1992. Campanian planktonic foraminifers and ostracodes from Hobetsu, Hokkaido, northern Japan. Part 1. Planktonic foraminifers. In Ishizaki, K. and Saito, T. (eds), *Centenary of Japanese Micropaleontology*, 317-325. Terra Scientific Publishing Company, Tokyo.

Korchagin, O.A. 2003. Mesozoic planktonic foraminifer classification (superfamilies Planomalinacea, Rotaliporacea and Globotruncanacea). *Trudy VNIGRI*, 547: 1–89. [in Russian]

Küpper, I. 1956. Mikropaläontologische Gliderung der Oberkreide der Beckenuntergrundes in den oberösterreichischen Molassebohrungen. *Mitteilungen der Geologischen Gessellschaft in Wien*, 56: 591–651.

Liszka, S. 1955. Foraminifera of the Lower Senonian in the vivinity of Cracow.
Rocznik Polskiego Towarzystwa Geologicznego, 23: 165–190.

Loeblich, A.R. Jr. and Tappan, H. 1988. *Foraminiferal genera and their classification.* Van Nostrand Reinhold Company, New York.

Majzon, L. 1963. *Globotruncana*-bearing sediments in Hungary. *Annales Instituti Geologici Publici Hungarici*, 49: 735–787. [in Russian]

Mancini, E.A., Puckett, T.M. and Tew, B.H. 1996. Integrated biostratigraphic and sequence stratigraphic framework for Upper Cretaceous strata of the eastern Gulf Coastal Plain, USA. *Cretaceous Research*, 17: 645–669.

Martin, L. 1964. Upper Cretaceous and Lower Tertiary Foraminifera from Fresno County, California. *Jahrbuch der Geologische Bundesanstaldt*, 9: 1–128.

Masters, B.A. 1977. Mesozoic planktonic Foraminifera. A world-wide review and analysis. In Ramsay, A.T.S. (ed.), *Oceanic Micropaleontology*, 1: 301-731. Academic Press, London-New York-San Francisco.

McGugan, A. 1957. Upper Cretaceous Foraminifera from Northern Ireland. *Journal of Paleontology*, 31: 329-348.

McNeil, D.H. 1984. The eastern facies of the Cretaceous System in the Canadian Western Interior. In Stott, D.F., Glass, D.J. (eds), *The Mesozoic of Middle North America.* Canadian Society of Petroleum Geologists Memoirs, 9: 145-171.

McNeil, D.H. and Caldwell, W.G.E. 1981. Cretaceous rocks and their Foraminifera in the Manitoba Escarpment. *The Geological Association of Canada Special Number*, 21: 1–439.

Moorkens, T.L. 1969. Quelques Globotruncanidae et Rotaliporidae du Cénomanien, Turonien et Coniacien de la Belgique. In Brönnimann, P., Renz, H.H., (eds), *Proceedings of the First International Conference on Planktonic Microfossils, Geneva 1967*, 2: 435-459. E.J. Brill, Leiden.

Nakkady, S.E. and Osman, A. 1954. The genus *Globotruncana* in Egypt, taxonomy and stratigraphical value. *19th International Geological Congress*, 13(15): 75–95.

Neagu, T. 1979. *Micropaleontologie*. Editura Tehnică, Bucureşti.

Neagu, T. 1987. White chalk foraminiferal fauna in southern Dobrogea (Romania). *Revista Española de Micropaleontología*, 19: 281–314.

Norris, R.D. 1992. Umbilical structures in Late Cretaceous planktonic Foraminifera. *Micropaleontology*, 38: 165–181.

Orbigny, A. d' 1840. Mémoire sur les foramnifères de la craie blanche du Bassin de Paris. *Mémoires de la Société Géologique de France*, 4: 1–51.

Pessagno, E.A. JR. 1960. Stratigraphy and micropaleontology of the Cretaceous and lower Tertiary of Puerto Rico. *Micropaleontology*, 6: 87–110.

Pessagno, E.A. JR. 1962. The Upper Cretaceous stratigraphy and micropaleontology of south-central Puerto Rico. *Micropaleontology*, 8: 349–368.

Pessagno, E.A. JR. 1967. Upper Cretaceous planktonic Foraminifera from the western Gulf coastal plain. *Palaeontographica Americana*, 5(37): 245–445.

Peryt, D. 1980. Planktic Foraminifera zonation of the Upper Cretaceous in the Middle Vistula River valley, Poland. *Palaeontologia Polonica*, 41: 3–101.

Petters, S.W., El-Nakhal, H.A. and Cifelli, R. 1983. *Costellagerina*, a new Late Cretaceous globigerine foraminiferal genus. *Journal of Foraminiferal Research*, 13: 247–251.

Plummer, H.J. 1926. Foraminifera of the Midway Formation in Texas. *University of Texas Bulletin*, 2644: 1–206. [published in 1927]

Postuma, J.A. 1971. *Manual of Planktonic Foraminifera.* Elsevier Publishing Company, Amsterdam-London-New York.

Premoli Silva, I. and Bolli, H.M. 1973. Late Cretaceous to Eocene planktonic Foraminifera and stratigraphy of Leg 15 Sites in the Caribbean Sea. In Edgar, N.T., Saunders, J.B. et al (eds), *Initial Reports of the Deep Sea Drilling Project*, 15: 499-547. United States Government Publishing Office, Washington, D.C.

Rao, B.R.J., Mamgain, V.D. and Sastry, M.V.A. 1968. *Globotruncana* in Ariyalur Group of Trichinopoly Cretaceous, South India. *Memoires of the Geological Society of India*, 2: 18–29.

Rasheed, D.A. and Govindan, A. 1968. Upper Cretaceous Foraminifera from Vridhachalam, South India. *Memoires of the Geological Society of India*, 2: 66–84.

Robaszynski, F. and Caron, M. 1995. Foraminifères planctoniques du Crétacé: commentaire de la zonation Europe-Mèditerranée. *Bulletin de la Société Géologique de France*, 6: 681–692.

Robaszynski, F., Caron, M. and The European Working Group on Planktonic Foraminifera 1979. Foramnifères planctoniques du Crétacé noyen (Mer Boréale et Téthys). *Cahiers de Micropaléontologie*, 2: 1–181.

Robaszynski, F., Caron, M., Gonzalez Donoso, J.M., Wonders, A.A.H. and The European Working group on Planktic Foraminifera 1984. Atlas of Late Cretaceous Globotruncanids. *Revue de Micropaléontologie*, 26: 145–305.

Robaszynski, F., González Donoso, J.M., Linares, D., Amédro, F., Caron, M., Dupuis, C., Dhont, A.V. and Gartner, S. 1998. The Upper Cretaceous of the Kalaat Senan region, central Tunisia. Integrated litho-biostratigraphy based on ammonites, planktonic foraminifera and nannofossil zones from upper Turonian to Maastrichtian. *Bulletin du Centre de Recherches Elf Exploration Production*, 22: 359–489.

Salaj, J. 1983. Quelques problèmes taxinomiques concernant les foramnifères planctoniques et al zonation du Sénonien supérieur d'El Kef. *Geologiký Zborník-Geologica Carpathica*, 34: 187–212.

Schijfsma, E. 1946. The Foraminifera from the Hervian (Campanian) of southern Limburg. *Mededeelingen van de Geologische Stichting*, 7: 1–174.

Seiglie, G.A. 1958. Notas sobre algunos foraminiferos planctonicos del Cretacico superior de la Cuenca de Jatibonico. *Memorias de la Sociedad Cubana de Historia Natural*, 14: 53–89.

Sliter, W.V. 1968. Upper Cretaceous Foraminifera from southern California and northwestern Baja California, Mexico. *The University of Kansas Paleontological Contributions*, 49: 1–141.

Sliter, W.V. 1973. Upper Cretaceous foraminifers from the Vancouver Island Area, British Columbia, Canada. *Journal of Foraminiferal Research*, 3: 167–186.

Solakius, N. 1983. Biozonation of the Maastrichtian of section II at Kef en Nsoura, Tunisia, and taxonomy if some Globotruncanidae. *Revista Española de Micropaleontología*, 15: 241–256.

Steineck, P.L. and Fleisher, R.L. 1978. Towards the classical evolutionary reclassification of Cenozoic Globigerinacea (Foraminiferida). *Journal of Paleontology*, 52: 618–635.

Takayanagi, Y. 1960. Cretaceous Foraminifera from Hokkaido, Japan. *Science Reports of the Tohoku University, Sendai, Japan, Second Series (Geology)*, 32: 1–154.

Tilev, N. 1951. Étude des Rosalines Maastrichtiennes (genre Globotruncana) du sud-est de la Turquie (Sondage de Ramandag). *Bulletin de la Laboratoires de Géologie, Minéralogie, Géophysique et Museum Géologique, Université du Lausanne*, 103: 1–101.

Walaszczyk, I., Wood, C.J., Less, J.A., Peryt, D., Voigt, S. and Wiese, F. 2010. The Sazgitter-Salder Quarry (Lower Saxony, Germany) and Słupia Nadbrzeżna river cliff section (central Poland): a proposed candidate composite Global Boundary Stratotype Section and Point for the base the Coniacian Stage (Upper Cretaceous). *Acta Geologica Polonica*, 60: 445–477.

Weidich, K.F. 1984. Feinstratigraphie, Taxonomie planktonischer Foraminiferen und Palökologie der Foraminiferenfauna der kalkalpinen tieferen Oberkreide (Untercenoman-Untercampan) der Bayerischen Alpen. *Bayerische Akademie der Wissenschaften, Matematisch-Naturwissenschaftliche Klasse, Abhandlungen*, 162: 1–151.

Weidich, K.F. 1997. Perforated portici and imperforated tegilla. On Upper Cretaceous planktonic foraminiferal taxonomy. *Revue de Micropaléontologie*, 30: 52–62.

Yassini, I. 1979. Maastrichtian-Lower Eocene biostratigraphy and the planktonic foraminiferal biozonation in Jordan. *Revista Española de Micropaleontología*, 11: 5–57.

Zapeda, M.A. 1998. Planktonic foraminiferal diversity, equitability and biostratigraphy of the uppermost Campanian-Maastrichtian, ODP Leg 122, Hole 762C, Exmouth Plateau, NW Australia, eastern Indian Ocean. *Cretaceous Research*, 19: 117–152.

Ziegler, J.H. 1957. Die Fauna der Cardientones der Oberpfalz und die Bedeutung der Foraminiferen für seine Altersbestimmung (Coniac). *Geologica Bavarica*, 30: 55–86.

In: Deep-Sea
Editors: D. R. Bailey and S. E. Howard

ISBN: 978-1-62257-173-4

Chapter 4

PROTEIN DEGRADATION OF DEEP-SEA SEDIMENT: COLLAGENOLYTIC ENZYMES PRODUCED FROM A NOVEL MARINE BACTERIUM

Atsushi Kurata[1,*], Tohru Kobayashi[2] and Koki Horikoshi[2]
[1]Department of Applied Biological Chemistry, Kinki University, Nara City, Nara, Japan
[2]Institute of Biogeosciences, Japan Agency for Marine-Earth Science and Technology (JAMSTEC), Yokosuka, Japan

ABSTRACT

Collagen, a well-characterized polypeptide, constitutes the extracellular matrix in higher organisms, such as whales. Deep-sea biological communities were found around whale bones near the summit of the Torishima Seamount (4,026 m deep). From the whale bone community, we obtained a novel collagenolytic enzyme-producing bacterium belonging to genus *Alkalimonas*. The bacterium produced two proteases, which degrades collagen, gelatin, and casein. The bacterial proteases could play an important role in the degradation of whale carcasses.

INTRODUCTION

In general, deep-sea environments (3,000-6,000 m deep) are characterized by low temperature, high hydrostatic pressure and the absence of solar radiation [1]. Thus, the abyssal environment is thought to be a stable system, but the outburst of activity in the locations of whale falls was known. A recent study of whale-falls has suggested various developments of the communities that colonize and devour whale carcasses. A four-stage model of succession was proposed [2]. According to this model, whale carcass degradation would begin with the stage of "mobile scavengers" which remove soft tissue. An "enrichment/opportunistic" stage would follow, which is characterized by aggregations of

---

* E-mail address: kurata090401@nara.kindai.ac.jp.

polychaetes and crustaceans. Next, a "sulfophilic" stage includes chemoautotrophic bacterial mats and animals that use chemoautotrophic endosymbionts such as *Idas washingtonia* and vesicomyid clams. Finally, a "reef" stage might occur after organic substances are exhausted and the remaining minerals were provided to suspension and filter feeder. In contrast, it was suggested that all stages do not occur at all whale-falls, and may overlap significantly in some cases [3]. Additionally, numerous aerobic and anaerobic bacteria have been isolated from the sediment adjacent to whale carcasses [4, 5]. Therefore, various animals and microorganisms were probably involved in the succession of the whale carcass.

Collagen, a well-characterized polypeptide containing numerous repeats of the tripeptide Gly-Pro-X. It is an insoluble structural protein found in animals, whales, fishes, sponge, sea urchin, and sea stars; in fact, collagen comprises more than 80% of total fish skin proteins [6]. Various microorganisms that produce collagenolytic proteases have been isolated [7]. Collagenolytic enzymes are divided into two groups: collagenases (E.C. 3.4.24.3), having a specific site for cleavage of collagen, and collagenolytic proteases (E.C. 3.4.21.62), which degrade various proteinaceous substrates including collagen. In marine environments, the collagen of higher organisms such as fish and whales [8, 9] could be important sources of nitrogen and participate in nitrogen recycling. However, there are few reports on marine collagenolytic enzyme-producing bacteria [6, 10]. Thus, new information about marine enzymes will be useful to elucidate the recycling of nitrogen derived from collagen.

A whale skeleton was discovered on the summit of the Torishima Seamount, 4,037 m deep, northwest Pacific Ocean [11]. The skeleton was encrusted with mussels and harbored galatheid crabs, echinoderms, sea anemones and tube worms, and various bacteria were found in the sediment near whale bones. We isolated a collagenolytic enzyme-producing bacterium, strain AC40, from the sediment around whale bones (4,026 m deep) [12]. The strain grows by utilizing collagen at initial pH values of 7.0 to 10.5, and it produces at least two extracellular proteases. In this chapter, we describe the isolation of bacterium and the biochemistry and genetics of proteases.

1. Collagenolytic Enzyme-Producing Bacterium

Some marine collagenolytic enzyme-producing bacteria belonging to genera Vibrio and *Pseudoalteromonas* have been reported [6, 10]. Studies on the collagenolytic proteases from marine bacteria are rather minor and preliminary.

In Section 1, we describe a psychrotolerant, obligatorily alkaliphilic bacterium that was isolated from a deep-sea sediment. Based on a phylogenetic analysis of its 16S rRNA gene sequence, DNA–DNA hybridization data and phenotypic characteristics, this organism is shown to belong to the genus *Alkalimonas*. At present, the genus *Alkalimonas* comprised just two species, *Alkalimonas amylolytica* N10^T and *Alkalimonas delamerensis* 1E1^T [13]. *A. amylolytica* N10^T was isolated from Lake Chahannor (39° 149' N 108° 049' E) located in Inner Mongolia Autonomous Region, China. *A. delamerensis* 1E1^T was isolated from Lake Elmenteita (0° 259' S 36° 159' E) in Kenya, East Africa. Both microorganisms are Gram-negative, slightly halophilic, alkaliphilic and mesophilic bacteria and belong to the class Gamma proteobacteria (formerly the gamma-3 subdivision of the Proteobacteria).

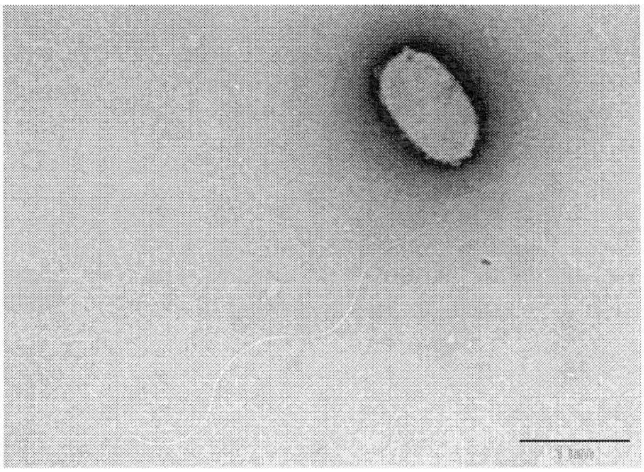

Bar, 1μm.

Figure 1. Transmission electron micrograph of a negatively stained cells of strain AC40 [12].

Thus, the bacterial isolate is the first novel species of the genus *Alkalimonas* to be isolated from a deep-sea sediment.

1.1. Isolation of Collagenolytic Enzyme-Producing Bacterium

Deep-sea sediments were collected around whale bones off Torishima Island (30° 55.0509' N 141° 48.9809' E), northwest Pacific Ocean, at a depth of 4,026 m [12]. An aliquot of the sample was spread on an alkaline plate consisting of (w/v) 1.0 % collagen from bovine Achilles tendon, type I (Sigma, St. Louis, MO), 0.1 % bonito extract (Wako Pure Chemical, Osaka, Japan), 0.5% yeast extract (Difco, Detroit, MI), 1.0% NaCl, 0.2 % $MgSO_4 \cdot 7H_2O$, 0.02 % K_2HPO_4, 1.0 % $NaHCO_3$ (sterilized separately) and 1.5 % agar (pH 9.0). After incubation at approximately 4°C for 1 month, bacteria having collagenolytic activity had grown on the plate with a zone of collagen dissolution around the colonies. After cultivation in an alkaline liquid medium at 4°C for 10 days, a bacterium showing apparent collagenolytic activity was finally selected. The bacterial isolate was subsequently purified three times on alkaline plates at 15°C. The isolate, designated strain AC40, was maintained on an alkaline plate without collagen at 4°C. Cells of strain AC40 were strictly aerobic, Gram-negative rods (0.6–0.86 × 1.1–2.2 mm) that were motile by means of a single polar flagellum (Figure 1).

1.2. Phylogenetic Position of Strain AC40

To determine phylogenetic relationships, 1,506 bp of the 16S rRNA gene sequence of strain AC40 was amplified by PCR with the eubacterial primers 27f and 1525r. The results of phylogenetic analysis based on 16S rRNA gene sequence data are shown in Figure 2. The nucleotide sequence of the 16S rRNA gene of strain AC40 was most closely related to those of *A. amylolytica* $N10^T$ and *A. delamerensis* $1E1^T$ with similarities at 96.7% and 97.9% identities, respectively. DNA–DNA hybridization experiments of strain AC40 with *A. delamerensis* $1E1^T$ were carried out as described by [14].

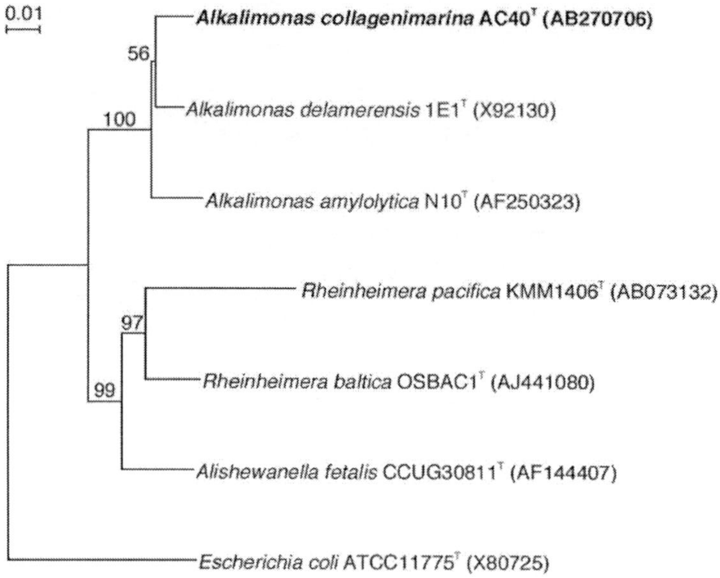

Figure 2. Neighbor-joining phylogenetic tree showing the relationships between strain AC40 and related bacteria based on 16S rRNA gene sequences [12].

Strain AC40 shared only 30% DNA–DNA relatedness at the whole genome level with *A. delamerensis* 1E1^T. This result suggested that strain AC40 represents a novel species of the genus *Alkalimonas*.

Bootstrap values were calculated from multiple re-samplings of the sequence data set, which are the basis for multiple tree topologies. GenBank accession numbers of 16S rRNA gene sequences are given in parentheses. Bar, 0.01 nucleotide substitutions per site.

1.3. Morphological and Physiological Properties of Strain AC40

Morphological and physiological tests with strain AC40 were performed according to Barrow and Feltham [15] and Baumann *et al.* [16] except regarding the use of alkaline liquid media. Strain AC40 grew over the temperature range 5–37°C with optimal growth at 33°C. It grew at initial pH values of 7.0–10.5 with optimal growth at pH 8.5–10.0. The optimum NaCl concentration for growth was 1% (w/v), although the strain was able to grow at concentrations from 0 to 10 % NaCl. Carbohydrate utilization was investigated based on acid production, which was assessed by using a modified OF basal medium [17] consisting of artificial seawater (1.5% NaCl, 0.035% KCl, 0.54% $MgCl_2 \cdot 6H_2O$, 0.27% $MgSO_4 \cdot 7H_2O$, 0.05% $CaCl_2 \cdot 2H_2O$), 0.05% $(NH_4)_2SO_4$, 0.01% yeast extract, 0.05% Tris, 1.0% carbohydrate (sterilized separately) and 0.003% bromothymol blue (pH 7.1) at 33°C. Strain AC40 utilized and produced acid from cellobiose, D-glucose, maltose, D-mannose and sucrose, but did not produce acid from L-arabinose, D-fructose, D-galactose, glycerol, myo-inositol, D-lactose, D-mannitol, D-raffinose, L-rhamnose, D-sorbitol, D-trehalose, xylose or palatinose. The strain was positive for hydrolysis of gelatin, casein, starch, tributyrin and Tweens 20, 40 and 80. It was positive for catalase and oxidase. It was negative for reduction of nitrate to nitrite and formation of indole and H_2S.

Whole-cell fatty acids were analyzed with a GC-MS system (GCMS-QP5050A; Shimadzu, Kyoto, Japan) according to the method of Komagata and Suzuki [18]. The cellular fatty-acid profile of strain AC40 was dominated by saturated and mono-unsaturated straight-chain components ($C_{16:0}$ and $C_{18:1}$), which is one of the distinctive features of the genus *Alkalimonas* [13].

Based on phylogenetic analysis of its 16S rRNA gene sequence, morphological and physiological features and predominant fatty acids, strain AC40 is considered to belong within the genus *Alkalimonas*. The genus *Alkalimonas* comprised just two species, *A. amylolytica* N10^T and *A. delamerensis* 1E1^T [13]. As described above, strain AC40 is a novel species of the genus *Alkalimonas*, which is probably involved in the degradation of whale carcasses in the deep-sea environment. Thus, strain AC40 was designated as *Alkalimonas collagenimarina* AC40^T.

2. Collagenolytic Proteases of *A. Collagenimarina* AC40^T

The bacterial isolate *A. collagenimarina* AC40^T can grow under utilization of collagen over the temperature range 5–37°C and can produce at least two extracellular proteases with molecular masses of 28 and 55 kDa, each of which has a different N-terminal amino acid sequence. It has been known that some Gram-negative bacteria produce alkaline serine proteases belonging to subtilase families A, B, and C [19]. We are trying to determine what kinds of protease are produced by genus *Alkalimonas* and also which proteases degrade primarily collagen because in marine environments, the collagen of higher organisms such as whales could be important sources of nitrogen and participate in nitrogen recycling.

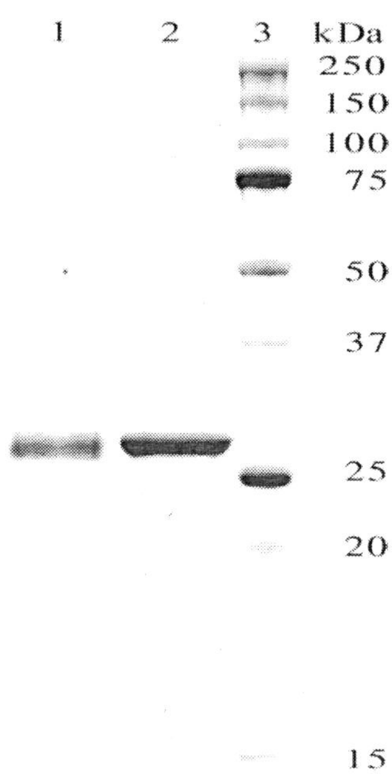

Lane 1, AcpI (1 µg); lane 2, rAcpI (16 µg); lane 3, molecular weight marker.

Figure 3. SDS-PAGE of AcpI and rAcpI [20].

In Section 2, we described enzymatic properties of two proteases, AcpI and AcpII. Furthermore, two genes coding for the proteases were cloned and deduced amino acid sequences were determined.

2.1. Purification of AcpI

In the culture supernatant of strain AC40^T, an alkaline protease (AcpI) was found and purified [20]. Strain AC40^T cells were grown aerobically at 15°C for 2 days in an alkaline medium containing 0.1% bonito extract (Wako Pure Chemical), 0.5% yeast extract (Difco), 1.0% NaCl, 0.2% $MgSO_4 \cdot 7H_2O$, 0.02% K_2HPO_4, and 1.0% $NaHCO_3$ (sterilized separately). The culture supernatant (300 ml) of strain AC40^T was sequentially concentrated and diluted with 100 mM glycine–NaOH buffer (pH 10.0) using an ultrafiltration system (Vivaflow 50, *Mr* cut-off 5 kDa; Vivascience, Hanover, Germany). AcpI was purified with a DEAE-Toyopearl 650M column (2.5× 16 cm; Tosoh, Tokyo, Japan) and a phenyl-Toyopearl column (2.5 × 6.5 cm; Tosoh) to homogeneity as judged by SDS-PAGE (Figure 3, Lane 1). Enzyme activity was measured with casein as the substrate [20].

2.2. Cloning of Alkaline Protease Gene

The *acpI* gene encoding AcpI was shotgun-cloned and sequenced [20]. Genomic DNA prepared from the strain was digested with restriction enzyme, *Eco*RI. The DNA fragment was ligated into the *Eco*RI-digested pHY300PLK plasmid (Takara Bio, Kyoto, Japan). The recombinant plasmid was introduced into *Bacillus subtilis* ISW1214 (Δ*apr*E Δ*npr*E). Transformants were inoculated on Luria-Bertani (LB) agar containing 2% (w/v) skim milk (Difco). The plate was incubated at 30°C for 5 days, and protease-producing clones formed clear halos around colonies. The recombinant plasmid containing a DNA fragment was subcloned with *Escherichia coli* cells. The nucleotide sequence of the DNA fragment encoding AcpI was determined. A single open reading frame (ORF) encoding AcpI, which started from an ATG codon at nucleotide +1 and ended with a TAA codon at nucleotide +1,615, was found in the sequence. The *acpI* gene encodes a protein of 538 amino acids with the calculated molecular mass of 56,810 Da.

The *acpI* gene product consisted of three parts, namely, an N-terminal prepropeptide, a catalytic domain, and a prepeptidase C-terminal domain (PPC domain), according to the prediction of a conserved domain search (http://www.ncbi.nlm.nih.gov/Structure/cdd/wrpsb.cgi). The determined N-terminal amino acid sequence of the purified AcpI from *A. collagenimarina* AC40^T, Ala-Gln-Thr-Thr-Pro-Tyr-Gly-Ile-Thr-Met-Val, was found in the initial Ala^{150} to Val^{160} [20]. Therefore, Met^1 through Met^{149} is the prepropeptide sequence. A prepeptide works as a signal peptide and a propeptide works as a molecular chaperone for a correctly folded protein; subsequently, it is cleaved off by autolysis to give an active protease.

The predicted mature AcpI showed the highest homology to AprP (AAA84886) from *Pseudomonas* sp. KFCC 10818 and an alkaline protease precursor of VapK (AAF23173) from *Vibrio metschnikovii* RH530 with 74 and 73% identities, respectively. As shown in Figure 4, amino acid sequences around the catalytic triads (Asp^{30}, His^{63}, and Ser^{219} by AcpI numbering), P1 cleft site (Ser^{127}-Leu^{128}-Gly^{129}), and oxyanion hole residue Asn^{157} are highly conserved in all proteases that belong to subtilase family A [19]. The putative Ca^{2+}-binding sites of AcpI are shown in Figure 4. The Ca^{2+}-binding sites of subtilisin Carlsberg were highly conserved in AcpI [21].

Amino acid sequences of AcpI (AB305158), AprP (AAA84886) from *Pseudomonas* sp. KFCC 10818, VapK (AAF23173) from *V. metschnikovii* RH530, and AprIV (BAB86297) from *Alteromonas* sp. O-7, and subtilisin Carlsberg (Carl) (P00780) from *Bacillus licheniformis* were aligned by the Clustal method using DNASTAR software (DNASTAR, Madison, WI). All proteases belong to subtilase family A [19]. Amino acid numbers of each sequence are shown on the left and right sides. The black and white boxes show conserved amino acid residues in proteases from both Gram-negative and Gram-positive bacteria and proteases from Gram-negative bacteria, respectively. The active sites of these proteases are marked by open circles. The Ca^{2+}-binding sites of subtilisin Carlsberg are indicated by arrowheads [21].

2.3. Purification and Characterization of Recombinant AcpI

To characterize of the *acpI* gene product, heterologous expression system of the gene was constructed with *E. coli* cells. A DNA fragment encoding AcpI was amplified by polymerase chain reaction (PCR) [20]. The amplified fragment was digested with *Xho*I and *Eco*RI and ligated into the corresponding restriction sites in a pRSET-A plasmid (Invitrogen, Carlsbad, CA). The resultant plasmid was introduced into *E. coli* BL21 (DE3) pLysS (Invitrogen). Recombinant AcpI (rAcpI) was produced in a culture of the recombinant *E. coli*. The enzyme was purified to homogeneity as judged by SDS-PAGE (Figure 3, Lane 2). The molecular mass of rAcpI was around 28 kDa, which is the same as that of AcpI.

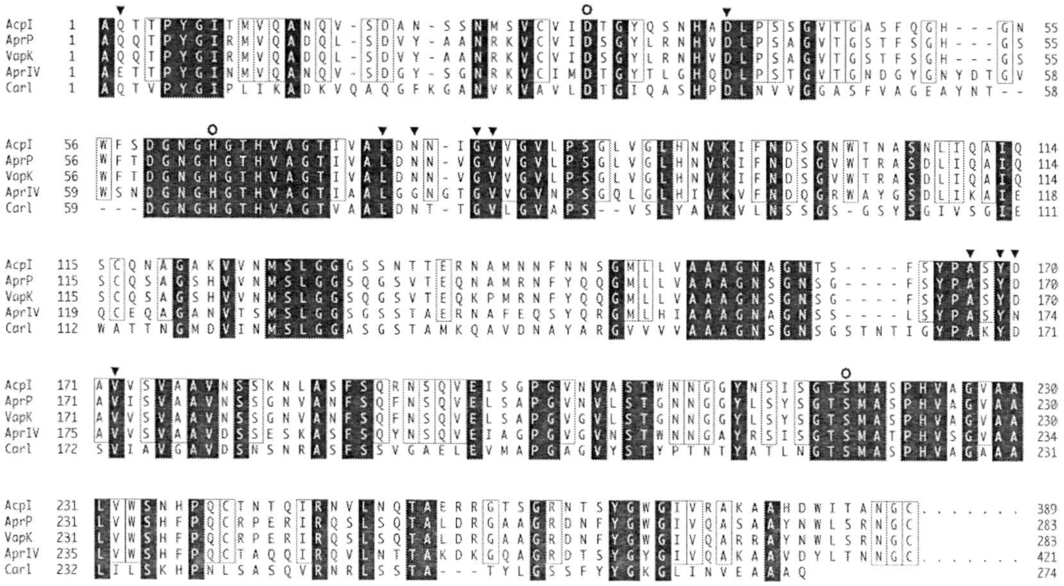

Figure 4. Multiple alignment of deduced amino acid sequence of catalytic domain of AcpI with those of other proteases from Gram-negative and Gram-positive bacteria [20].

Table 1. Hydrolysis of proteinaceous substrates by rAcpI [20]

Substrate	Relative activity (%)
Casein	64
Collagen	24
Gelatin	100
Elastin	4.5
Keratin	1.8

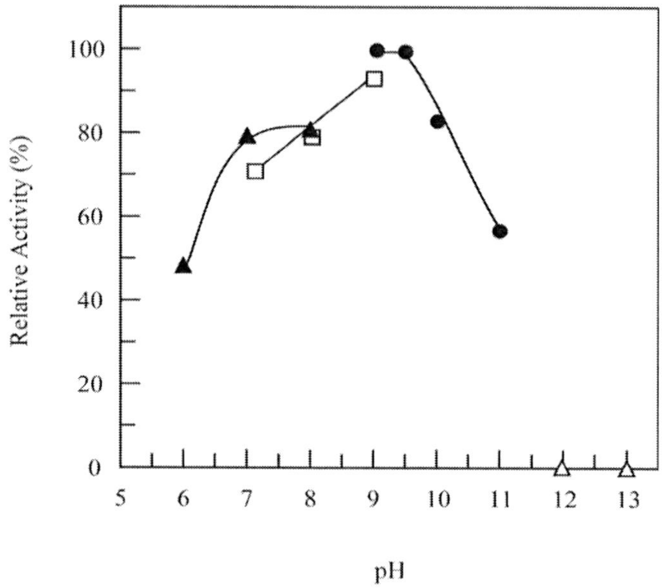

Figure 5. Effect of pH on enzymatic activity [20].

The reduced molecular masses of AcpI and rAcpI are probably caused by removal of a PPC domain from the secreted AcpI and rAcpI molecules. N-terminal amino acid sequence, Ala^{150}-Gln-Thr-Thr-Pro-Tyr-Gly-Ile-Thr-Met-Val^{160}, is completely identical to that of AcpI [20].

Among natural proteinous substrates, casein was found to be a suitable substrate for rAcpI, although rAcpI showed the highest activity toward gelatin (Table 1). Additionally, the enzyme showed some activity toward collagen but less toward other insoluble proteins, keratin and elastin.

The effect of pH on rAcpI activity was examined over the pH range from 6 to 13. The optimal pH was found to be 9.0–9.5 in 100 mM glycine–NaOH buffer and the enzyme did not work at a pH higher than 12 in 100 mM KCl–NaOH buffer (Figure 5).

The activity was measured at 50°C for 15 min in a total volume of 100 μl that contained 0.28 μg rAcpI in the following buffers (100 mM): closed triangles, phosphate (pH 6.0–8.0); open squares, Tris–HCl (pH 7.1–9.0); closed circles, glycine–NaOH (pH 9.0–11.0); open triangles, KCl–NaOH (pH 12.0– 13.0). The activity in glycine–NaOH buffer (pH 9.0) is taken as 100%.

The optimal temperature for the activity of rAcpI was around 45°C at pH 9.5 in 100 mM glycine–NaOH buffer, and the addition of 1 mM $CaCl_2$ shifted it from 45 to 50°C (Figure 6).

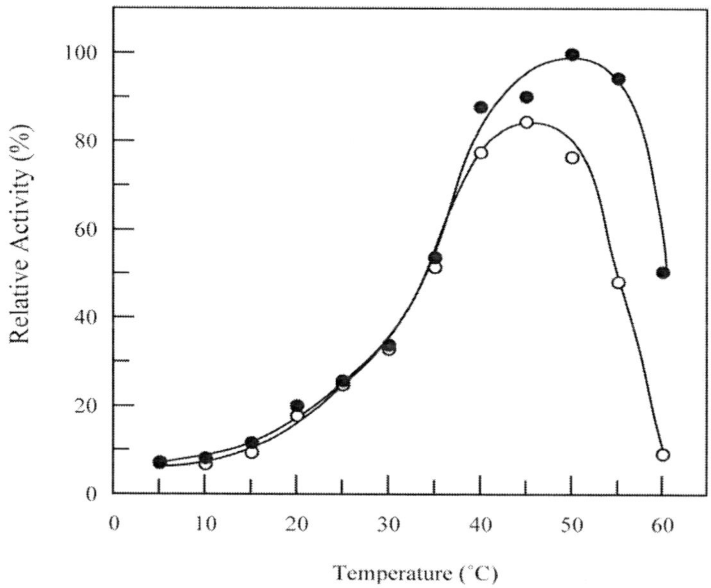

Figure 6. Effect of temperature on enzyme activity [20].

rAcpI (0.28 μg) was added to 0.5% (w/v) casein in 100 mM glycine–NaOH buffer (pH 9.5) in a total volume of 100 μl. The reaction was carried out for 15 min at various temperatures. The activities are expressed as a percentage of the activity at 50°C in the presence of 1 mM $CaCl_2$. The relative activities in the presence of $CaCl_2$ are indicated as closed circles and, in the absence of $CaCl_2$, as open circles.

rAcpI was incubated at 15°C for 20 min in 100 mM Tris–HCl buffer (pH 7.0) in the presence of the several metal ions (2 mM each), and the residual caseinolytic activity was measured under the standard condition of the enzyme assay. The enzyme activity was slightly increased to 9% by Ca^{2+} ions. Mg^{2+}, Co^{2+}, Mn^{2+}, Fe^{2+}, Zn^{2+}, Ni^{2+}, and Cu^{2+} ions did not significantly affect the enzyme activity.

rAcpI was treated with 1 mM phenylmethylsulfonyl fluoride (PMSF) or 5 mM ethylene diamine tetraacetate (EDTA) in 100 mM Tris–HCl buffer (pH 7.0) at 15°C for 20 min, and the residual activity was measured. PMSF completely abolished the enzyme activity, consistent with the classification of AcpI as a serine protease. EDTA inhibited the enzyme activity by 51%, suggesting that part of the enzyme activity and/or stability depends on the presence of Ca^{2+} ions near the surface of the enzyme. AprI from *Alteromonas* sp. strain O-7 was also inhibited by both PMSF and EDTA [22].

2.4. Purification and Properties of AcpII

In the culture supernatant of strain $AC40^T$, we found another protease, AcpII. To characterize AcpII, purification was carried out at 4°C. *A. collagenimarina* $AC40^T$ cells were grown aerobically at 15°C for 2 days in an alkaline medium. AcpII was purified from the culture supernatant of strain $AC40^T$ with a DEAE-Toyopearl 650 M column (2.5 × 16 cm; Tosoh) and a Butyl-Toyopearl column (2.5 × 6.5 cm; Tosoh). The enzymatic activity of AcpII was measured with collagen as the substrate. The acid-soluble peptides produced were quantified by a DC protein assay kit (Bio-Rad, Hercules, CA) [23].

As shown in Figure 7 (Lane 2), AcpII migrated as two protein bands with molecular masses of around 55 and 65 kDa on SDS–PAGE. The N-terminal amino acid sequences of both proteins were completely the same, Ala-Gln-Gln-Thr-Pro-Tyr-Gly-Tyr-Thr-Met-Val-Gln-Ala-Asp-Gln-Val-Ser-Asp-Gln. We suppose that both proteins had the same origin, AcpII. In fact, the 55-kDa protein was derived from a deletion of one of the PPC domains during secretion of an AcpII precursor as described below. In the course of two purification steps, a 364-fold purification was obtained with a specific activity of 115.9 U/mg toward collagen and a recovery of 24.9% of the original activity.

Among the proteinaceous substrates tested, AcpII preferred hydrolysis of gelatin and collagen to casein (Table 2). Additionally, the enzyme showed a bit of activity toward other insoluble proteins, such as keratin and elastin.

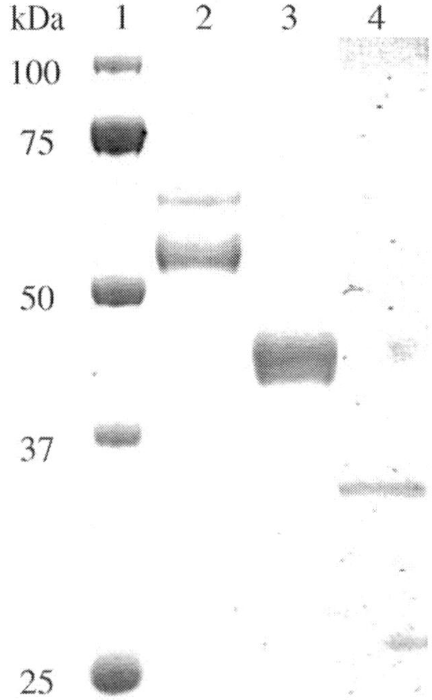

Lane 1; molecular weight marker, lane 2; AcpII (1.0 μg), lane 3; AcpII-C (2.5 μg), and lane 4; AcpII-C-ΔPA (1.6 μg).

Figure 7. SDS-PAGE of AcpII and AcpII derivatives [23].

Table 2. Substrate specificity of AcpII [23]

Substrate	Relative activity (%)
Casein	27
Collagen	81
Gelatin	100
Elastin	7.0
Keratin	1.5

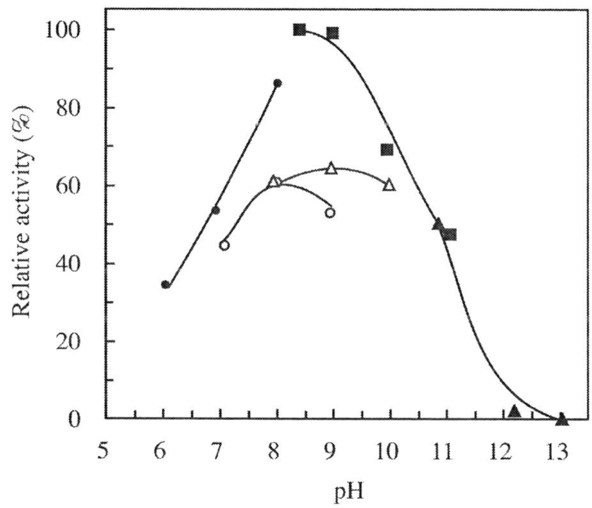

Figure 8. Effect of pH on AcpII activity [23].

The effects of pH on AcpII activity were examined over the pH range of 6 to 13. The optimal pH was 8.5–9.0 in 100 mM glycine–NaOH buffer. In the same pH range, the maximal enzyme activity was reduced to around 40% to 60% in Tris–HCl buffer and borate–KCl–NaOH buffer (Figure 8). The optimal temperature for the activity of AcpII was around 45°C at pH 9.5 in 100 mM glycine–NaOH buffer. To assess the thermal stability of AcpII, the enzyme was incubated at various temperatures for 15 min in 100 mM Tris–HCl buffer (pH 8.0) with or without 1 mM $CaCl_2$, and the residual collagenolytic activity was measured. AcpII was stable up to 40°C, and the residual activity was 80%, 60%, or 3% of the original activity after 15 min incubation at 45°C, 50°C, or 55°C, respectively. $CaCl_2$ slightly increased the thermal stability of AcpII up to 45°C, and the residual activity was decreased to 80% after incubation at 50°C for 15 min.

Enzyme activity was measured at 45°C for 25 min in a total volume of 100 μl that contained 1.9 μg AcpII in the following buffers (100 mM): closed circles MOPS (pH 6.0–8.0), open circles Tris–HCl (pH 7.1–9.0), open triangles borate–KCl–NaOH (pH 8.0–10.0), closed square glycine–NaOH (pH 8.5–11.1), closed triangles KCl–NaOH (pH 10.9–13.0). The activity in glycine–NaOH buffer (pH 9.0) is taken as 100%.

AcpII was treated with 1 mM PMSF or 5 mM EDTA in 100 mM Tris–HCl buffer (pH 7.0) at 15°C for 20 min, and the residual activity was measured. PMSF completely abolished the enzyme activity, which is consistent with the classification of AcpII as a serine protease. EDTA inhibited the enzyme activity by 35%, suggesting that part of the enzyme activity and/or stability depends on the presence of divalent cations, probably Ca^{2+} ions, near the surface of the enzyme. AprI from *Alteromonas* sp. strain O-7 [22] and recombinant AcpI from *A. collagenimarina* AC40^{T} were also inhibited by both PMSF and EDTA [20].

2.6. Cloning and Sequencing of *acpII* Gene

A part of the gene for AcpII (*acpII* gene) was amplified by degenerate PCR with a sense primer, 5'-TAYGGNTAYACNATGGT-3', an antisense primer, 5'-

TCNACYTGNGCRTTNCKYTG-3', and the genomic DNA of *A. collagenimarina* AC40^{T} as the template using Ex Taq DNA polymerase (Takara Bio). Based on the peptide fragment derived from purified AcpII, the degenerate primers were prepared. The 5' region and 3' region of the gene were amplified by an LA PCR *in vitro* Cloning kit (Takara Bio) according to the manufacturer's instructions.

The nucleotide sequence of the *acpII* gene and its flanking regions (2,520 bp) was sequenced as shown in Figure 9. We found one ORF with an ATG initiation codon at nucleotide 1 and a TAA termination codon at nucleotide 2,281 that encodes a protein of 760 amino acids with a calculated molecular mass of 79,012 Da. The G + C content of the ORF was 51%. A possible ribosome-binding site, with the sequence 5'-GGAGGGT-3', was located seven bases upstream from the initiation codon. Putative promoter sequences, with 5'-TTGAAT-3' as the potential −35 region and 5'-TAAGTT-3' as the potential −10 region, separated by 17 bp, were located 135 bp upstream from the initiation codon. An inverse-repeat sequence was found 19 bp downstream of the TAA stop codon.

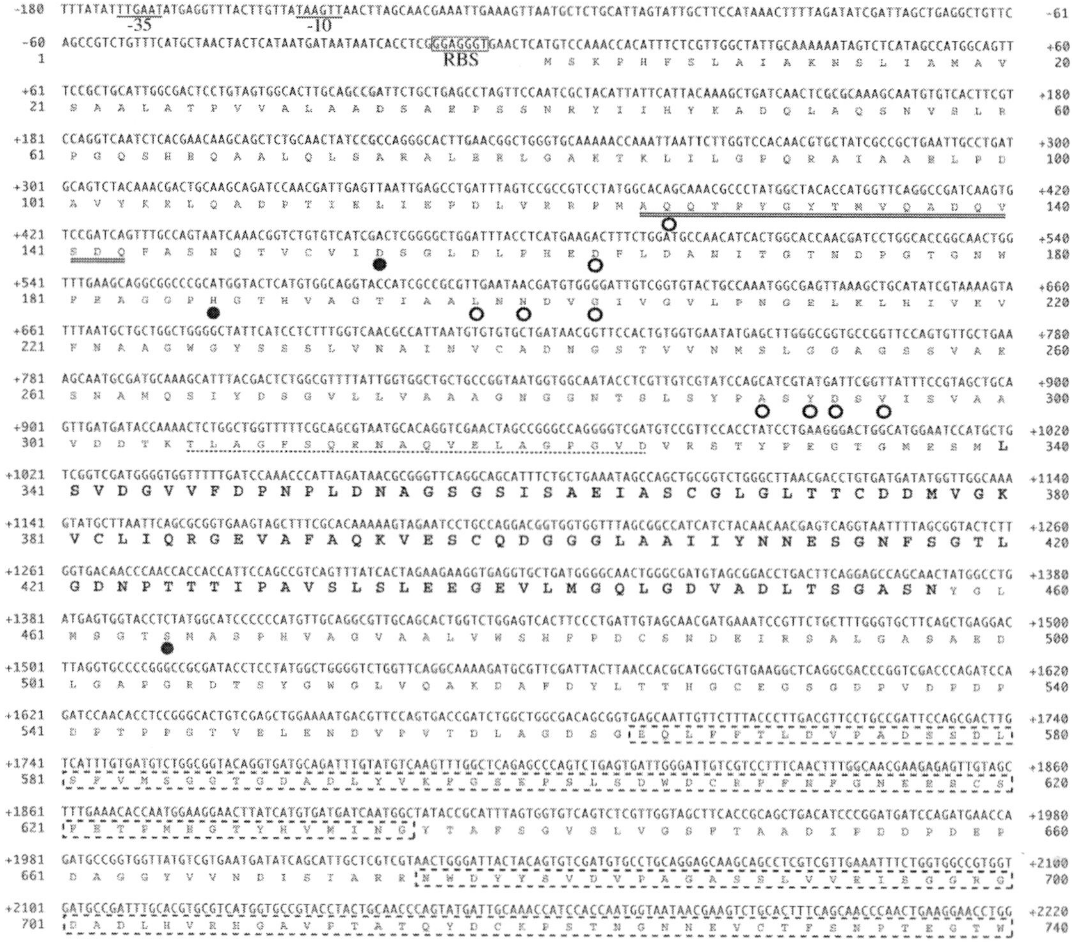

Figure 9. Complete nucleotide sequence and deduced amino acid sequences of AcpII [23].

In the 2,520 bp determined, the nucleotide sequence of the *acpII* gene and its flanking regions are shown. The possible −35 and −10 consensus promoter sequences are underlined. The putative ribosome-binding site is boxed. The deduced amino acids are indicated by the single-letter codes under the nucleotide sequence. The N-terminal amino acid sequence (Ala^{125} to Gln^{143}) and the internal one (Thr^{306} to Asp^{325}) are double underlined and dash underlined, respectively. Inverted repeats downstream from the stop codon TAA (asterisk) of the ORF are designated by convergent arrows. A possible catalytic triad, Asp^{154}, His^{187}, and Ser^{465}, is indicated by filled circles beneath the deduced amino acid sequences. Possible Ca^{2+} binding sites, Gln^{126}, Asp^{163}, Leu^{198}, Asn^{200}, Gly^{203}, Ala^{290}, Tyr^{292}, Asp^{293}, and Val^{295}, are indicated by open circles. The putative PA domain is indicated by boldface type (Leu^{340} to Asn^{457}). The putative PPC domains 1 and 2 are surrounded by dash boxes (PPC domain 1; Asn^{565} to Glu^{635}, PPC domain 2; Asn^{676} to Glu^{760}).

The overall deduced amino acid sequence of AcpII was compared with available protein sequences in the Basic Local Alignment Search Tool program (http://blast.ncbi.nlm.nih.gov/). AcpII exhibited the closest sequence similarity to the cold-active alkaline serine proteases (YP_269576 and ZP_01131839) of *Colwellia psychrerythraea* 34H and *Pseudoalteromonas tunicata* D2 with 49% identities. The next closest level of similarity was a cold-active alkaline serine protease (YP_267500) of *C. psychrerythraea* 34H with 43% identity.

AcpII consisted of four domains, namely the N-terminal prepropeptide, catalytic domain, which includes the protease-associated domain (PA domain), and tandem repeat PPC domains, as was predicted by a conserved domain search (http://www.ncbi.nlm.nih.gov/Structure/cdd/wrpsb.cgi). The determined N-terminal amino acid sequence of the purified AcpII from *A. collagenimarina* $AC40^T$ was found in the internal Ala^{125} to Gln^{143}, as shown in Figure 9. Therefore, Met^1 through Met^{124} may work as a prepropeptide sequence. The prepeptide is thought to be cleaved between Ala^{32} and Ala^{33} according to the SignalP 3.0 program (http://www.cbs.dtu.dk/services/SignalP/), and a propeptide would be further cleaved between Met^{124} and Ala^{125}. Thus, the propeptide of AcpII consists of 92 amino acids and probably works as a molecular chaperone for a correctly folded protein, and then is subsequently cleaved off by autolysis to give an active protease molecule.

Predicted tandem repeat PPC domains (Pfam family PF04151) were present in the C-terminal region of AcpII. The PPC domain is normally found in the C-terminal region of Gram-negative bacteria secretory peptidases [24]. A tandem repeat C-terminal extension like AcpII has is also reported in other peptidases [25, 26]. As described above, the purified AcpII from *A. collagenimarina* $AC40^T$ migrated as two protein bands with molecular masses of around 55 and 65 kDa on SDS-PAGE. The difference between the molecular masses is around 10 kDa, which is inferred to be almost the same molecular mass as the one PPC domain closest to the C terminus (9 kDa). It suggests that one of the PPC domains is cleaved from a mature AcpII (65 kDa) to yield the 55-kDa protease.

The predicted catalytic domain (Ala^{125} to Gly^{513}) showed the closest similarity to VapT (CAA82213) of *V. metschnikovii* RH530 with 46% identity. The next similarity was found in Apa1 (BAD51450) of *Pseudoalteromonas* sp. AS-11, Asp (ABA28307) of *Vibrio alginolyticus* HY9901, and SapSh (AAC04871) of *Shewanella* sp. Ac10 with 42% to 44% identities. The sequence similarity to AcpI of *A. collagenimarina* $AC40^T$ [20] was less than 40% identity. The catalytic triads (Asp^{154}, His^{187}, and Ser^{465} in AcpII numbering), P1 cleft site (Ser^{250}-Leu^{251}-Gly^{252}), and oxyanion hole residue Asn^{280} in the catalytic domain of AcpII were highly conserved, suggesting that it belongs to subtilase family A [19].

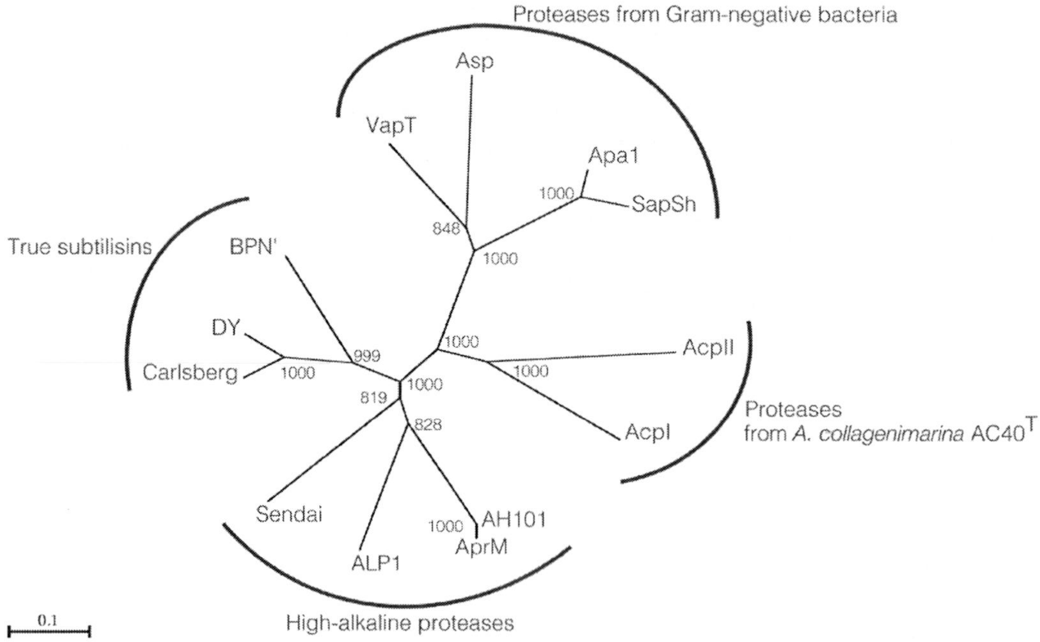

Figure 10. Unrooted phylogenetic tree of subtilisins based on the amino acid sequence alignment.

The putative Ca^{2+}-binding residues of AcpII (Gln^{126}, Asp^{163}, Leu^{198}, Asn^{200}, Gly^{203}, Ala^{290}, Tyr^{292}, Asp^{293}, and Val^{295}) were also found in subtilisin Carlsberg [21] and other Gram-negative bacteria proteases [27, 28] including AcpI [20].

The predicted PA domain (Leu^{340} to Asn^{457}) was inserted into the catalytic domain of AcpII. The closest sequence similarity to the PA domain of AcpII was found in another PA domain of a cold-active alkaline serine protease (YP_929185) from *Shewanella amazonensis* SB2B, serine protease (YP_271274) from *C. psychrerythraea* 34H, and cold-active alkaline serine protease (ZP_01131839) from *P. tunicata* D2 with 41% identities.

A phylogenetic tree was inferred by the neighbor-joining method in the Clustal X program. Bar represents *k*nuc unit. Source of sequences aligned: AcpI (AB305158) and AcpII (AB505451) from *A. collagenimarina* $AC40^T$; VapT (CAA82213) from *V. metschnikovii* strain RH530; Apa1 (BAD51450) from *Pseudoalteromonas* sp. AS-11; Asp (ABA28307) of *V. alginolyticus* HY9901; SapSh (AAC04871) from *Shewanella* sp. Ac10; Carlsberg (P00780) from *B. licheniformis*; BPN' (Q44684) from *Bacillus amyloliquefaciens*; DY (P00781) from *B. subtilis* DY; AH101 (D13158) from *Bacillus* sp. strain AH101; AprM (Q45521) from *Bacillus* sp. strain B18; ALP1 (Q45523) from *Bacillus* sp. strain NHS21; Sendai (Q45522) from *Bacillus* sp. strain G-825-6.

A phylogenetic tree of alkaline proteases belonging to subtilase family A is shown in Figure 10. Phylogenetic clusters of true subtilisins and high-alkaline proteases that are derived from Gram-positive bacteria are clearly located on different branches, as reported previously [29]. The proteases from *A. collagenimarina* $AC40^T$ and other Gram-negative bacteria, *V. metschnikovii* RH530, *Pseudoalteromonas* sp. AS-11, *V. alginolyticus* HY9901, *Shewanella* sp. Ac10, formed individual clusters. This suggests that AcpI and AcpII are characteristically distinguished from other enzymes of Gram-negative bacteria as well as those of Gram-positive bacteria on substrate specificity, especially collagenolytic activity.

3. ROLE OF INSERTION DOMAIN OF ACPII

In Section 2, we showed AcpII preferred hydrolysis of gelatin and collagen to casein (Table 2) and found that the catalytic domain of AcpII includes an insertion, namely, PA domain (Figure 9). It has been reported that several proteases contain a PA domain in their catalytic domains, namely the cell-envelope proteinase from *Lactococcus lactis* [30], C5a peptidase from *Streptococcus agalactiae* [31], VapT of *V. metschnikovii* strain RH530 [27], SapSh of *Shewanella* sp. Ac10 [28], and Apa1 (BAD51450) of *Pseudoalteromonas* sp. As-11. The PA domain would be involved in protein–protein interaction [32, 33]. Additionally, it is speculated that the PA domain participates in substrate binding or promoting conformational changes of the enzyme itself, which affect the accessibility of substrates [30, 31]. In Section 3, the role of a PA domain of AcpII was described.

3.1. Construction of Heterologous Expression System of AcpII Derivatives

To evaluate a role of the PA domain, we constructed the genes for AcpII, one of the PPC domains, truncated AcpII (AcpII-tr: Met^{1}-Arg^{675}), the catalytic domain of AcpII (AcpII-C: Met^{1}- Gly^{513}), and AcpII-C with the PA domain deleted (AcpII-C-ΔPA: Met^{1}- Gly^{513} without Leu^{340}- Asn^{457}; see Figure 11). To construct the genes encoding three AcpII derivatives, the directions of PCR primers were designed to start from the N or C terminus of each derivative. The resultant expression plasmids, designated pRSET-acpII, pRSET-acpII-tr, pRSET-acpII-C, and pRSET-acpII-C-ΔPA, were introduced into *E. coli* BL21(DE3) pLysS (Invitrogen) to produce each AcpII derivative.

	Molecular mass (kDa)	
	Calculated	Actual
AcpII	65.7	65 and 55
AcpII-tr	56.5	42
AcpII-C	44.6	42
AcpII-C-ΔPA	32.8	35

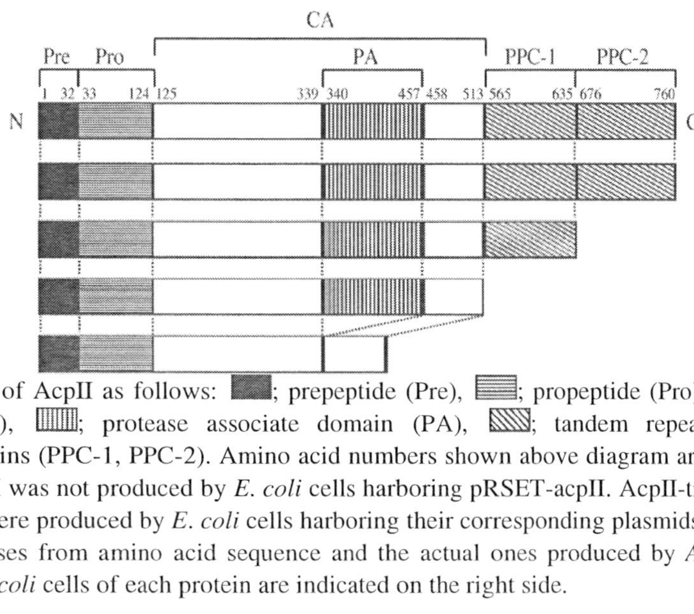

Diagrams of the domain structure of AcpII as follows: ▪; prepeptide (Pre), ▤; propeptide (Pro), ☐; catalytic domain (CA), ▥; protease associate domain (PA), ▧; tandem repeat prepeptidase C-terminal domains (PPC-1, PPC-2). Amino acid numbers shown above diagram are the same as in Figure 9. AcpII was not produced by *E. coli* cells harboring pRSET-acpII. AcpII-tr, AcpII-C, and AcpII-C-ΔPA were produced by *E. coli* cells harboring their corresponding plasmids. The predicted molecular masses from amino acid sequence and the actual ones produced by *A. collagenimarina* AC40^{T} or *E. coli* cells of each protein are indicated on the right side.

Figure 11. Domain structures of AcpII and its derivatives [23].

Table 3. Substrate specificity of AcpII derivatives [23]

	Relative activity (%)		AcpII-C-ΔPA /AcpII-C
	AcpII-C	AcpII-C-ΔPA	(fold)
Gelatin	100	253	2.5
Casein	25	88	3.5
Collagen	63	85	1.3

3.2. Purification of AcpII Derivatives and Comparison of Substrate Specificity

For the production of recombinant AcpII, *E. coli* BL21(DE3) pLysS was transformed with pRSET-acpII. However, host strains did not grow at all. Although the reason is not clear, the complete AcpII might be a strong stress on the host strain. Next, we introduced expression plasmid, pRSET-acpII-tr, pRSET-acpII-C, and pRSET-acpII-C-ΔPA, into *E. coli* BL21(DE3) pLysS cells. The recombinant enzymes were successfully produced extracellularly by each host strain. AcpII-C and AcpII-C-ΔPA were purified to almost homogeneity as judged by SDS–PAGE (Figure 7, Lanes 3 and 4). AcpII-tr had a molecular mass of around 42 kDa, which is smaller than the predicted molecular mass of 56.5 kDa (data not shown). The molecular masses of AcpII-C and AcpII-C-ΔPA were around 42 and 35 kDa, respectively, which are in agreement with the calculated molecular masses from the amino acid sequence.

The relative specific activities toward gelatin, casein, and collagen of the purified AcpII-C and AcpII-C-ΔPA were determined (Table 3). They acted well on gelatin, casein, and collagen, but the specific activities of AcpII-C-ΔPA were greater than those of AcpII-C. The results indicate that the PA domain of AcpII-C regulates the access of substrates to the active site. Meanwhile, the specific activity ratio between both enzymes for collagen (1.3) was lower than that for gelatin (2.5) or casein (3.5). This suggests that the PA domain relatively enhances the affinity of the catalytic domain of AcpII for collagen. The PA domain of C5a peptidase is positioned as a lid covering the active site [31]. Likewise, the PA domain of AcpII might be located as a lid to restrict the accessibility of the proteinaceous substrate.

Next, we examined the binding of both enzymes to collagen. The collagen-binding ability of AcpII derivatives was examined using the batch method [34]. First, 5 mg of collagen was added to 0.2 ml of chilled 10 mM glycine-NaOH buffer (pH 10.0), and the mixture was kept in an ice bath for 30 min. After centrifugation (1,000×*g* for 5 min at 4°C), the precipitated collagen was washed three times with the same buffer. Approximately 8 μg of each AcpII derivative, AcpII-C and AcpII-C-ΔPA, was added to 10 mM glycine-NaOH buffer (pH 10.0) containing 5 mg washed collagen and 400 mM NaCl. The mixture (100 μl) was kept in an ice bath for 180 min. During incubation, samples (5, 30, 75, and 180 min) were withdrawn, and the centrifuged supernatant (5 μl) was subjected to enzyme assay. As a result, during a 180-min incubation with collagen at 4°C, both AcpII-C and AcpII-C-ΔPA activities were detected in supernatants without any loss of the original activities. The results indicate that collagen is not bound to either derivative, which means that the PA domain is not responsible for binding to collagen.

CONCLUSION

A psychrotolerant, obligatorily alkaliphilic bacterium, *A. collagenimarina* AC40^T, was isolated from deep-sea sediment around whale bones near the summit of the Torishima Seamount (4,026 m deep). The strain AC40^T produces two alkaline proteases, AcpI and AcpII. AcpI prefers casein to collagen as the substrate, while AcpII degrades well collagen. Also, both enzymes hydrolyze well gelatin which is a denatured collagen molecule. According to pH- and temperature-activity profiles of both enzymes, it is supposed that they work well in deep-sea environment. AcpI and AcpII belong to subtilase family A based on their deduced amino acid sequences. In the catalytic domain of AcpII, we found an unidentified insertion, namely, a PA domain. From the substrate specificity of several AcpII derivatives, the PA domain seems to relatively enhance hydrolytic activity toward collagen. Hence, AcpII as well as AcpI of the strain AC40^T can participate fully in collagen degradation and accordingly contribute to nitrogen cycle in the deep-sea ecosystem.

REFERENCES

[1] Bartlett, D.H. *Sci. Prog.*, 76, 479 (1992).
[2] Smith, C.R., Baco, A.R. *Oceanography and Marine Biol.*, 41, 311 (2003).
[3] Braby, C.E., Rouse, G.W., Johnson, S.B., Jones, W.J., Vrijenhoek, R.C. *Deep-Sea Research I.*, 54, 1773 (2007).
[4] Azam, F., Malfatti, F. *Nat. Rev. Microbiol.*, 5, 782 (2007).
[5] Miyazaki, M., Nagano, Y., Fujiwara, Y., Hatada, Y., Nogi, Y. *Int. J. Syst. Evol. Microbiol.*, 60, 2298 (2010).
[6] Kang, S.I., Jang, Y.B., Choi, Y.J., Kong, J.Y. *B. B. E.*, 10, 593 (2005).
[7] Watanabe, K. *Appl. Microbiol. Biotechnol.*, 63, 520 (2004).
[8] Ishikawa, H., Amasaki, H., Dohguchi, H., Furuya, A., Suzuki, *J. Vet. Med. Sci.*, 61, 227 (1999).
[9] Le Guellec, D., Morvan-Dubois, G., Sire, J.Y. *Int. J. Dev. Biol.*, 48, 217 (2004).
[10] Zhao, G.Y., Chen, X.L., Zhao, H.L., Xie, B.B., Zhou, B.C., Zhang, Y.Z. *J. Biol. Chem.*, 283, 36100 (2008).
[11] Naganuma, T., Wada, H., Fujioka, K. *J. Oceanogr.*, 52, 1 (1996).
[12] Kurata, A., Miyazaki, M., Kobayashi, T., Nogi, Y., Horikoshi, K. *Int. J. Syst. Evol. Microbiol.*, 57, 1549 (2007).
[13] Ma, Y., Xue, Y., Grant, W.D., Collins, N.C., Duckworth, A.W., Van Steenbergen, R.P., Jones, B.E. *Extremophiles.*, 8, 193 (2004).
[14] Ezaki, T., Hashimoto, Y., Yabuuchi, E. *Int. J. of Syst. Bacteriol.*, 39, 224 (1989).
[15] Barrow G.I., Feltham, R.K.A. *Cowan and Steel's Manual for the Identification of Medical Bacteria 3rd edn.*, Cambridge University Press., Cambridge, (1993).
[16] Baumann, L., Baumann, P., Mandel, M., Allen, R.D. *J. Bacteriol.*, 110, 402 (1972).
[17] Hugh R., Leifson, E. *J. Bacteriol.*, 66, 24 (1953).
[18] Komagata K., Suzuki, K. *Methods in Microbiology*, 19, 161 (1987).
[19] Siezen R.J., Leunissen, J.A.M. *Protein Sci.*, 6, 501 (1997).

[20] Kurata, A., Uchimura, K., Shimamura, S., Kobayashi, T., Horikoshi, K. *Appl. Microbiol. Biotechnol.*, 77, 311 (2007).

[21] Neidhart D.J., Petsko, G.A. *Protein Eng.*, 2, 271 (1988).

[22] Tsujibo, H., Miyamoto, K., Tanaka, K., Kaidzu, Y., Imada, C., Okami, Y., Inamori,Y. *B. B. B.*, 60, 1284 (1996).

[23] Kurata, A., Uchimura, K., Kobayashi, T., Horikoshi, K. *Appl. Microbiol. Biotechnol.*, 86, 589 (2010).

[24] Yeats, C., Bentley, S., Bateman, A. *BMC Microbiol.*, 3, 1 (2003).

[25] Miyamoto, K., Tsujibo, H., Nukui, E., Itoh, H., Kaidzu, Y., Inamori, Y. *B. B. B.*, 66, 416 (2002).

[26] Helland, R., Larsen, A.N., Smalas, A. O., Willassen, N.P. *FEBS J.*, 273, 61 (2006).

[27] Kwon, Y.T., Kim, J.O., Moon, S.Y., Yoo, Y.D., Rho, H.M. *Gene.*, 152, 59 (1995).

[28] Kulakova, L., Galkin, A., Kurihara, T., Yoshimura, T., Esaki, N. *A. E. M.*, 65, 611 (1999).

[29] Kobayashi, T., Lu, J. Li, Z., Hung, V.S., Kurata, A., Hatada, Y., Takai, K., Ito, S., Horikoshi, K. *Appl. Microbiol. Biotechnol.*, 75, 71 (2007).

[30] Bruinenberg, P.G., Doesburg, P., Alting, A.C., Exterkate, F.A., de Vos, W.M., Siezen, R.J. *Protein Eng*, 7, (1994).

[31] Brown, C.K., Gu, Z.Y., Matsuka, Y.V., Purushothaman, S.S., Winter, L.A., Cleary, P.P., Olmsted, S.B., Ohlendorf, D.H., Earhart, C.A. *Proc Natl Acad Sci USA*, 102, (2005).

[32] Mahon, P., Bateman, A. *Protein Sci.*, 9, (2000).

[33] Luo, X., Hofmann, K. *Trends Biochem Sci.*, 26, (2001).

[34] Itoi, Y., Horinaka, M., Tsujimoto, Y., Matsui, H., Watanabe, K. *J.Bacteriol.*, 188, (2006).

[35] Okamoto, M., Yonejima, Y., Tsujimoto, Y., Suzuki, Y., Watanabe, *Appl. Microbiol. Biotechnol.*, 57, 103 (2001).

In: Deep-Sea
Editors: D. R. Bailey and S. E. Howard ISBN: 978-1-62257-173-4

Chapter 5

PRIMARY PRODUCTION AND CARBON CYCLING IN THE DEEP SEA: ARCHAEAL CARBON FIXATION AND METHANE OXIDATION*

Nianzhi Jiao and Chuanlun Zhang
National Key Laboratory for Marine Environmental Sciences,
University of Xiamen, Xiamen, Fujian, China

INTRODUCTION

The discovery of nonthermophilic archaea (crenarchaeota and euryarchaeota) in the open ocean and coastal water using culture-independent molecular technologies was reported in two seminal papers in the early 1990's (DeLong, 1992; Fuhrman et al., 1992), and this corrected our previous biased view that crenarcheota only existed in extreme environments such as hydrothermal vents or terrestrial thermal springs.Since then, research on nonthermophilic archaea has been intensely pursued intensely and it is now clear that these organisms occur widely in low temperature marine and terrestrial environments (see reviews by DeLong, 1998; Schleper et al., 2005). Overall, in marine water columns, nonthermophilic archaea are dominated by Group I crenarchaeota and Group II euryarchaeota. In marine sediments, not only Group I crenarchaeota and Group II euryarchaeota are present but also a variety of other archaeal species, including the Marine Benthic Group crenarchaeota, Miscellaneous Crenarchaeotal Group, and an unidentified euryarchaeota group (Vetriani et al., 1999; Inagaki et al., 2006; Biddle et al., 2006).

In addition, anaerobic methane-oxidizing archaea (ANME groups) widely occur in association with gas hydrates, cold seeps, and organic rich thermal sediments (e.g., Boetius et al., 2000; Orphan et al., 2001; Mills et al., 2003; Knittel et al., 2005) as well as in anoxic

* A version of this chapter also appears in *Archaea: Structure, Habitats, and Ecological Significance*, edited by Sakura Y. Kato, published by Nova Science Publishers, Inc. It was submitted for appropriate modifications in an effort to encourage wider dissemination of research.

water bodies like the Black Sea and Cariaco Basin (Madrid et al., 2001; Durisch-Kaiser et al., 2005; Schubert et al., 2006).

This section reviews recent observations of the occurrence and diversity, abundance, metabolic pathways, and contributions to carbon and nitrogen cycles of nonthermophilic archaea in the deep oceans and marine sediments.

Occurrence and Diversity of Nonthermophilic Archaea in Water Column and Marine Sediments

The first detection of PCR-based archaeal 16S rRNA genes in the marine plankton included a few sequences that are closely related to known hyperthermophilic species (Furman et al., 1992; DeLong, 1992). This observation was initially uncertain as some of these archaea were questioned as being allochthonous thermophiles transported from a putative hydrothermal vent.

Soon after, however, DeLong and co-workers discovered high abundance of archaea in cold aerobic Arctic water (DeLong et al., 1994) and also a crenarchaeal symbiont in a cold marine sponge (Preston et al., 1996), crystallizing the fact that the newly discovered archaea are native to the low-temperature marine environment. Since then, a large number of novel archaeal phylotypes have been detected in almost every niche of the marine biotope, ranging from surface to abyssal depths of the Pacific and Atlantic Oceans as well as in deep marine sediments (see review by Schelper et al., 2005). So far, the vast majority of planktonic archaea fall into the group I. 1A crenarchaeota and group II euryarchaeota. It has been observed that group I. 1A crenarcheota generally increase in abundance at greater depths (>200 m); whereas, the group II euryarchaeota are predominant in the photic zone (<200 m) but also are found in significant abundance in deeper waters (Massana et al., 1997; Karner et al., 2001; Herndl et al., 2005; DeLong et al., 2006; Teira et al., 2006). Nonthermophilic archaea in marine sediments are mostly reported from cold-seep and gas hydrate-related environments where methane serves as an abundant energy source fueling the anaerobic oxidation of methane (AOM) coupled to sulfate reduction (Hinrichs et al., 1999; Orphan et al., 2001).

In diffusion prone settings of deep sea marine sediments, where methane and sulfate fluxes are low, the AOM process is identified in a sediment horizon, namely the sulfate-methane transition zone, into which both methane and sulfate diffuse (Biddle et al., 2006). In cold seeps and gas hydrates, the archaeal community is dominated by the uncultured ANME groups of euryarchaeota that are known to mediate the AOM (*e.g.,* Inagaki et al., 2003; Biddle et al., 2006; Inagaki et al., 2006; Kendall et al., 2007).

In particular, ANME-1 and ANME-2 phylotypes are most abundant although their relative distributions varies from location to location. For example, ANME- 1 is predominant in the Black Sea, whereas ANME-2 is predominant at Hydrate Ridge (Knittel et al., 2005; Nauhaus et al., 2005). In the Gulf of Mexico, both ANME-1 and ANME-2 are conspicuous (Lanoil et al., 2001; Mills et al., 2003; Lloyd et al., 2006; Reed et al., 2006; Pi et al., in rivision). In the diffusion prone underlying the highly productive surface water off the coast of Peru, however, the sedimentary archaeal communities are dominated by the Marine Benthic Group B and the Miscellaneous Crenarchaeotal Group, whereas methanotrophic archaea such as ANME-1 and ANME-2 are not detected (Biddle et al., 2006). Other non-cold seep or non-gas hydrate deep-sea marine sediments are also dominated by nonthermophilic crenarchaeota and nonANME groups of euryarchaeota (e.g., Vetriani et al., 1999; Inagaki et al., 2003; Lanoil et al.,2005; Wang et al., 2005; Inagaki et al., 2006). In particular, marine benthic group B and Thermoplasmatales appear to be cosmopolitan in normal marine sediments (Orphan et al., 2001; Teske, 2006).

ABUNDANCE OF NONTHERMOPHILIC ARCHAEA IN THE WATER COLUMN AND MARINE SEDIMENTS

The observation of very large numbers of nonthermophilic archaea in the open ocean adds another dimension to the importance of these organisms in the marine ecosystem. An early estimate of the abundance of planktonic archaea was made in temperate coastal marine waters and polar seas, which ranged from about 10 to 30% of the total prokaryotic biomass(DeLong, 1992). In 2001, a major study was undertaken in the Pacific Ocean, which demonstrated the dominance of archaea in the mesopelagic zone of this water body (Karner et al., 2001). Monthly sampling within a 12-month period showed that pelagic crenarchaeota comprised a large fraction of the total marine picoplankton below 150 m (the bottom of the euphotic zone) and increased with depth, reaching about 40% of the total DNA-containing picoplankton detected using DAPI staining and fluorescent isothiocyanate multiple labeled probes (Karner et al., 2001).

Estimates based on this study indicate that the world's oceans contain approximately 1.3 x 1028 archaeal cells, which is of the same order of magnitude as bacteria (3.1 x 1028 cells) (Karner et al., 2001). Another comprehensive study in the Atlantic Ocean, demonstrated the consistently higher abundance of archaea than bacteria below the 100-m depth, using an improved catalyzed reporter deposition-fluorescence in situ hybridization method and specific oligonucleotide probes (Herndl et al., 2005). Estimates of the archaeal production in the mesopelagic and bathypelagic North Atlantic showed 13-27% contributions to the total prokaryotic production in the oxygen minimum layer, 41-84% in Labrador sea water, and 10-20% in the North Atlantic deep water (Herndl et al., 2005).

While group I crenarchaeota commonly comprise the majority of deep water planktonic archaea, it was observed in one instance that a deeply branching crenarchaeal group related to a hot spring clade (pSL 12) was significantly more abundant than the group I crenarchaeota below the euphotic zone in the North Pacific subtropical gyre (Mincer et al., 2007), suggesting that certain environmental conditions may stimulate the growth of multiple species of crenarchaeota.

Another observation involves the unusual abundance of group I crenarchaeota in the suboxic layers of the Black Sea, which makes up to 98% of the total archaeal 16S rRNA gene copies (Coolen et al., 2007), suggesting that these archaeal organisms may be microaerophilic and may be able to tolerant suboxic conditions. Estimates of archaea in marine sediments are complicated by the heterogeneous distribution of these organisms in the porous medium. Geochemical proxies, however, have indicated that 90% of the methane produced in the marine sediments is consumed before it reaches the water column (Valentine and Reeburgh, 2000), suggesting that a large population of methane-oxidizing archaea exists in the sediments and plays a significant role in controlling the flux of methane into the atmosphere. Other studies have estimated that the abundance of archaea in marine sediments varies between 0.01% and 30% of total biomass with high values often occurring near the seafloor (Vetriani et al., 1999; Inakagi et al., 2006).

METABOLIC PATHWAYS OF NONTHERMOPHILIC ARCHAEA IN THE WATER COLUMN AND MARINE SEDIMENTS

The physiology and biochemistry of nonthermophilic archaea have also begun to be elucidated as a result of developments in environmental genomics (Venter et al., 2004; Schleper et al., 2005; Hallam et al., 2006a) and the genomic analysis of a single crenarchaeal phylotype Cenarchaeum symbiosium in a marine sponge (Preston et al., 1996; Hallam et al., 2006b). The recently isolated pure culture of the mesophilic crenarchaeon, Nitrosopumilus maritimus, adds new insight into the physiology of non-thermophilic crenarchaeota (Könneke et al., 2005). Mounting evidence indicates that some of the nonthermophilic crenarchaeota use ammonia as a major energy source for autotrophic growth (Francis et al., 2005; Könneke et al., 2005; Nicol and Schleper, 2006; Wuchter et al., 2006). Other molecular and geochemical studies show that nonthermophilic crenarchaeota can take up amino acids and organic carbon, indicating that some of them are heterotrophs or mixotrophs (Biddle et al., 2006; Herndl et al., 2005; Ingalls et al., 2006; Ouverney et al., 2000; Teira et al., 2006). One recent study showed that some of the benthic crenarchaeota may oxidize methane using it as an energy source and other organic carbon as the carbon source (Biddle et al., 2006).

On the other hand, ANME groups of the euryarchaeota oxidize methane using a reverse methanogenesis pathway (Boetius et al., 2000; Hallam et al., 2003, 2004; Valentine, 2000). While no isolate is available of any of the ANME group members for detailed biochemical studies, genomic analysis of the AOM community in Eel River Basin sediment could distinguish the ANME-1 group from the ANME-2 group based on G+C content and depth of the whole genome shotgun coverage, Importantly, Hallam et al., (2004) demonstrated that nearly all the genes typically associated with methane production are present in the ANME-1 group, except a step encoded by the mer gene. It was proposed that the loss of mcr activity in ANME-1 could promote AOM by increasing the activation barrier for conversion of methylene- H4MPT to methyl-H4MPT. Overall, the association of most of the genes for methanogenesis in the ANME groups strongly supports the hypothesis that AOM proceeds as a reversed pathway of methane production.

The Group II euryarchaeota are known to contain light-harvesting proteins (PRs), suggesting that these novel euryarchaeota may perform light-dependent cellular energy metabolism (DeLong et al., 2006; Frigaard et al., 2006).

CONTRIBUTION OF NONTHERMOPHILIC ARCHAEA TO CARBON AND NITROGEN CYCLES IN THE DEEP OCEAN

A picture is emerging that the majority of the nonthermophilic archaea in the deep ocean are chemoautotrophs and may use ammonia as a major energy source. Herndl et al., (2005) presented leucine-incorporation evidence that the actively growing archaea in the oxygenated water column, which numerically dominate the prokaryotic community in the North Atlantic below a depth of 100 m to 2435-2790 m in the North Atlantic Deep Water, use bicarbonate or CO2 as a carbon source.

Independently, Ingalls et al., (2006) quantified archaeal autotrophy in the subtropical North Pacific gyre using natural radiocarbon and demonstrated, based on an isotopic mass balance model, that 83% of total archaeal metabolism is autotrophy that fixes ^{14}C-depleted dissolved inorganic carbon in the deep water; whereas, the remaining metabolism may contribute to heterotrophic consumption of the ^{14}C-enriched modern organic carbon. Based on the depth-integrated mean C fixation rate (0.014 fmol C archaeon^{-1} day^{-1}) and on the assumption that the global ocean contains approximately 1.3 x 10^{28} cells of archaea, Herndl et al., (2005) derived a global inorganic carbon fixation rate of 6.55 x 10^{13} mol C year^{-1}. This global carbon fixation rate is compared with that derived from archaeal nitrification, which is estimated to be 3.3 x 10^{13} mol C year^{-1} (assuming all of the ammonium would be oxidized by crenarchaeota fixing one carbon atom for every 10 nitrogen molecules oxidized) (Wuchter et al., 2006).

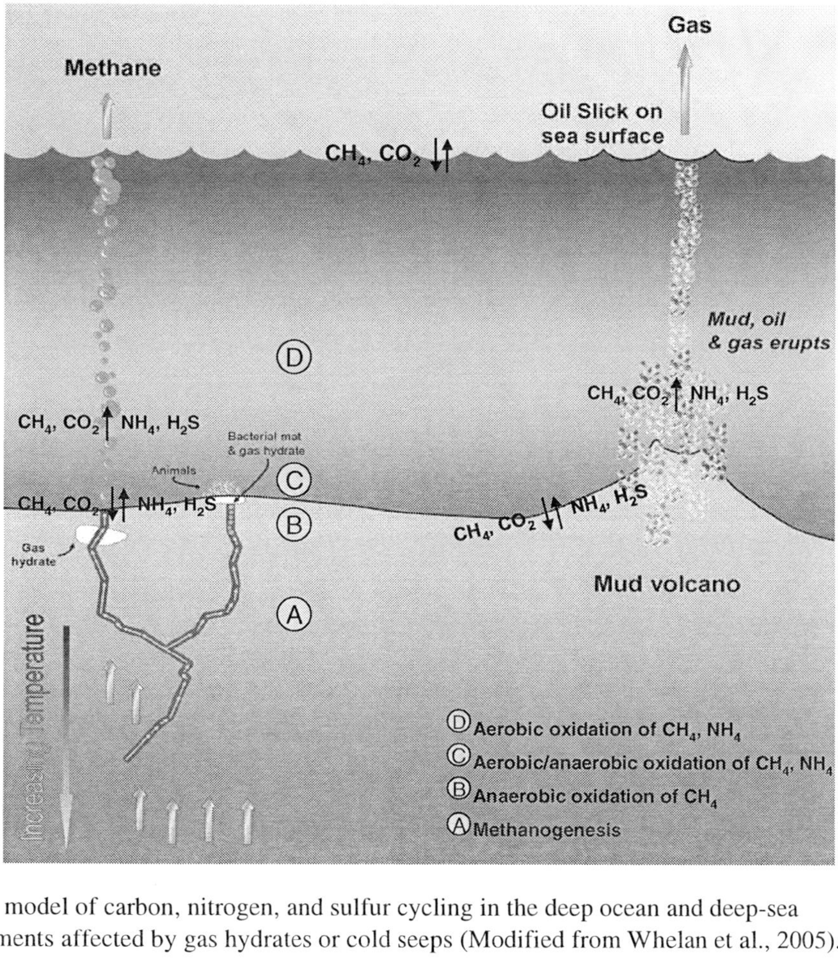

Figure 1. A model of carbon, nitrogen, and sulfur cycling in the deep ocean and deep-sea marinesediments affected by gas hydrates or cold seeps (Modified from Whelan et al., 2005).

The agreement between these estimated rates suggest that archaeal nitrification coupled to inorganic carbon fixation may contribute profoundly to the biogeochemical cycles of carbon and nitrogen in the global ocean (Figure 1). Methane formation by methanogens in the dark ocean is also a major sink of CO_2; however, its contribution to carbon fixation is two orders of magnitude lower (~8×10^{11} mol C $year^{-1}$) than that from crenarchaeota in the global ocean (Herndl et al., 2005). In marine sediments, methanogens are the sole source of biogenic methane below the sulfate reduction zone; whereas within the sulfate reduction zone, methane-oxidizing archaea in consortia with sulfate-reducing bacteria are mainly responsible for methane consumption (Figure 1).

Although heterotrophic crenarchaeota are known to perform oxidation of methane (Biddle et al., 2006), their relative contribution to methane consumption is unknown.

In either case, CO_2 from anaerobic methane oxidation is largely sequestrated into carbonate minerals that constitute a major fraction of the marine sediment or form carbonate mounds above the ocean floor. This has been observed frequently in the Gulf of Mexico (Formolo et al., 2004; Sassen et al., 2004) as well as other methane-rich marine sediments (e.g., Aloisi et al., 2002), where gas hydrates and cold seeps are found abundantly. Globally, a huge amount of methane (> 10 trillion tons; Zimmer, 2001) may be sequestrated in the hydrate structure in marine sediment or hydrate mounds, which are known as an attractive alternative energy resource beyond petroleum hydrocarbons and coals.

REFERENCES

Aloisi, G., Bouloubassi, I., Heijs, S. K., Pancost, R. D., Pierre, C., Damste, J. S. S., Gottschal, J. C., Forney, L. J., and Rouchy, J. M. (2002). CH_4-consuming microorganisms and the formation of carbonate crusts at cold seeps. *Earth and Planetary Science Letters*, 203, 195-203.

Anbar, A. D. and Knoll, A. H. (2002). Proterozoic ocean chemistry and evolution: A bioinorganic bridge? *Science*, 297, 1137-1142.

Anderson, N. L. and Anderson, N. G. (1998). Proteome and proteomics: New technologies, new concepts, and new words. *Electrophoresis*, 19, 1853-1861.

Angly, F. E., Felts, B., Breitbart, M., Salamon, P., Edwards, R. A., Carlson, C., Chan, A. M., Haynes, M., Kelley, S., Liu, H., Mahaffy, J. M., Mueller, J. E., Nulton, J., Olson, R., Parsons, R., Rayhawk, S., Suttle, C. A., and Rohwer, F. (2006). The marine viromes of four oceanic regions. *Plos Biology*, 4, 2121-2131.

Arce, F. T., Avci, R., Beech, I. B., Cooksey, K. E., and Wigglesworth-Cooksey, B. (2004). live bioprobe for studying diatom-surface interactions. *Biophysical Journal*, 87, 4284-4297.

Auerbach, I.D., Sorensen, C., Hansma, H. G., and Holden, P. A. (2000). Physical morphology and surface properties of unsaturated *Pseudomonas putida* biofilms. *Journal of Bacteriology*, 182, 3809-3815.

Azam, F., Fenchel, T., Field, J. G., Gray, J. S., Meyer-Reil, L. A., and Tingstad, F. (1983). The ecological role of water-column microbes in the sea. *Marine Ecology Progress Series*, 10, 257-263.

Bahatyrova, S., Frese, R. N., Siebert, C. A., Olsen, J. D., van der Werf, K. O., van Grondelle, R., Niederman, R. A., Bullough, P. A., and Hunter, C. N. (2004). The native architecture of a photosynthetic membrane. *Nature*, 430, 1058-1062.

Bauer, J. E., Williams, P. M., and Druffel, E. R. M. (1992). ^{14}C activity of dissolved organic carbonfractions in the northcentral Pacific and Sargasso Sea. *Nature*, 357, 667-670.

Beatty, J. T. (2002). On the natural selection and evolution of the aerobic phototrophic bacteria. *Photosynthesis Research*, 73, 109-114.

Béjà, O., Aravind, L., Koonin, E. V., Suzuki, M. T., Hadd, A., Nguyen, L. P., Jovanovich, S., Gates, C. M., Feldman, R. A., Spudich, J. L., Spudich, E. N., and DeLong, E. F. (2000). Bacterial rhodopsin: Evidence for a new type of phototrophy in the sea. *Science*, 289, 1902-1906.

Béjà, O., Spudich, E. N., Spudich, J. L., Leclerc, M., and DeLong, E. F. (2001). Proteorhodopsin phototrophy in the ocean. *Nature*, 411, 786-789.

Béjà, O., Suzuki, M. T., Heidelberg, J. F., Nelson, W. C., Preston, C. M., Hamada, T., Eisen, J. A., Fraser, C. M., and DeLong, E. F. (2002). Unsuspected diversity among marine aerobic anoxygenic phototrophs. *Nature*, 415, 630-633.

Bergh, O., Borsheim, K. Y., Bratbak, G., and Heldal, M. (1989). High abundance of viruses found in aquatic environments. *Nature,* 340, 467–468.

Beveridge, T. J. and Graham, L. L. (1991). Surface layers of bacteria. *Microbiology and Molecular Biology Reviews*, 55, 684-705.

Beveridge, T. J., Makin, S. A., Kadurugamuwa, J. L., and Li, Z. (1997). Interactions between biofilms and the environment. *FEMS Microbiology Reviews*, 20, 291-303.

Beveridge, T. J. (1999). Structures of Gram-Negative Cell Walls and Their Derived Membrane Vesicles. *Journal of Bacteriology*, 181, 4725-4733.

Biddle, J. F., Lipp, J. S., Lever, M. A., Lloyd, K. G., Sorensen, K. B., Anderson, R., Fredricks, H. F., Elvert, M., Kelly, T. J., Schrag, D. P., Sogin, M. L., Brenchley, J. E., Teske, A., House, C. H., and Hinrichs, K. U. (2006). Heterotrophic Archaea dominate sedimentary subsurface ecosystems off Peru. *Proceedings of the NationalAcademy of Sciences of the United States of America*, 103, 3846-3851.

Binder, B. J., Chisholm, S. W., Olson, R. J., Frankel, S. L., and Worden, A. Z. (1996). Dynamics of picophytoplankton, ultraphytoplankton and bacteria in the central equatorial Pacific. *Deep-Sea Research II*, 43, 907-931.

Binnig, G., Quate, C. F., and Gerber, C. (1986). Atomic force microscope. *Physical Review Letters*, 56, 930 - 933.

Blackstock, W. P. and Weir, M. P. (1999). Proteomics: quantitative and physical mapping of cellular proteins. *Trends in Biotechnology*, 17, 121-127.

Boetius, A., Ravenschlag, K., Schubert, C. J., Rickert, D., Widdel, F., Gieseke, A., Amann, R., Jorgensen, B. B., Witte, U., and Pfannkuche, O. (2000). A marine microbial consortium apparently mediating anaerobic oxidation of methane. *Nature*, 407, 623-626.

Boschker, H. T. S., Nold, S. C., Wellsbury, P., Bos, D., de Graaf, W., Pel, R., Parkes, R. J., and Cappenberg, T. E. (1998). Direct linking of microbial populations to specific biogeochemical processes by C-13-labelling of biomarkers. *Nature*, 392, 801-805.

Bralower, T. J., Silva, I. P., and Malone, M. J. (2002). New evidence for abrupt climate change in the Cretaceous and Paleogene: An Ocean Drilling Program expedition to Shatsky rise, northwest Pacific. *GSA Today*, 12, 4-10.

Brocks, J. J. and Pearson, A. (2005) Building the biomarker tree of life. In *Molecular Geomicrobiology*. Chantilly: Mineralogical Soc America, pp. 233-258.

Brussow, H., Canchaya, C., and Hardt, W. D. (2004). Phages and the evolution of bacterial pathogens: From genomic rearrangements to lysogenic conversion. *Microbiology and Molecular Biology Reviews*, 68, 560-602.

Buck, K. R., Chavez, F. P., and Campbell, L. (1996). Basin-wide distributions of living carbon components and the inverted trophic pyramid of the central gyre of the north Atlantic Ocean, summer 1993. *Aquatic Microbial Ecology*, 10, 283-298.

Butt, H. J., Jaschke, M., and Ducker, W. (1995). Measuring surface forces in aqueous electrolyte solution with the atomic force microscope. *Bioelectrochemistry and Bioenergetics*, 38, 191-201.

Camesano, T. A., Natan, M. J., and Logan, B. E. (2000). Observation of changes in bacterial cell morphology: using tapping mode atomic force microscopy. *Langmuir*, 16, 4563-4572.

Campbell, L., Liu, H. B., Nolla, H. A., and Vaulot, D. (1997). Annual variability of phytoplankton and bacteria in the subtropical North Pacific Ocean at Station ALOHA during the 1991-1994 ENSO event. *Deep-Sea Research Part I-Oceanographic Research Papers,* 44, 167-and.

Canchaya, C., Proux, C., Fournous, G., Bruttin, A., and Brussow, H. (2003). Prophage genomics. *Microbiology and Molecular Biology Reviews*, 67, 473-473.

Chada, V. G. R., Sanstad, E. A., Wang, R., and Driks, A. (2003). Morphogenesis of *Bacillus* spore surfaces. *Journal of Bacteriology*, 185, 6255-6261.

Chen, F., Lu, J. R., Binder, B. J., Liu, Y. C., and Hodson, R. E. (2001). Application of digital image analysis and flow cytometry to enumerate marine viruses stained with SYBR gold. *Applied and Environmental Microbiology*, 67, 539-545.

Chen, F., Wang, K., Kan, J. J., Suzuki, M. T., and Wommack, K. E. (2006). Diverse and unique picocyanobacteria in Chesapeake Bay, revealed by 16S-23S rRNA internal transcribed Spacer sequences. *Applied and Environmental Microbiology*, 72, 2239-2243.

Chen, F., Wang, K., Stewart, J., and Belas, R. (2006). Induction of multiple prophages from a marine bacterium: a genomic approach. *Applied and Environmental Microbiology*, 72, 4995-5001.

Cochran, P. K., Kellogg, C. A., and Paul, J. H. (1998). Prophage induction of indigenous marine lysogenic bacteria by environmental pollutants. *Marine Ecology-Progress Series*, 164, 125-133.

Cochran, P. K. and Paul, J. H. (1998). Seasonal abundance of lysogenic bacteria in a subtropical estuary. *Applied and Environmental Microbiology*, 64, 2308-2312.

Comita, P. B. and Gagosian, R. B. (1983). Membrane lipid from deep sea hydrothermal vent methanogen: A new macrocyclic glycerol diether. *Science*, 222, 1329-1331.

Coolen, M. J. L., Abbas, B., van Bleijswijk, J., Hopmans, E. C., Kuypers, M. M. M., Wakeham, S. G., and Damste, J. S. S. (2007). Putative ammonia-oxidizing Crenarchaeota in suboxic waters of the Black Sea: a basin-wide ecological study using 16S ribosomal and functional genes and membrane lipids. *Environmental Microbiology*, 9, 1001-1016.

Curtis, T. P., Sloan, W. T., and Scannell, J. W. (2002). Estimating prokaryotic diversity and its limits. *Proceedings of the NationalAcademy of Sciences of the United States of America*, 99, 10494-10499.

Damste, J. S. S., Rijpstra, W. I. C., Hopmans, E. C., Prahl, F. G., Wakeham, S. G., and Schouten, S. (2002). Distribution of membrane lipids of planktonic Crenarchaeota in the Arabian seat. *Applied and Environmental Microbiology*, 68, 2997-3002.

Damste, J. S. S., Strous, M., Rijpstra, W. I. C., Hopmans, E. C., Geenevasen, J. A. J., van Duin, A. C. T., van Niftrik, L. A., and Jetten, M. S. M. (2002). Linearly concatenated cyclobutane lipids form a dense bacterial membrane. *Nature*, 419, 708-712.

Delong, E. F. (1992). Archaea in coastal marine environments. *Proceedings of the National Academy of Sciences of the United States of America*, 89, 5685-5689.

DeLong, E. F., Wu, K. Y., Prezelin, B. B., and Jovine, R. V. (1994). High abundance of Archaea in Antarctic marine picoplankton. *Nature*, 371, 695-697.

DeLong, E. F., King, L. L., Massana, R., Cittone, H., Murray, A., Schleper, C., and Wakeham, S. G. (1998). Dibiphytanyl ether lipids in nonthermophilic crenarchaeotes. *Applied and Environmental Microbiology*, 64, 1133-1138.

DeLong, E. F., Preston, C. M., Mincer, T., Rich, V., Hallam, S. J., Frigaard, N. U., Martinez, A., Sullivan, M. B., Edwards, R., Brito, B. R., Chisholm, S. W., and Karl, D. M. (2006). Community genomics among stratified microbial assemblages in the ocean's interior. *Science*, 311, 496-503.

Doktycz, M. J., Sullivan, C. J., Hoyt, P. R., Pelletier, D. A., Wu, S., and Allison, D. P. (2003). AFM imaging of bacteria in liquid media immobilized on gelatin coated mica surfaces. *Ultramicroscopy*, 97, 209-216.

Dufrêne, Y. F. (2002). Atomic Force Microscopy, a Powerful Tool in Microbiology. *Journal of Bacteriology*, 184, 5205-5213.

Dufrêne, Y. F. (2004). Using nanotechniques to explore microbial surfaces. *Nature*, 2, 451-460.

Dufrêne, Y. F. (2004). Refining our perception of bacterial surfaces with the atomic force microscope. *Journal of Bacteriology*, 186, 3283-3285.

Dufresne, A., Salanoubat, M., Partensky, F., Artiguenave, F., Axmann, I. M., Barbe, V., Duprat, S., Galperin, M. Y., Koonin, E. V., Le Gall, F., Makarova, K. S., Ostrowski, M., Oztas, S., Robert, C., Rogozin, I. B., Scanlan, D. J., de Marsac, N. T., Weissenbach, J., Wincker, P., Wolf, Y. I., and Hess, W. R. (2003). Genome sequence of the cyanobacterium *Prochlorococcus marinus* SS120, a nearly minimal oxyphototrophic genome. *Proceedings of the NationalAcademy of Sciences of the United States of America*, 100, 10020-10025.

Dumitrescu, M. and Brassell, S.C. (2003) Biomarkers confirm cyanobacteria as a major source of planktonic organic matter during the Early Aptian. Geological Society of America (GSA) Seattle Annual Meeting (November 2-5, 2003, Abstracts with Programs, pp. 391). URL: http://gsa.confex.com/gsa/2003AM/finalprogram/abstract_67241.htm.

Durisch-Kaiser, E., Klauser, L., Wehrli, B., and Schubert, C. (2005). Evidence of intense archaeal and bacterial methanotrophic activity in the black sea water column. *Applied and Environmental Microbiology*, 71, 8099-8106.

Emerson, R. J. and Camesano, T. A. (2004). Nanoscale investigation of pathogenic microbial adhesion to a biomaterial. *Applied and Environmental Microbiology*, 70, 6012-6022.

Fang, H. H. P., Chan, K. Y., and Xu, L. C. (2000). Quantification of bacterial adhesion forces using atomic force microscopy (AFM). *Journal of Microbiological Methods*, 40, 89-97.

Feely, R. A., Sabine, C. L., Takahashi, T., and Wanninkhof, R. (2001). Uptake and Storage of carbon dioxide in the Ocean: the Global CO2 Survey. *Oceanography*, 14, 18-32.

Fenchel, T. (2001). Ecology - Marine bugs and carbon flow. *Science,* 292, 2444-2445.

Fichmann, J. and Westermeier, R. (1999) *2-D proteome analysis protocols.* New Jersey: Humana Press.

Fletcher, M., 1996. Bacterial attachment in aquatic environments: a diversity of surfaces and adhesion strategies. In: Fletcher, M. (Eds.), *Bacterial Adhesion: Molecular and Ecological Diversity* (pp. 1–24), New York: John Wiley and Sons.

Formolo, M. J., Lyons, T. W., Zhang, C. L., Kelley, C., Sassen, R., Horita, J., and Cole, D. R. (2004). Quantifying carbon sources in the formation of authigenic carbonates at gas hydrate sites in the Gulf of Mexico. *Chemical Geology*, 205, 253-264.

Francis, C. A., Roberts, K. J., Beman, J. M., Santoro, A. E., and Oakley, B. B. (2005). Ubiquity and diversity of ammonia-oxidizing archaea in water columns and sediments of the ocean. *Proceedings of the NationalAcademy of Sciences of the United States of America*, 102, 14683-14688.

Freeman, K. H., Hayes, J. M., Trendel, J. M., and Albrecht, P. (1990). Evidence from carbon isotope measurements for diverse origins of sedimentary hydrocarbons. *Nature*, 343, 254-256.

Friedrich, M. W. (2006). Stable-isotope probing of DNA: insights into the function of uncultivated microorganisms from isotopically labeled metagenomes. *Current Opinion in Biotechnology*, 17, 59-66.

Frigaard, N. U., Martinez, A., Mincer, T. J., and DeLong, E. F. (2006). Proteorhodopsin lateral gene transfer between marine planktonic Bacteria and Archaea. *Nature*, 439, 847-850.

Fuhrman, J. A., McCallum, K., and Davis, A. A. (1992). Novel major archaebacterial group from marine plankton. *Nature*, 356, 148-149.

Fuhrman, J. A. and Suttle, C. A. (1993). Viruses in marine planktonic systems. *Oceanography*, 6, 50-62.

Fuhrman, J. A. (1999). Marine viruses and their biogeochemical and ecological effects. *Nature*, 399, 541-548.

Gaboriaud, F., Bailet, S., Dague, E., and Jorand, F. (2005). Surface Structure and Nanomechanical Properties of *Shewanella putrefaciens* Bacteria at Two pH values (4 and 10) Determined by Atomic Force Microscopy. *Journal of Bacteriology*, 187, 3864-3868.

Gad, M. and Ikai, A. (1995). Method for immobilizing microbial cells on gel surface for dynamic AFM studies. *Biophysical Journal*, 69, 2226-2233.

Georicke, R. and Welschmeyer, N. A. (1993). The marine prochlorophyte *Prochlorococcus* contributes significantly to phytoplankton biomass and primary production in the Sargasso Sea. *Deep-Sea Research*, 2283-2294.

Giovannoni, S. J., Bibbs, L., Cho, J. C., Stapels, M. D., Desiderio, R., Vergin, K. L., Rappe, M. S., Laney, S., Wilhelm, L. J., Tripp, H. J., Mathur, E. J., and Barofsky, D. F. (2005). Proteorhodopsin in the ubiquitous marine bacterium SAR11. *Nature*, 438, 82-85.

Giovannoni, S. J., Tripp, H. J., Givan, S., Podar, M., Vergin, K. L., Baptista, D., Bibbs, L., Eads, J., Richardson, T. H., Noordewier, M., Rappe, M. S., Short, J. M., Carrington, J. C., and Mathur, E. J. (2005). Genome streamlining in a cosmopolitan oceanic bacterium. *Science*, 309, 1242-1245.

Glaeser, J. and Overmann, J. (2003). Characterization and in situ carbon metabolism of phototrophic consortia. *Applied and Environmental Microbiology*, 69, 3739-3750.

Gobler, C. J., Hutchins, D. A., Fisher, N. S., Cosper, E. M., and Sanudo-Wilhelmy, S. A. (1997). Release and bioavailability of C, N, P, Se, and Fe following viral lysis of a marine chrysophyte. *Limnology and Oceanography*, 42, 1492-1504.

Hafenbradl, D., Keller, M., Thiericke, R., and Stetter, K. O. (1993). A novel unsaturated archaeal ether core lipid from the hyperthermophile *Methanopyrus kandleri*. *Systematics and Applied Microbiology*, 16, 165-169.

Hallam, S. J., Girguis, P. R., Preston, C. M., Richardson, P. M., and DeLong, E. F. (2003). Identification of methyl coenzyme M reductase A (mcrA) genes associated with methane-oxidizing archaea. *Applied and Environmental Microbiology*, 69, 5483-5491.

Hallam, S. J., Putnam, N., Preston, C. M., Detter, J. C., Rokhsar, D., Richardson, P. M., and DeLong, E. F. (2004). Reverse methanogenesis: Testing the hypothesis with environmental genomics. *Science*, 305, 1457-1462.

Hallam, S. J., Konstantinidis, K. T., Putnam, N., Schleper, C., Watanabe, Y., Sugahara, J., Preston, C., de la Torre, J., Richardson, P. M., and DeLong, E. F. (2006). Genomic analysis of the uncultivated marine crenarchaeote Cenarchaeum symbiosum. *Proceedings of the NationalAcademy of Sciences of the United States of America*, 103, 18296-18301.

Hallam, S. J., Mincer, T. J., Schleper, C., Preston, C. M., Roberts, K., Richardson, P. M., and DeLong, E. F. (2006). Pathways of carbon assimilation and ammonia oxidation suggested by environmental genomic analyses of marine Crenarchaeota. *Plos Biology*, 4, 520-536.

Hayes, J. M., Freeman, K. H., Popp, B. N., and Hoham, C. H. (1990). Compound-specific isotopic analyses: A novel tool for reconstruction of ancient biogeochemical processes. *Organic Geochemistry*, 16, 1115-1128.

Hedges, J. I. (1992). Global biogeochemical cycles: progress and problems. *Marine Chemistry*, 39, 67-93.

Hedrick, D. B., Gucker, J. B., and White, D. C. (1991). Archaebacterial ether lipid diversity analyzed by supercritical fluid chromatography: integration with a bacterial lipid protocol. *Journal of Lipid Research*, 32, 659-666.

Herndl, G. J., Reinthaler, T., Teira, E., van Aken, H., Veth, C., Pernthaler, A., and Pernthaler, J. (2005). Contribution of Archaea to total prokaryotic production in the deep Atlantic Ocean. *Applied and Environmental Microbiology*, 71, 2303-2309.

Higgins, M. J., Sader, J. E., Mulvaney, P., and Wetherbee, R. (2003). Probing the surface of living diatoms with atomic force microscopy: the nanostructure and nanomechanical properties of the mucilage layer. *Journal of Phycology*, 39, 722-734.

Hinrichs, K. U., Hayes, J. M., Sylva, S. P., Brewer, P. G., and DeLong, E. F. (1999). Methane-consuming archaebacteria in marine sediments. *Nature*, 398, 802-805.

Hinterdorfer, P., Baumgartner, W., Gruber, H. J., Schilcher, K., and Schindler, H. (1996). Detection and localization of individual antibody-antigen recognition events by atomic force microscopy. *Proceedings of the NationalAcademy of Sciences of the United States of America*, 93, 3477-3481.

Hopmans, E. C., Schouten, S., Pancost, R. D., van der Meer, M. T. J., and Damste, J. S. S. (2000). Analysis of intact tetraether lipids in archaeal cell material and sediments by high performance liquid chromatography/atmospheric pressure chemical ionization mass spectrometry. *Rapid Communications in Mass Spectrometry*, 14, 585-589.

House, C. H., Schopf, J. W., and Stetter, K. O. (2003). Carbon isotopic fractionation by Archaeans and other thermophilic prokaryotes. *Organic Geochemistry*, 34, 345-356.

Huber, R., Wilharm , T., Huber, D., Trincone, A., Burgraf, S., Konig, H., Rachel, R., Rochinger, I., Fricke, H., and Stetter, K. O. (1992). *Aquifex pyrophilus* gen. nov. sp. nov., represents a novel group of marine hyperthermophilic hydrogen-oxidizing bacteria. *Systematics and Applied Microbiology*, 15, 340-351.

Inagaki, F., Suzuki, M., Takai, K., Oida, H., Sakamoto, T., Aoki, K., Nealson, K. H., and Horikoshi, K. (2003). Microbial communities associated with geological horizons in coastal subseafloor sediments from the Sea of Okhotsk. *Applied and Environmental Microbiology*, 69, 7224-7235.

Inagaki, F., Nunoura, T., Nakagawa, S., Teske, A., Lever, M., Lauer, A., Suzuki, M., Takai, K., Delwiche, M., Colwell, F. S., Nealson, K. H., Horikoshi, K., D'Hondt, S., and Jorgensen, B. B. (2006). Biogeographical distribution and diversity of microbes in methane hydrate-bearing deep marine sediments, on the Pacific Ocean Margin. *Proceedings of the NationalAcademy of Sciences of the United States of America*, 103, 2815-2820.

Ingalls, A. E., Shah, S. R., Hansman, R. L., Aluwihare, L. I., Santos, G. M., Druffel, E. R. M., and Pearson, A. (2006). Quantifying archaeal community autotrophy in the mesopelagic ocean using natural radiocarbon. *Proceedings of the NationalAcademy of Sciences of the United States of America*, 103, 6442-6447.

Ivanova, E. P., Alexeeva, Y. V., Pham, D. K., Wright, J. P., and Nicolaul, D. V. (2006). ATP level variations in heterotrophic bacteria during attachment on hydrophilic and hydrophobic surfaces. *International Microbiology*, 9, 37-46.

Jahnke, L. L., Summons, R. E., Dowling, L. M., and Zahiralis, K. D. (1995). Identification of methanotrophic lipid biomarkers in cold-seep mussel gills-Chemical and isotopic analysis. *Applied and Environmental Microbiology*, 61, 576-582.

Jahnke, L. L., Eder, W., Huber, R., Hope, J. M., Hinrichs, K. U., Hayes, J. M., Des Marais, D. J., Cady, S. L., and Summons, R. E. (2001). Signature lipids and stable carbon isotope analyses of octopus spring hyperthermophilic communities compared with those of Aquificales representatives. *Applied and Environmental Microbiology*, 67, 5179-5189.

Jiang, S. C. and Paul, J. H. (1996). Occurrence of lysogenic bacteria in marine microbial communities as determined by prophage induction. *Marine Ecology Progress Series*, 142, 27-28.

Jean, W., Lorraine, E., Lawrence, C., Steven, L., Harry, R. (2005). Surface and subsurface manifestations of gas movement through a N-S stransect of the Gulf of Mexico, *Marine and Petroleum Geology*, 22, 479-497.

Jiao, N. Z., Yang, Y. H., Koshikawa, H., and Watanabe, M. (2002). Influence of hydrographic conditions on picoplankton distribution in the East China Sea. *Aquatic Microbial Ecology*, 30, 37-48.

Jiao, N. Z., Sieracki, M. E., Zhang, Y., and Du, H. L. (2003). Aerobic anoxygenic phototrophic bacteria and their roles in marine ecosystems. *Chinese Science Bulletin*, 48, 1064-1068.

Jiao, N. Z., Yang, Y. J., and Luo, T. W. (2004). Membrane potential based characterization by flow cytometry of physiological states in an aerobic anoxygenic phototrophic bacterium. *Aquatic Microbial Ecology*, 37, 149-158.

Jiao, N. Z., Yang, Y. H., Hong, N., Ma, Y., Harada, S., Koshikawa, H., and Watanabe, M. (2005). Dynamics of autotrophic picoplankton and heterotrophic bacteria in the East China Sea. *Continental Shelf Research*, 25, 1265-1279.

Jiao, N. Z., Feng, F. Y., and Wei, B. (2006). Proteorhodopsin - A new path for biological utilization of light energy in the sea. *Chinese Science Bulletin*, 51, 889-896.

Jiao, N. Z., Zhang, Y., and Chen, Y. (2006). Time series observation based InfraRed Epifluorescence Microscopic (TIREM) approach for accurate enumeration of bacteriochlorophyll-containing microbes in marine environments. *Journal of Microbiological Methods*, 65, 442-452.

Jiao N. Z. (2006). *Marine Microbial Ecology*. Beijing, China: Science Press.

Jiao, N. Z., Yang, Y. H., Zeng, Y. H., Hong, N., Liu, R. L. Chen, F., and Wang, P. X. (2007). Distinct Distribution Pattern of Abundance and Diversity of Aerobic Anoxygenic Phototrophic Bacteria in the Global Ocean. *Environmental Microbiology*, 9(12):3091-3099.

Kan, J., Hanson, T. E., Ginter, J. M., Wang, K., and Chen, F. (2005). Metaproteomic analysis of Chesapeake Bay microbial communities. *Saline Systems*, 1, 7.

Karner, M. B., DeLong, E. F., and Karl, D. M. (2001). Archaeal dominance in the mesopelagic zone of the Pacific Ocean. *Nature*, 409, 507-510.

Kasas, S., Gotzos, V., and Celio, M. R. (1993). Obeservation of living cells using the atomic force microscope. *Biophysical Journal*, 64, 539-544.

Kendall, M. M., Wardlaw, G. D., Tang, C. F., Bonin, A. S., Liu, Y. T., and Valentine, D. L. (2007). Diversity of Archaea in marine sediments from Skan Bay, Alaska, including cultivated methanogens, and description of *Methanogenium boonei* sp nov. *Applied and Environmental Microbiology*, 73, 407-414.

Kerr, R. A. (2002). Evolution - Could poor nutrition have held life back? *Science*, 297, 1104-1105.

Knittel, K., Losekann, T., Boetius, A., Kort, R., and Amann, R. (2005). Diversity and distribution of methanotrophic archaea at cold seeps. *Applied and Environmental Microbiology*, 71, 467-479.

Koblizek, M., Béjà, O., Bidigare, R. R., Christensen, S., Benitez-Nelson, B., Vetriani, C., Kolber, M. K., Falkowski, P. G., and Kolber, Z. S. (2003). Isolation and characterization of *Erythrobacter* sp strains from the upper ocean. *Archives of Microbiology*, 180, 327-338.

Kolber, Z. S., Van Dover, C. L., Niederman, R. A., and Falkowski, P. G. (2000). Bacterial photosynthesis in surface waters of the open ocean. *Nature*, 407, 177-179.

Kolber, Z. S., Plumley, F. G., Lang, A. S., Beatty, J. T., Blankenship, R. E., VanDover, C. L., Vetriani, C., Koblizek, M., Rathgeber, C., and Falkowski, P. G. (2001). Contribution of aerobic photoheterotrophic bacteria to the carbon cycle in the ocean. *Science*, 292, 2492-2495.

Könneke, M., Bernhard, A. E., de la Torre, J. R., Walker, C. B., Waterbury, J. B., and Stahl, D. A. (2005). Isolation of an autotrophic ammonia-oxidizing marine archaeon. *Nature*, 437, 543-546.

Kuznetsov, Y. G., Malkin, A. J., Lucas, R. W., Plomp, M., and McPherson, A. (2001). Imaging of viruses by atomic force microscopy. *Journal of General Virology*, 82, 2025-2034.

Kuznetsov, Y. G., Gurnon, J. R., Etten, J. L., and McPherson, A. (2005). Atomic force microscopy investigation of a chlorella virus, PBCV-1. *Journal of Structural Biology*, 149, 256-263.

Lang, A. S. and Beatty, J. T. (2007). Importance of widespread gene transfer agent genes in alpha-proteobacteria. *Trends in Microbiology*, 15, 54-62.

Langworthy, T. A., Holzer, G., Zeikus, J. G., and Tornabene, T. G. (1983). Iso-and anteiso-branched glycerol diethers of the thermophilic anaerobes *Thermo- desulfotobacterium commune*. *Systematics and Applied Microbiology*, 4, 1-17.

Lanoil, B. D., Sassen, R., La Duc, M. T., Sweet, S. T., and Nealson, K. H. (2001). Bacteria and Archaea physically associated with Gulf of Mexico gas hydrates. *Applied and Environmental Microbiology*, 67, 5143-5153.

Lanoil, B. D., La Duc, M. T., Wright, M., Kastner, M., Nealson, K. H., and Bartlett, D. (2005). Archaeal diversity in ODP legacy borehole 892b and associated seawater and sediments of the Cascadia Margin. *FEMS Microbiology Ecology*, 54, 167-177.

Lead, J. R., Muirhead, D., and Gibson, C. T. (2005). Characterization of freshwater natural aquatic colloids by atomic force microscopy (AFM). *Environmental Science and Technology*, 39, 6930 -6936.

Li, W. K. W. and Wood, A. M. (1988). Vertical distribution of North Atlantic ultraphytoplankton: analysis by flow cytometry and epifluorescence microscopy. *Deep-Sea Research*, 35, 1615-1638.

Lloyd, K. G., Lapham, L., and Teske, A. (2006). Anaerobic methane-oxidizing community of ANME-1b archaea in hypersalineGulf of Mexico sediments. *Applied and Environmental Microbiology*, 72, 7218-7230.

Lower, S. K., Tadanier, C. J., and Hochella, M. F. (2000). Measuring interfacial and adhesion forces between bacteria and mineral surfaces with biological force microscopy. *Geochimica. Et. Cosmochimica. Acta.*, 64, 3133-3139.

Lower, S. K., Hochella, M. F., and Beveridge, T. J. (2001). Bacterial Recognition of Mineral Surfaces: Nanoscale Interactions Between *Shewanella* and α-FeOOH. *Science*, 292, 1360 - 1363.

Lu, J., Chen, F., and Hodson, R. E. (2001). Distribution, isolation, host specificity, and diversity of cyanophages infecting marine *Synechococcus* spp. in river estuaries. *Applied and Environmental Microbiology*, 67, 3285-3290.

Macko, S. A., Engel, M. H., and Qian, Y. (1994). Early diagenesis and organic matter preservation—A molecular stable carbon isotope perspective. *Chemical Geology*, 114, 365-379.

Madigan M. T., and Martinko J. M. (2006). *Brock Biology of Microorganisms* (11th Ed.). New Jersey: Prentice Hall.

Madrid, V. M., Taylor, G. T., Scranton, M. I., and Chistoserdov, A. Y. (2001). Phylogenetic diversity of bacterial and archaeal communities in the anoxic zone of the CariacoBasin. *Applied and Environmental Microbiology*, 67, 1663-1674.

Man, D. L., Wang, W. W., Sabehi, G., Aravind, L., Post, A. F., Massana, R., Spudich, E. N., Spudich, J. L., and Béjà, O. (2003). Diversification and spectral tuning in marine proteorhodopsins. *Embo Journal*, 22, 1725-1731.

Man-Aharonovich, D., Sabehi, G., Sineshchekov, O. A., Spudich, E. N., Spudich, J. L., and Béjà, O. (2004). Characterization of RS29, a blue-green proteorhodopsin variant from the Red Sea. *Photochemical and Photobiological Sciences*, 3, 459-462.

Marie, D., Brussaard, C. P. D., Thyrhaug, R., Bratbak, G., and Vaulot, D. (1999). Enumeration of marine viruses in culture and natural samples by flow cytometry. *Applied and Environmental Microbiology*, 65, 45-52.

Massana, R., Murray, A. E., Preston, C. M., and DeLong, E. F. (1997). Vertical distribution and phylogenetic characterization of marine planktonic Archaea in the Santa Barbara Channel. *Applied and Environmental Microbiology*, 63, 50-56.

McElfresh, M., Baesu, E., Balhorn, R., Belak, J., and Allen, M. J. (2002). Combining constitutive materials modeling with atomic force microscopy to understand the mechanical properties of living cells. *Proceedings of the NationalAcademy of Sciences of the United States of America*, 99, 6493-6497.

McFeters, G. A., Yu, F. P., Pyle, B. H., and Stewart, P. S. (1995). Physiological assessment of bacteria using fluorochromes. *Journal of Microbiological Methods*, 21, 1-13.

Michel, H. and Oesterhelt, D. (1980). Electrochemical proton gradient across the cell membrane of *Halobacterium halobium*: effect of N,N9-dicyclohexylcarbodiimide, relation to intracellular adenosine triphosphate, adenosine diphosphate, and phosphate concentration, and influenceof the potassium gradient. *Biochemistry*, 19, 4607~4614.

Miller, W. and Moran, M. A. (1997). Interaction of photochemical and microbial processes in the degradation of refractory dissolved organic matter from a coastal marine environment. *Limnology and Oceanography*, 42, 1317-1324.

Mills, H. J., Hodges, C., Wilson, K., MacDonald, I. R., and Sobecky, P. A. (2003). Microbial diversity in sediments associated with surface-breaching gas hydrate mounds in the Gulf of Mexico. *FEMS Microbiology Ecology*, 46, 39-52.

Mincer, T. J., Church, M. J., Taylor, L. T., Preston, C., Kar, D. M., and DeLong, E. F. (2007). Quantitative distribution of presumptive archaeal and bacterial nitrifiers in MontereyBay and the North Pacific Subtropical Gyre. *Environmental Microbiology*, 9, 1162-1175.

Mopper, K., Zhou, X., Kieber, R. J., Kieber, D. J., Sikorski, R. J., and Jones, R. D. (1991). Photochemical degradation of dissolved organic carbonand its impact on the oceanic carbon cycle. *Nature*, 353, 60-62.

Mozes, N., Handley, P. S., Busscher, H. J., and Rouxhet, P. G. (1991). *Microbial cell surface analysis: structural and physicochemical methods*. New York: VCH Publishers.

Müller, D. J., Baumeister, W., and Engel, A. (1999). Controlled unzipping of a bacterial surface layer with atomic force microscopy. *Proceedings of the NationalAcademy of Sciences of the United States of America*, 96, 13170-13174.

Müller, S., Ullrich, S., Lösche, A., Loffhagen, N., and Babel, W. (2000). Flow cytometric techniques to characterise physiological states of *Acinetobacter calcoaceticus*. *Journal of Microbiological Methods*, 25, 79-86.

Nauhaus, K., Treude, T., Boetius, A., and Kruger, M. (2005). Environmental regulation of the anaerobic oxidation of methane: a comparison of ANME-I and ANME-II communities. *Environmental Microbiology*, 7, 98-106.

Nicol, G. W. and Schleper, C. (2006). Ammonia-oxidising Crenarchaeota: important players in the nitrogen cycle? *Trends in Microbiology*, 14, 207-212.

Noble, R. T. and Fuhrman, J. A. (1998). Use of SYBR Green I for rapid epifluorescence counts of marine viruses and bacteria. *Aquatic Microbial Ecology*, 14, 113-118.

Nogawa, H. and Tanoue, E. (2003). Dissolved organic matter in oceanic waters. *Journal of Oceanography*, 59, 129-147.

Novo, D., Perlmutter, N. G., Hunt, R. H., and Shapiro, H. M. (1999). Accurate flow cytometric membrane potential measurement in bacteria using diethyloxacarbocyanine and a ratiometric technique. *Cytometry*, 35, 55-63.

Novo, D. J., Perlmutter, N. G., Hunt, R. H., and Shapiro, H. M. (2000). Multiparameter flow cytometric analysis of antibiotic effects on membrane potential, membrane permeability, and bacterial counts of Staphylococcus aureus and Micrococcus luteus. *Antimicrobial Agents and Chemotherapy*, 44, 827-834.

O'Farrell, P. H. (1975). High resolution two-dimensional electrophoresis of proteins. *Journal of Biological Chemistry*, 250, 4007-4021.

Orphan, V. J., Hinrichs, K. U., Ussler, W., Paull, C. K., Taylor, L. T., Sylva, S. P., Hayes, J. M., and Delong, E. F. (2001). Comparative analysis of methane-oxidizing archaea and sulfate-reducing bacteria in anoxic marine sediments. *Applied and Environmental Microbiology*, 67, 1922-1934.

Orphan, V. J., House, C. H., Hinrichs, K. U., McKeegan, K. D., and DeLong, E. F. (2001). Methane-consuming archaea revealed by directly coupled isotopic and phylogenetic analysis. *Science*, 293, 484-487.

Orphan, V. J., House, C. H., Hinrichs, K. U., McKeegan, K. D., and DeLong, E. F. (2002). Multiple archaeal groups mediate methane oxidation in anoxic cold seep sediments. *Proceedings of the NationalAcademy of Sciences of the United States of America*, 99, 7663-7668.

Ouverney, C. C. and Fuhrman, J. A. (2000). Marine planktonic Archaea take up amino acids. *Applied and Environmental Microbiology*, 66, 4829-4833.

Palenik, B., Brahamsha, B., Larimer, F. W., Land, M., Hauser, L., Chain, P., Lamerdin, J., Regala, W., Allen, E. E., McCarren, J., Paulsen, I., Dufresne, A., Partensky, F., Webb, E. A., and Waterbury, J. (2003). The genome of a motile marine *Synechococcus*. *Nature*, 424, 1037-1042.

Palenik, B., Ren, Q. H., Dupont, C. L., Myers, G. S., Heidelberg, J. F., Badger, J. H., Madupu, R., Nelson, W. C., Brinkac, L. M., Dodson, R. J., Durkin, A. S., Daugherty, S. C., Sullivan, S. A., Khouri, H., Mohamoud, Y., Halpin, R., and Paulsen, I. T. (2006). Genome sequence of *Synechococcus* CC9311: Insights into adaptation to a coastal environment. *Proceedings of the NationalAcademy of Sciences of the United States of America*, 103, 13555-13559.

Pancost, R. D., Freeman, K. H., Patzkowsky, M. E., Wavrek, D. A., and Collister, J. W. (1998). Molecular indicators of redox and marine photoautotroph composition in the late Middle Ordovician of Iowa, USA. *Organic Geochemistry*, 29, 1649-1662.

Pancost, R. D., Freeman, K. H., and Wakeham, S. G. (1999). Controls on the carbon-isotope compositions of compounds in Peru surface waters. *Organic Geochemistry*, 30, 319-340.

Pancost, R. D., Bouloubassi, I., Aloisi, G., and Damste, J. S. S. (2001). Three series of non-isoprenoidal dialkyl glycerol diethers in cold-seep carbonate crusts. *Organic Geochemistry*, 32, 695-707.

Pancost, R. D., Hopmans, E. C., and Damste, J. S. S. (2001). Archaeal lipids in Mediterranean cold seeps: Molecular proxies for anaerobic methane oxidation. *Geochimica Et Cosmochimica Acta*, 65, 1611-1627.

Pancost, R. D. and Damste, J. S. S. (2003). Carbon isotopic compositions of prokaryotic lipids as tracers of carbon cyclingin diverse settings. *Chemical Geology*, 195, 29-58.

Pearson, A., McNichol, A. P., Benitez-Nelson, B. C., Hayes, J. M., and Eglinton, T. I. (2001). Origins of lipid biomarkers in Santa MonicaBasin surface sediment: A case study using compound-specific Delta C-14 analysis. *Geochimica. Et. Cosmochimica. Acta.*, 65, 3123-3137.

Pearson, A., Huang, Z., Ingalls, A. E., Romanek, C. S., Wiegel, J., Freeman, K. H., Smittenberg, R. H., and Zhang, C. L. (2004). Nonmarine crenarchaeol in Nevada hot springs. *Applied and Environmental Microbiology*, 70, 5229-5237.

Pelling, A. E., Li, Y., Shi, W., and Gimzewski, J. K. (2005). Nanoscale visualization and characterization of Myxococcus xanthus cells with atomic force microscopy. *Proceedings of the NationalAcademy of Sciences of the United States of America*, 102, 6484-6489.

Pi Y., Ye Q., Jiang H., Pearson A., Li S., Noakes J., Culp R., Dong H., and Zhang C. L. (2007). Lipids and phylogenetic characterization of Archaea associated with gas hydrates in the Gulf of Mexico. *FEMS Microbiology Ecology*, In revision.

Pradet-Balade, B., Boulme, F., Beug, H., Mullner, E. W., and Garcia-Sanz, J. A. (2001). Translation control: bridging the gap between genomics and proteomics? *Trends in Biochemical Sciences*, 26, 225-229.

Preston, C. M., Wu, K. Y., Molinsky, T. F., and DeLong, E. F. (1996). A psychrophilic crenarchaeote inhabits a marine sponge: *Cenarchaeum symbiosum* gen. nov. *Proceedings of the NationalAcademy of Sciences of the United States of America*, 93, 6241-6246.

Proctor, L. M. and Fuhrman, J. A. (1990). Viral mortality of marine bacteria and cyanobacteria. *Nature*, 343, 60–62.

Radajewski, S., McDonald, I. R., and Murrell, J. C. (2003). Stable-isotope probing of nucleic acids: a window to the function of uncultured microorganisms. *Current Opinion in Biotechnology*, 14, 296-302.

Ram, R. J., VerBerkmoes, N. C., Thelen, M. P., Tyson, G. W., Baker, B. J., Blake, R. C., Shah, M., Hettich, R. L., and Banfield, J. F. (2005). Community proteomics of a natural microbial biofilm. *Science*, 308, 1915-1920.

Razatos, A., Ong, Y. L., Sharma, M. M., and Georgiou, G. (1998). Molecular determinants of bacterial adhesion monitored by atomic force microscopy. *Proceedings of the NationalAcademy of Sciences of the United States of America*, 95, 11059-11064.

Reed, A. J., Lutz, R. A., and Vetriani, C. (2006). Vertical distribution and diversity of bacteria and archaea in sulfide and methane-rich cold seep sediments located at the base of the Florida Escarpment. *Extremophiles*, 10, 199-211.

Richardson, T. L., and Jackson, G. A. (2007). Small phytoplankton and carbon export from the surface ocean. *Science*, 315, 838-840.

Rief, M., Oesterhelt, F., Heymann, B., and Gaub, H. E. (1997). Single molecule force spectroscopy on polysaccharides by atomic force microscopy. *Science*, 275, 1295 - 1297.

Ringelberg, D. B., Davis, J. D., Smith, G. A., Pfiffner, S. M., Nichols, P. D., Nickels, J. S., Henson, J. M., Wilson, J. T., Yates, M., Kampbell, D. H., Read, H. W., Stocksdale, T. T., and White, D. C. (1989). Validation of signature polarlipid fatty acid biomarkers for alkane-utilizing bacteria in soils and subsurface aquifer materials. *FEMS Microbiology Ecology*, 62, 39-50.

Rocap, G., Larimer, F. W., Lamerdin, J., Malfatti, S., Chain, P., Ahlgren, N. A., Arellano, A., Coleman, M., Hauser, L., Hess, W. R., Johnson, Z. I., Land, M., Lindell, D., Post, A. F., Regala, W., Shah, M., Shaw, S. L., Steglich, C., Sullivan, M. B., Ting, C. S., Tolonen, A., Webb, E. A., Zinser, E. R., and Chisholm, S. W. (2003). Genome divergence in two *Prochlorococcus* ecotypes reflects oceanic niche differentiation. *Nature*, 424, 1042-1047.

Rodriguez, G. G., Phipps, D., Ishiguro, K., and Ridgway, H. F. (1992). Use of a fluorescent redox probe for direct visualization of actively respiring bacteria. *Applied and Environmental Microbiology*, 58, 1801-1808.

Roszack, D. B. and Colwell, R. R. (1987). Survival strategies of bacteria in the natural environment. *Microbiology and Molecular Biology Reviews*, 51, 365-379.

Rotsch, C. and Radmacher, M. (1997). Mapping local electrostatic forces with the atomic force microscope. *Langmuir*, 13, 2825-2832.

Rusch, D. B., Halpern, A. L., Sutton, G., Heidelberg, K. B., Williamson, S., Yooseph, S., Wu, D. Y., Eisen, J. A., Hoffman, J. M., Remington, K., Beeson, K., Tran, B., Smith, H., Baden-Tillson, H., Stewart, C., Thorpe, J., Freeman, J., Andrews-Pfannkoch, C., Venter, J. E., Li, K., Kravitz, S., Heidelberg, J. F., Utterback, T., Rogers, Y. H., Falcon, L. I., Souza, V., Bonilla-Rosso, G., Eguiarte, L. E., Karl, D. M., Sathyendranath, S., Platt, T., Bermingham, E., Gallardo, V., Tamayo-Castillo, G., Ferrari, M. R., Strausberg, R. L., Nealson, K., Friedman, R., Frazier, M., and Venter, J. C. (2007). The Sorcerer II Global Ocean Sampling expedition: Northwest Atlantic through Eastern Tropical Pacific. *Plos Biology*, 5, 398-431.

Sabehi, G., Loy, A., Jung, K. H., Partha, R., Spudich, J. L., Isaacson, T., Hirschberg, J., Wagner, M., and Béjà, O. (2005). New insights into metabolic properties of marine bacteria encoding proteorhodopsins. *Plos Biology*, 3, 1409-1417.

Santschi, P. H., Balnois, E., Wilkinson, K. J., Zhang, J., and Buffle, J. (1998). Fibrillar polysaccharides in marine macromolecular organic matter as imaged by atomic force microscopy and transmission electron microscopy. *Limnology and Oceanography*, 43, 896-908.

Sassen, R., Roberts, H. H., Carney, R., Milkov, A. V., DeFreitas, D. A., Lanoil, B., and Zhang, C. L. (2004). Free hydrocarbon gas, gas hydrate, and authigenic minerals in chemosynthetic communities of the northern Gulf of Mexico continental slope: relation to microbial processes. *Chemical Geology*, 205, 195-217.

Scheuring, S., Seguin, J., Marco, S., Levy, D., Robert, B., and Rigaud, J. L. (2003). Nanodissection and high-resolution imaging of the *Rhodopseudomonas viridis* photosynthetic core complex in native membranes by AFM. *Proceedings of the NationalAcademy of Sciences of the United States of America*, 100, 1690-1693.

Scheuring, S. and Sturgis, J. N. (2005). Chromatic Adaptation of Photosynthetic Membranes. *Science*, 309, 484-487.

Schleper, C., Jurgens, G., and Jonuscheit, M. (2005). Genomic studies of uncultivated archaea. *Nature Reviews Microbiology*, 3, 479-488.

Schouten, S., Hopmans, E. C., Pancost, R. D., and Damste, J. S. S. (2000). Widespread occurrence of structurally diverse tetraether membrane lipids: Evidence for the ubiquitous presence of low-temperature relatives of hyperthermophiles. *Proceedings of the NationalAcademy of Sciences of the United States of America*, 97, 14421-14426.

Schouten, S., Hopmans, E. C., Schefuss, E., and Damste, J. S. S. (2002). Distributional variations in marine crenarchaeotal membrane lipids: a new tool for reconstructing ancient sea water temperatures? *Earth and Planetary Science Letters*, 204, 265-274.

Schubert, C. J., Coolen, M. J. L., Neretin, L. N., Schippers, A., Abbas, B., Durisch-Kaiser, E., Wehrli, B., Hopmans, E. C., Damste, J. S. S., Wakeham, S., and Kuypers, M. M. M. (2006). Aerobic and anaerobic methanotrophs in the Black Sea water column. *Environmental Microbiology*, 8, 1844-1856.

Schulze, W. X., Gleixner, G., Kaiser, K., Guggenberger, G., Mann, M., and Schulze, E. D. (2005). A proteomic fingerprint of dissolved organic carbonand of soil particles. *Oecologia*, 142, 335-343.

Schwalbach, M. S. and Fuhrman, J. A. (2005). Wide-ranging abundances of aerobic anoxygenic phototrophic bacteria in the world ocean revealed by epifluorescence microscopy and quantitative PCR. *Limnology and Oceanography*, 50, 620-628.

Seki, M. P., Polovina, J. J., Brainard, R. E., Bidigare, R. R., Leonard, C. L., and Foley, D. G. (2001). Biological enhancement at cyclonic eddies tracked with GOES thermal imagery in Hawaiian waters. *Geophysical Research Letters*, 28, 1583-1586.

Shalapyonok, A., Olson, R. J., and Shalapyonok, L. S. (1998). Ultradian growth in *Prochlorococcus* spp. *Applied and Environmental Microbiology*, 64, 1066-1069.

Shepard, J. L., Olsson, B., Tedengren, M., and Bradley, B. P. (2000). Protein expression signatures identified in Mytilus edulis exposed to PCBs, copper and salinity stress. *Marine Environmental Research*, 50, 337-340.

Sherr, B. F., A., d. G. P., and Sherr, E. B. (1999). Estimating abundance and single-cell characteristics of respiring bacteria via the redox dye, CTC. *Aquatic Microbial Ecology*, 18, 117-131.

Shibata, A., Goto, Y., Saito, H., Kikuchi, T., Toda, T., and Taguchi, S. (2006). Comparison of SYBR Green I and SYBR Gold stains for enumerating bacteria and viruses by epifluorescence microscopy. *Aquatic Microbial Ecology*, 43, 223-231.

Sieracki, M. E., Gilg, I. C., Thier, E. C., Poulton, N. J., and Goericke, R. (2006). Distribution of planktonic aerobic anoxygenic photoheterotrophic bacteria in the northwest Atlantic. *Limnology and Oceanography*, 51, 38-46.

Sullivan, M. B., Waterbury, J. B., and Chisholm, S. W. (2003). Cyanophages infecting the oceanic cyanobacterium *Prochlorococcus*. *Nature*, 424, 1047-1051.

Sullivan, M. B., Coleman, M. L., Weigele, P., Rohwer, F., and Chisholm, S. W. (2005). Three *Prochlorococcus* cyanophage genomes: Signature features and ecological interpretations. *Plos Biology*, 3, 790-806.

Summons, R. E., Jane, L. L., and Roksandic, Z. (1994). Carbon isotopic fractionation in lipids from metamorphic bacteria: Relevance for interpretation of the geochemical record of biomarkers. *Geochimica. Et. Cosmochimica. Acta.*, 58, 2853-2863.

Summons R. E., Jahnke L. L., and Simoneit R. T. (1996). Lipid biomarkers for bacterial ecosystems: Studies of cultured organisms, hydrothermal environments and ancient sediments. In G. R. Bock and J. A. Goode (Eds.), *Evolution of Hydrothermal Ecosystems on Earth (and Mars?)* (pp. 174-194). New York: John Wiley and Sons.

Suttle, C. A., Chan, A. M., and Cottrell, M. T. (1990). Infection of phytoplankton by viruses and reduction of primary productivity. *Nature*, 347, 467-469.

Suttle, C. A., and Chan, A. M. (1993). Marine cyanophages infecting oceanic and coastal strains of *Synechococcus*: abundance, morphology, cross-reactivity and growth characteristics. *Marine Ecology Progress Series*, 92, 99-109.

Suttle, C. A. a. C. A. M. (1993). Marine cyanophages infecting oceanic and coastal strains of Synechococcus: abundance, morphology, cross-reactivity and growth characteristics. *Marine Ecology Progress Series*, 92, 99-109.

Suttle, C. A. (1994). The significance of viruses to mortality in aquatic microbial communities. *Microbial Ecology*, 28, 237-243.

Suttle, C. A., and Chan, A. M. (1994). Dynamics and distribution of cyanophages and their effect on marine *Synechococcus* spp. *Applied and Environmental Microbiology*, 60, 3167-3174.

Suttle, C. A. (2005). Viruses in the sea. *Nature*, 437, 356-361.

Suyama, T., Shigematsu, T., Suzuki, T., Tokiwa, Y., Kanagawa, T., Nagashima, K. V. P., and Hanada, S. (2002). Photosynthetic apparatus in *Roseateles depolymerans*61A is transcriptionally induced by carbon limitation. *Applied and Environmental Microbiology*, 68, 1665-1673.

Teira, E., Lebaron, P., van Aken, H., and Herndl, G. J. (2006). Distribution and activity of Bacteria and Archaea in the deep water masses of the North Atlantic. *Limnology and Oceanography*, 51, 2131-2144.

Teske, A. P. (2006). Microbial communities of deep marine subsurface sediments: Molecular and cultivation surveys. *Geomicrobiology Journal*, 23, 357-368.

Thingstad, T. F., Heldal, M., Bratbak, G., and Dundas, I. (1993). Are viruses important partners in pelagic food webs. *Trends in Ecology and Evolution*, 8, 209-213.

Touhami, A., Hoffmann, B., Vasella, A., Denis, F. A., and Dufrêne, Y. F. (2003). Aggregation of yeast cells: direct measurement of discrete lectin–carbohydrate interactions *Microbiology*, 149, 2873-2878.

Tringe, S. G., von Mering, C., Kobayashi, A., Salamov, A. A., Chen, K., Chang, H. W., Podar, M., Short, J. M., Mathur, E. J., Detter, J. C., Bork, P., Hugenholtz, P., and Rubin, E. M. (2005). Comparative metagenomics of microbial communities. *Science*, 308, 554-557.

Tsutomu, M. and Koichi, K. (1992). Quantum conversion and image detection by a bacteriorhodopsin-based artificial photoreceptor. *Science*, 255, 342~344.

Tyson, G. W., Chapman, J., Hugenholtz, P., Allen, E. E., Ram, R. J., Richardson, P. M., Solovyev, V. V., Rubin, E. M., Rokhsar, D. S., and Banfield, J. F. (2004). Community structure and metabolism through reconstruction of microbial genomes from the environment. *Nature*, 428, 37-43.

Valentine, D. L. and Reeburgh, W. S. (2000). New perspectives on anaerobic methane oxidation. *Environmental Microbiology*, 2, 477-484.

Vandevivere, P. and Kirchman, D. L. (1993). Attachment stimulates synthesis by a bacterium exopolysaccharide synthesis by a bacterium. *Applied and Environmental Microbiology*, 59, 3280-3286.

Vaulot, D., Marie, D., Olson, R. J., and Chisholm, S. W. (1995). Growth of *Prochlorococcus*, a photosynthetic prokaryote, in the Equatorial Pacific Ocean. *Science*, 8, 1480-1482.

Velegol, S. B., Pardi, S., Li, X., Velegol, D., and Logan, B. E. (2003). AFM Imaging Artifacts due to Bacterial Cell Height and AFM Tip Geometry. *Langmuir*, 19, 851-857.

Venter, J. C., Remington, K., Heidelberg, J. F., Halpern, A. L., Rusch, D., Eisen, J. A., Wu, D. Y., Paulsen, I., Nelson, K. E., Nelson, W., Fouts, D. E., Levy, S., Knap, A. H., Lomas, M. W., Nealson, K., White, O., Peterson, J., Hoffman, J., Parsons, R., Baden-Tillson, H., Pfannkoch, C., Rogers, Y. H., and Smith, H. O. (2004). Environmental genome shotgun sequencing of the Sargasso Sea. *Science*, 304, 66-74.

Vetriani, C., Jannasch, H. W., MacGregor, B. J., Stahl, D. A., and Reysenbach, A. L. (1999). Population structure and phylogenetic characterization of marine benthic archaea in deep-sea sediments. *Applied and Environmental Microbiology*, 65, 4375-4384.

Wakeham, S. G. (1995). Lipid biomarkers for heterotrophic alteration of suspended particulate organic matter in oxygenated and anoxic water columns of the ocean. *Deep-Sea ResearchPart I-Oceanographic Research Papers*, 42, 1749-1771.

Whelan, J., Eglinton, L., Cathles III, L., Losh, S., Roberts, H. (2005). Surface and subsurface manifestations of gas movement through a N–S transect of the Gulf of Mexico. *Marine and Petroleum Geology* 22, 479–497.

Wang, W. W., Sineshchekov, O. A., Spudich, E. N., and Spudich, J. L. (2003). Spectroscopic and photochemical characterization of a deep ocean proteorhodopsin. *Journal of Biological Chemistry*, 278, 33985-33991.

Wang, P., Xiao, X., and Wang, F. P. (2005). Phylogenetic analysis of Archaea in the deep-sea sediments of west Pacific Warm Pool. *Extremophiles*, 9, 209-217.

Wang, K. and F., Chen. (2007). Prevalence of highly host-specific cyanophages in the estuarine environment. *Environmental Microbiology*, In press.

Wasinger, V. C., Cordwell, S. J., Cerpa Poljak, A., Yan, J. X., Gooley, A. A., wilkins, M. R., Duncan, M. W., Harris, R., Williams, K. L., and Humphery-Smith, I. (1995). Progress with gene product mapping of Moli cutes: Mycoplasma genitalium. *Electrophoresis*, 16, 1090-1094.

Waterbury, J. B. and Valois, F. W. (1993). Resistance to co-occuring phages enables marine *Synechococcus* communities to coexist wth cyanophages abundant in seawater. *Applied and Environmental Microbiology*, 59, 3393-3399.

Weijers, J. W. H., Schouten, S., Hopmans, E. C., Geenevasen, J. A. J., David, O. R. P., Coleman, J. M., Pancost, R. D., and Sinninghe Damste, J. S. (2006). Membrane lipids of mesophilic anaerobic bacteria thriving in peats have typical archaeal traits. *Environmental Microbiology*, 8, 648-657.

Weinbauer, M. G. and Suttle, C. A. (1996). Potential significance of lysogeny to bacteriophage production and bacterial mortality in coastal waters of the Gulf of Mexico. *Applied and Environmental Microbiology*, 62, 4374–4380.

Weinbauer, M. G. and Rassoulzadegan, F. (2004). Are viruses driving microbial diversification and diversity? *Environmental Microbiology*, 6, 1-11.

Wilcox, R. M. and Fuhrman, J. A. (1994). Bacterial viruses in coastal seawater: lytic rather than lysogenic production. *Marine Ecology Progress Series*, 114, 35-45.

Wilhelm, S. W., Weinbauer, M. G., Suttle, C. A., and Jeffrey, W. H. (1998). The role of sunlight in the removal and repair of viruses in the sea. *Limnology and Oceanography*, 43, 586-592.

Wilhelm, S. W. and Suttle, C. A. (1999). Viruses and Nutrient Cycles in the Sea - Viruses play critical roles in the structure and function of aquatic food webs. *Bioscience*, 49, 781-788.

Williams, P. M. and Druffel, E. R. M. (1987). Radiocarbon in dissolved organic matter in the central North Pacific Ocean. *Nature*, 330, 246-248.

Williams, S. C., Hong, Y., Danavall, D. C. A., Howard-Jones, M. H., Gibson, D., Frischer, M. E., and Verity, P. G. (1998). Distinguishing between living and nonliving bacteria: Evaluation of the vital stain propidium iodide and its combined use with molecular probes in aquatic samples. *Journal of Microbiological Methods*, 32, 225-236.

Wilmes, P. and Bond, P. L. (2004). The application of two-dimensional polyacrylamide gel electrophoresis and downstream analyses to a mixed community of prokaryotic microorganisms. *Environmental Microbiology*, 6, 911-920.

Wilmes, P. and Bond, P. L. (2006). Metaproteomics: studying functional gene expression in microbial ecosystems. *Trends in Microbiology*, 14, 92-97.

Wilson, W. H., Joint, I. R., Carr, N. G., and Mann, N. H. (1993). Isolation and molecular characterization of five marine cyanophages propagated on *Synechococcus* sp. strain WH7803. *Applied and Environmental Microbiology*, 59, 3736-3742.

Wommack, K. E. and Colwell, R. R. (2000). Virioplankton: Viruses in aquatic ecosystems. *Microbiology and Molecular Biology Reviews*, 64, 69-114.

Wuchter, C., Abbas, B., Coolen, M. J. L., Herfort, L., van Bleijswijk, J., Timmers, P., Strous, M., Teira, E., Herndl, G. J., Middelburg, J. J., Schouten, S., and Damste, J. S. S. (2006). Archaeal nitrification in the ocean. *Proceedings of the NationalAcademy of Sciences of the United States of America*, 103, 12317-12322.

Xiong, J., Fischer, W. M., Inoue, K., Nakahara, M., and Bauer, C. E. (2000). Molecular evidence for the early evolution of photosynthesis. *Science*, 289, 1724-1730.

Xu, W., Mulhern, P. J., Blackford, B. L., Jericho, M. H., and Templeton, I. (1994). A new atomic force microscopy technique for the measurement of the elastic properties of biological materials. *Scanning Microscopy*, 8, 499-506.

In: Deep-Sea
Editors: D. R. Bailey and S. E. Howard
ISBN: 978-1-62257-173-4

Chapter 6

THE DEEP-MEDITERRANEAN SEA: HABITATS, BIODIVERSITY AND ECOSYSTEM FUNCTIONING*

Roberto Danovaro†
Department of Life and Environmental Sciences,
Polytechnic University of Marche, Ancona, Italy

ABSTRACT

The deep-Mediterranean basin is a peculiar system characterized by high species richness. The unexpectedly high biodiversity of the deep-sea benthos is apparently related to the presence of strong environmental gradients and the richness of deep-sea habitats. Despite the increasing sampling effort and the advancements made in the last three decades, the knowledge of the deep Mediterranean Sea and its biodiversity is still poor, as highlighted by recent estimates reporting that approximately 75% of Mediterranean deep-sea biodiversity is still undiscovered. Here is provided an overview of the main deep-sea habitat types, their distribution, ecological characteristics and biodiversity. These habitats include: open slope systems, submarine canyons, deep basins, seamounts, deep-water corals, cold seeps, hydrothermal vents and permanent anoxic systems. Most of these habitats are extremely vulnerable and under serious threats related to the increasing direct and indirect human impact of deep-sea ecosystems.

INTRODUCTION

Deep-sea ecosystems include the waters and sediments beneath approximately 200 m depth. They represent the world's largest biome, covering more than 65% of the world surface and including more than 95% of the global biosphere (Ramirez-Llodra et al. 2010). Traditionally the Mediterranean basin is one of the most intensively investigated areas of the

* A version of this chapter also appears in *Life in the Mediterranean Sea: A Look at Habitat Changes*, edited by Noga Stambler, published by Nova Science Publishers, Inc. It was submitted for appropriate modifications in an effort to encourage wider dissemination of research.

† E-mail: r.danovaro@univpm.it

world in terms of both terrestrial and coastal marine biodiversity. The Mediterranean basin is indeed reported as a hot spot of biodiversity with a uniquely high percentage of endemic species (Myers et al. 2000). Despite its small dimensions (ca 0.82% of the ocean surface), this basin hosts more than 7.5% of global biodiversity (Bianchi and Morri 2000). However, this information is almost completely confined to coastal ecosystems and studies on the deep-sea fauna remain limited, especially when compared to the rest of the world (CIESM 2003). The average depth of the Mediterranean basin is ca 1,450 m, much shallower than the average depth of the world oceans (ca 3,680 m, Charette and Smith 2010). This has several implications for the deep-water turnover (ca 50 years) and the vulnerability to climate change and deep-water warming (Danovaro et al. 2010a). The Mediterranean Sea has been considered as a "miniature ocean" that can be used as a model to anticipate the response of the global oceans to various kinds of pressures. The deepest depth of the Mediterranean Sea (5,121 m) is reported for the Matapan-Vavilov Trench, Ionian Sea. The main environmental characteristics of the Mediterranean Sea, which can influence also deep-sea biodiversity are: a) high homeothermy from ca 300-500 m to the bottom, and bottom temperatures of ca 12.8 to 13.5 °C in the Western basin and 13.5 to 15.5 °C in the Eastern basin (i.e., there are no thermal boundaries, whereas in the Atlantic Ocean temperature decreases with depth; b) high salinity, from about 38 to 39.5 by the stratification of the water column; c) limited freshwater inputs (the freshwater deficit is equivalent to ca 0.5-0.9 m y-1, compensated by the Atlantic inflow of surface water); d) high oxygen concentrations; e) oligotrophic conditions, with a strong decreasing gradient moving Eastward (Danovaro et al. 1999a). The Eastern basin is considered to be one of the most oligotrophic areas of the world (Psarra et al. 2000; Tselepides et al. 2000). Inputs of organic carbon are 15-80 times lower than in the Western basin and there are extremely low concentrations of chlorophyll-a in surface offshore waters (ca 0.05 μg L^{-1}) (Krom et al. 1991;Yacobi et al. 1995) as well as of the potentially limiting organic nutrients (e.g., proteins and lipids) that sharply decline with increasing distance from the coast and depth within the sediment.

The composition and bioavailability of the food sources is likely to influence also the biodiversity and biochemical composition of the benthic species. Such findings implement previous studies on both planktonic (Bamstedt 1978; Hopkins et al. 1993) and benthic organisms (Lehtonen 1996; Miliou et al. 1992) that emphasised the influence of seasonal fluctuations in physical (e.g., temperature) and biological constraints on the biochemical compositions, energy content and calorific value of marine invertebrates (Falk-Petersen 1981; Lehtonen 1996).

Different trophic conditions (as those reported in the Eastern and North-Western Mediterranean Sea) determine different ecological and bioenergetic strategies. A comparative study carried out on biochemical composition of nematodes (the dominant metazoan taxon of deep-sea benthos) from the Eastern and Western basin reported significantly higher lipid concentrations in deep-sea specimens from the Eastern Mediterranean Sea (Danovaro et al. 1999b). Deep-sea benthic organisms from regions characterised by highly variable food inputs must have efficient energy storage systems to be able to cope with long periods of very low food availability (Sargent and Whittle 1981). The differences in lipid level between nematodes from the two study areas of the Mediterranean Sea might be explained with the need to accumulate, opportunistically, lipids during short periods of food availability to be able to survive long periods of indigence. Bioenergetic strategies similar to those reported above have been reported for deposit-feeding amphipods and other deep-sea species living at high latitudes which are known to be characterised by high temporal variability with short periods of extremely large food availability (e.g., in the Arctic; Clarke 1983; Sargent and Henderson 1986). Therefore, it is possible that the unpredictability of the food supply rather than the oligotrophy of the environment could determine the bioenergetic strategy of the benthic components.

The extreme oligotrophy of certain systems has been demonstrated to play an important role in the community structure and, for instance can result in a diminished average size of the benthic organisms. The combined effects of food limitation and temporal variability in food supply might have important implication for bioenergetic strategies at ecosystem level. In the deep-Cretan Sea a close coupling has been observed between particle fluxes, sedimentary organic matter composition and concentration, individual nematode energy content and biomass (Danovaro et al. 1999a). Calculations of the energy content of nematodes suggest that an apparent balance between diminished size and increased energy content (in terms of lipid storage) is a possible mechanism to optimise resource partitioning and exploitation.

In order to define the efficiency of the benthic metazoans in exploiting the available organic matter inputs from the photic layers, ratios between labile organic carbon fluxes (as the sum of protein, carbohydrate and lipid carbon content of the settling particles) and the number and biomass of organisms supported by the food input were calculated (Danovaro 2003). Meiofaunal abundance in the NW-Mediterranean Sea is typically higher than in the deep Eastern Mediterranean Sea but each meiofaunal organism in the NW-Mediterranean Sea has a larger nutrient availability. For instance, about 0.14 mg C of C flux are supplied every daily per each individual of meiofauna in the Western Mediterranean compared to the 0.07 mg C in the Eastern Mediterranean Sea (Danovaro 2003). The same applies to benthic prokaryotes. The ratio between organic carbon fluxes and bacterial biomass indicates that 3.3 mg of C flux are supplied daily per 1 mg C of prokaryotic biomass in the Western Mediterranean Sea, compared to 0.07 in the Eastern Mediterranean Sea. These data, confirm that quantitative differences between the two areas are related to the different organic carbon inputs from the photic layer but also allow to infer about the efficiency of the deep-sea ecosystems of the two basins in exploiting the nutrient inputs. In fact, if we hypothesise that a ratio of 0.14 mg C ind.-1 is required by meiofauna in the NW-Mediterranean Sea, but meiofauna from the Western basin would receive the organic carbon fluxes of the Eastern Mediterranean Sea, the meiofauna would be so much nutrient limited to be reduced to be about half of those observed. With a similar approach it might be estimated that bacterial biomass in the NW-Mediterranean Sea in the scenario of food limiting conditions would be 50 times lower. Data for meiofaunal comparison at biomass level are not available but calculation based on bacterial biomass reported above would suggest a significant functional difference (Danovaro et al. 1999c).

History of Research in the Deep-Mediterranean Sea

The Mediterranean Sea (and particularly the deep-Cretan Sea) has seen the first investigations on the deep-sea biodiversity, where Forbes noticed that biodiversity extinguished altogether by 500 m depth (Anderson and Rice 2006). Benthic deep-sea fauna in the Tyrrhenian Sea was investigated also during the Washington expedition (1881-83) down to 3,115 m depth and subsequently by the Hirondelle and Princess Alice expeditions (1888-1922), and the voyages of the Pola (1890-1893). The Danish oceanographic cruises of the Thor (1908) and Dana (1928-29) also reported deep-sea fishes at depths greater than 1,000 m in the Mediterranean Sea. The Polymède campaign (N/O Jean Charcot) in the Western basin (Geistdoerfer and Rannou 1972) and the German Meteor expedition in the Eastern basin (Klausewitz 1989) increased significantly the knowledge of the deep-sea biota. Studies based on the use of samplings adequate for quantitative studies started in the late 1950s in both the Western and Eastern basins and extended in the 1970s, 80s, and 90s (De Stigter 1996; Jorrisen 1988; Massiotta et al. 1976; Wright and Rupert 1981) down to 4,523 m depth, but the study of the diversity of benthic prokaryotic assemblages (Bacteria and Archaea) in deep-sea sediments of the Mediterranean Sea was initiated only after 2000 (Luna et al. 2004; Polymenakou et al. 2005).

Deep-Sea Habitats and Their Biodiversity

A central paradigm of marine diversity is that species richness increases with increasing water depth to a maximum at mid-slopes (around 2,000 m) and thereafter decreases (Levin et al. 2001; Rex 1981).

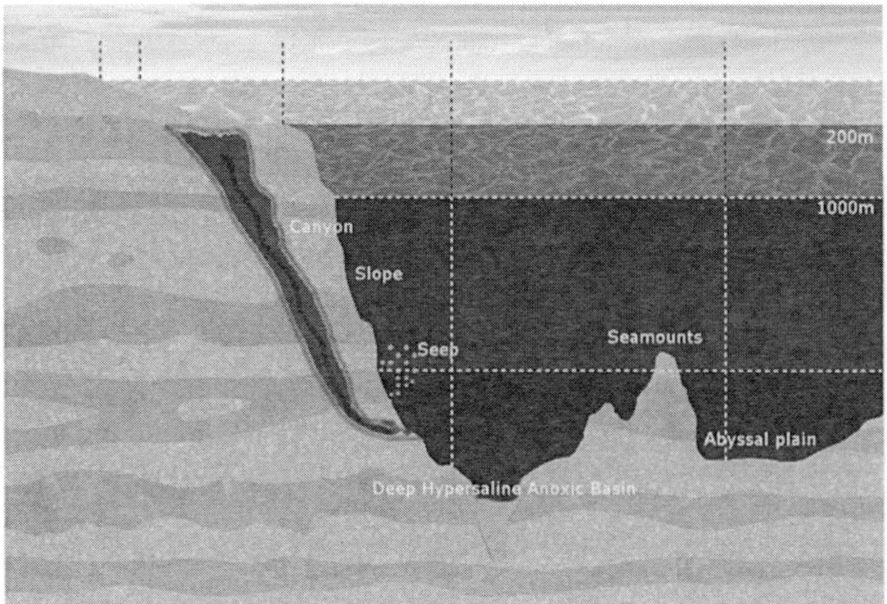

Figure 1. A conceptual scheme of the location of the different deep-sea habitats in the Mediterranean Sea; reported are: slope, canyon, seamounts, cold seep, abyssal plain and deep hypersaline anoxic basin. Deep-water corals are typically located at the canyon head, or along the slopes or on the top of seamounts.

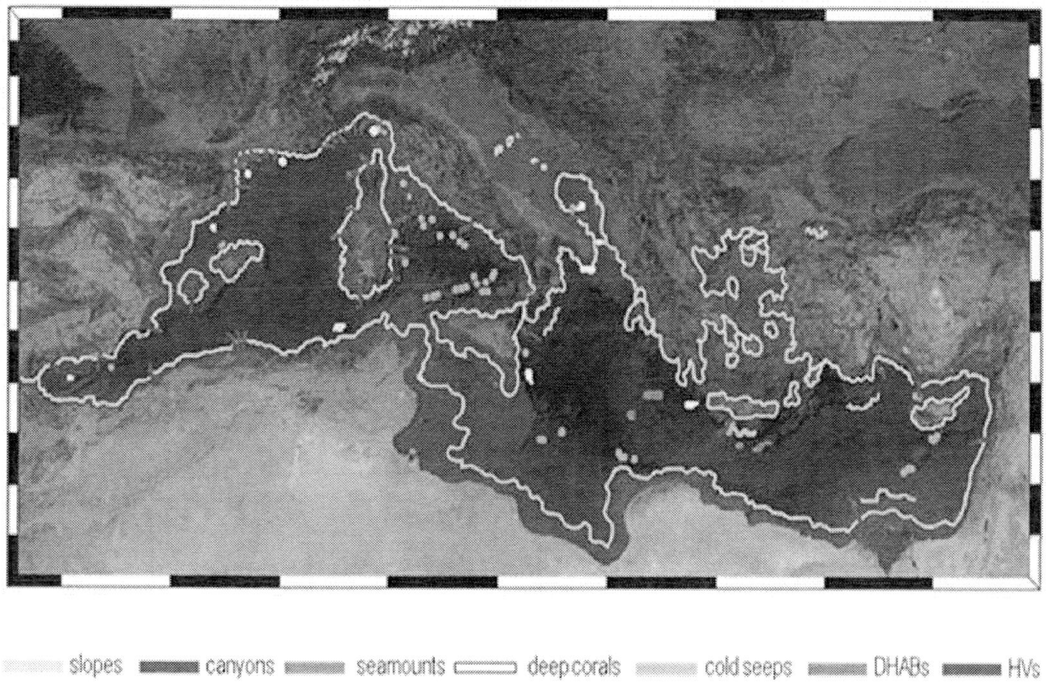

Figure 2. Mediterranean basin and topography of the deep-sea floor.Deep Basins.

The enhanced levels of biodiversity along slopes are possibly a source for biodiversity for deeper basins and shelves, through radiation and dispersal processes, closely coupled with benthic topography and the hydrodynamic, physical, and biogeo-chemical characteristics of the deep sea. However, different benthic components display different spatial patterns with increasing depth. For instance number of bacterial and archaeal OTUs (operational taxonomic units) in the deep-Mediterranean Sea did not change significantly with increasing water depth (Danovaro unpublished data).

The biodiversity of all other benthic components decreased significantly with increasing water depth. These results indicate that the patterns of biodiversity could be dependent on the size of the organisms and, likely, on the greater ability to move and disperse across different bathymetric ranges among larger organisms, which can be crucial for shaping biodiversity patterns.

The deep-sea floor of the Mediterranean basin includes regions characterised by complex sedimentological and structural features. Therefore, the Mediterranean basin contains, over relatively limited spatial scales, a number of habitats, which can represent potential "hotspots" of biodiversity. A tentative, possibly not exhaustive list of these systems includes: a) deep basins; b) open slope systems; c) submarine canyons, d) seamounts; e) deep-water coral systems; f) cold seeps and carbonate mounds; g) hydrothermal vents; and h) permanently anoxic systems. The conceptual scheme of the location of these habitats across the Mediterranean basin is reported in Figure 1 and specific distribution is reported in Figure 2. Here I will present an in-depth analysis on the main types of these deep-sea habitats.

Deep Basins

The abyssal basins of the Mediterranean Sea are extremely unusual deep-sea systems. With water temperatures at 4,000 m in excess of ca 14 °C (rather than < 4 °C for the deep oceanic basins) the entire benthic environment is a warm deep-sea ecosystem, with temperatures resembling those of the water around a hydrothermal vent system, but lacking the vents' rich chemical energy supply. The deep-sea basin is the largest habitat of the dark portion of the Mediterranean Sea. Bathyal/abyssal plains cover a large portion of the deep Western Mediterranean basin (Acosta et al. 2002), these having a triangular shape and an overall area of ca. 240,000 km2. In the Western basin the 2,600-2,700 m isobaths have been used as the upper limit of the bathyal/abyssal plain, which has a maximum depth of ca 3,050 m. Conversely the deepest part of the Tyrrhenian Basin exceeds 3,600 m depth (Vanney and Gennesseaux 1985). Sediments filling the Mediterranean abyssal plains are dominated by the deposition of turbidites and are characterized by the presence of seafloor features up to 35 m in height (Acosta et al. 2002). Barriers to colonization of the Mediterranean Sea include the differences in temperature, salinity, and food supply between the Atlantic Ocean and Mediterranean Sea, as well as the existence of shallow sills in the Strait of Gibraltar and Strait of Sicily.

The deep Mediterranean plains differ from other bathyal/abyssal plains for the different species composition. Typical deep-water groups, such as echinoderms, glass sponges, and macroscopic foraminifera (Xenophyophora) are scarce or absent in the deep basins of the Mediterranean Sea, while other faunistic groups (fishes, decapods crustaceans, mysids, and gastropods) are poorly represented compared to the NE Atlantic. In the deep-Mediterranean basin the deep-water Grenadier fish Coryphaenoides armatus and the amphipod Eurythenes gryllus (Figure 3) are absent. Instead, Acanthephyra eximia appears to have functionally replaced E. gryllus, the dominant deep-sea scavenging crustacean throughout most of the world's oceans (Christiansen 1989). The Eastern deep basins had a complex geological history, which precluded the colonisation by deep-sea fauna until less than 6,000 years ago (Galil and Goren 1994). A certain degree of eurythermy may have allowed A. eximia to become a dominant member of the Mediterranean abyssal community in the absence of the stenothermal amphipod E. gryllus. Despite these inferences and the relative youth of the system, a deep-sea fauna has developed, although it is depauperate compared to that of the oceans (Jones et al. 2003).

Open Slopes

The continental slopes are, after the bathyal/abyssal plains the largest deep-sea habitats of the deep-Mediterranean Sea and represent the connection between the shelf and basin plain. All the studied margins show that the flux of particles increases with depth owing to the presence of lateral inputs, ranging from 50% in the Gulf of Lions to 80% in the Cretan Sea. The analysis of benthic biodiversity along the Mediterranean slopes has revealed a clear decrease of the benthic abundance and biomass with increasing depth. Danovaro et al. (2009a) investigated the biodiversity of meiofaunal (as richness of higher taxa) and nematode (as species richness) assemblages along the continental margins at large spatial scales and reported that open slopes display a species richness similar or higher than that reported for bathyal and abyssal plain ecosystems.

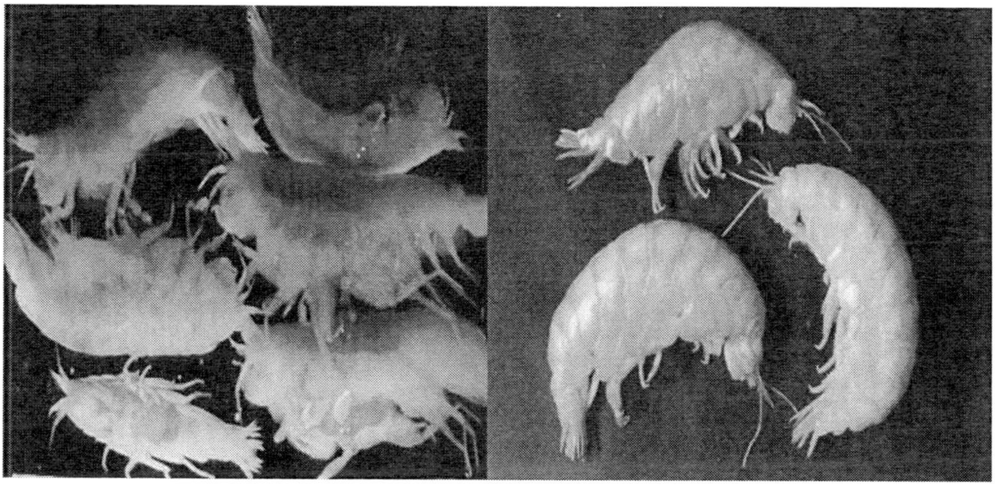

Figure 3. Specimens of Eurythenes gryllus collected at ca 8,000 m depth from the Atacama trench in the South Pacific Ocean. This species, which is common and dominant in most abyssal habitats of the world is absent in the Mediterranean Sea.

However, a unique, general driver capable of explaining the spatial patterns of biodiversity was not identified.

This result is not surprising, considering the multiplicity of interactions among "local" ecological characteristics, environmental factors, and topographic and textural conditions in each specific slope environment. This complexity probably has considerable influence on the conditions, allowing settlement of a high number of species. The patterns of deep-sea biodiversity along the slope of the Mediterranean Sea are different from those hypothesized so far (i.e., reflecting a hump shaped curve), and reflect a mosaic of life more complex and varied than previously imagined (Danovaro et al. 2010a). A recent study (Bianchelli et al. 2010) has demonstrated the presence of significant differences in the composition of the metazoan meiofauna between slopes and adjacent canyons apparently related to the pronounced local variability (Soltwedel 2000). When the analyses of the structure of meiofaunal assemblages are performed excluding the three most abundant taxa (Nematoda, Copepoda and Polychaeta), i.e., considering only the rare taxa, some taxa of the temporary meiofauna were exclusively present in open slopes, such as Priapulida, Holothuroidea, Ascidiacea and Cnidaria. If confirmed, the apparent preference of some rare taxa for a specific habitat within the same region could be responsible for the high dissimilarity (high turnover diversity) reported in most deep-sea habitats. These results suggest that habitat characteristics could play a key role in the supply and settlement of the temporary meiofaunal taxa (Levin et al. 2001).

Overall the environmental variables investigated (organic matter, depth, grain size, temperature, salinity) explained only 23% of the variance encountered in the meiofaunal assemblages, suggesting that most of the variance depends on additional physical variables, (e.g., slope inclination) and hydrodynamic features (e.g., bottom-current speeds, sediment resuspension and mineralogy) and biotic interactions with other benthic components. Understanding the mechanisms controlling deep-sea biodiversity within and across these attractive environments will open new perspectives for the conservation and sustainable management of open slope systems crucial to the functioning of the global ecosystem.

Canyons

Submarine canyons are major topographic systems that enhance the heterogeneity of continental slopes (Levin et al. 2010). Submarine valleys are mostly incised on the continental slope and form part of the drainage system of continental margins. Their cross-sections tend to be V-shaped along the upper course and U-shaped in the lower course. Submarine canyons are widespread on many continental margins of the Mediterranean basin, but their extension is likely comprising 10 - 20% of the extension of the continental margins. Complex canyon networks (e.g., the Gulf of Lions) are sometimes adjacent to sections of the margin with only linear canyons (e.g., the Catalonia margin), or no canyons at all (e.g., the North Balearic margin). Most canyons are relatively inactive, but others are characterized by an important sediment transport (Puig et al. 2000). They are major pathways for the transportation and burial of organic carbon, and fast-track corridors for material transported from the land to the deep sea (Puig et al. 2003) and act as temporary buffers for sediment and carbon storage. Rapid, episodic flushing of canyons may at times transport large amounts of sediment to the deep basin (Canals et al. 2006). Several submarine canyons cross the continental slope of the Western and Central Mediterranean Sea. They represent hotspots of species diversity and endemism (Gili et al. 1999; 2000) and are preferential areas for the recruitment of megafaunal species (Sardà et al. 2004). Canyons play an important role in structuring the populations and life cycles of both deep-sea plankton and meroplankton of the benthic megafauna (Gili et al. 2000). Canyons are also important habitats for fisheries (e.g., the hake Merluccius merluccius and the rose shrimp Aristeus antennatus; Capezzuto et al. 2010; Sardà et al. 2009; Sardà and Cartes 1997; Stanley and Wezel 1985; Stefanescu et al. 1994; Tudela et al. 2003). The presence and/or higher abundance of sessile taxa such as corals and sponges is explained by the intensified hydrodynamics associated with the proximity of the canyon (Ramirez-Llodra et al. 2008). Faunal abundance and biomass are usually higher inside the canyons when compared to similar depths in the surrounding habitat. These habitats can support the presence of suspension feeders (that benefit from accelerated currents - Zúñiga et al. 2009 - and the presence of hard substrate), demersal planktivores that exploit the krill concentrated in the canyons (Greene et al. 1988), and detritivores that can profit of the high sedimentation rates and accumulation of macrophytic debris (Harrold et al. 1998; Vetter 1994; Vetter and Dayton 1998). Because of their characteristics, the biodiversity of faunal assemblages can be markedly different from that on the adjacent open slopes (Ramirez-Llodra et al. 2009). Their biomass and abundance can be 2 to 15-fold higher compared to the surrounding areas at similar depths (Vetter and Dayton 1998). Species composition within canyons is also different from that found on the surrounding slopes. Canyon assemblages generally display lower diversity for the meiofaunal components, because of the high dominance of a few species and the lower evenness (Bianchelli et al. 2008). On the other hand, certain canyons may contain a higher diversity of megafauna than the slopes and can be considered hotspots of diversity as they may display high rates of endemism (Gage and Tyler

1991; Gili et al. 2000). The higher habitat heterogeneity and higher organic matter availability can partly explain the higher biodiversity of megafauna in the upper part of the Mediterranean canyons (Ramirez-Llodra et al. 2008). A comparative analysis of nematodes at similar depths in four deep-sea canyons and on adjacent open slopes in the Western and Central Mediterranean Sea suggested that species richness changed significantly with increasing water depth only in about half of the investigated systems. The analysis of nematode biodiversity revealed the presence of significant differences in species composition also among different canyons. The topographic features can contribute to the observed differences; for example, at 500 m depth in the Central Mediterranean Sea (South Adriatic margin), the lower nematode species richness in canyons could be related to the presence of hard substrates (Trincardi et al. 2007).

Seamounts

Seamounts are like islands separated by great depths and can play an important role in the connectivity among deep-sea ecosystems and their species. Seamounts can also serve as refuges for relict populations of species that have been disappeared from other areas. A complete and detailed map of all Mediterranean seamounts is not available yet. In the Western Mediterranean Sea, the Tyrrhenian bathyal plain is characterized by a large number of seamounts (Magnaghi, Vavilov, Marsili, Vercelli, Cassinis; Wezel 1985). The Eastern Mediterranean basin is characterized by a higher topographic heterogeneity than the Western sector and a high number of seamounts that include some impressive geological structures (e.g., Eratosthenes Seamount), but their biology is almost unknown. At 800 m depth a diverse fauna was found including two species of scleractinian corals (Caryophylla calveri and Desmophyllum cristagalli, now D. dianthus), two types of encrusting foraminiferans, two species of encrusting poriferans, abundant scyphozoan polyps, many individuals of the small actiniarian Kadophellia bathyalis, seven species of bivalves, one sipunculan, one asteroid and one fish (Galil and Zibrowius 1998). Recent studies have demonstrated that also soft sediments surrounding the seamounts (e.g., the Marsili and Palinuro Seamounts) are enriched in terms of biodiversity, likely as an effect of the presence of these topographic structures (Danovaro et al. 2009b; Pusceddu et al. 2009). The complex hydrodynamic conditions associated to the seamounts can also result in a lower number of meiofaunal taxa and low nematode species richness, when compared with non-seamount sediments. However, some families and species were apparently exclusively present in sediments close to the seamounts and absent in adjacent sediments. These findings suggest that the deep-sea benthic assemblages associated or close to Mediterranean seamounts can be highly site-specific and support the hypothesis that seamounts contribute to increase the turnover diversity of this basin (Danovaro et al. 2008). Current research also involves other seamounts, such as the Vercelli and the Dauno Seamounts and seamounts in the Alboran Sea.

Deep-Water Coral Habitats

A deep-water coral reef is formed by living animals and coral rubbles, which provide hard substrate for the colonization of several organisms (Hovland 2008). The colonial stone corals Lophelia pertusa and Madrepora oculata occur along the North-Western European continental margin, the deep shelves and in Scandinavian fjords, and are present also in different sectors of the deep-Mediterranean Sea (Tursi et al. 2004), but the distribution of these habitats along Mediterranean margins is still incomplete. At present, a total of 14 coral bank areas have been censused and include the areas from the Gibraltar sill to the Gulf of Lions canyons, from the Ligurian Sea to the Sicilian Channel, and from the Apulian margin to the trough off Tassos in the Aegean Sea (Freiwald et al. 2009).

The depth distribution of the corals ranges from 150 m (Strait of Gibraltar) to 1,100 m (Santa Maria di Leuca). In the Mediterranean Sea, cold-water corals generally occur along the edge of the continental shelf, on offshore submarine banks and canyons. These coral communities share a set of common characteristics, including hydrographic conditions and food supply within a complex local topographic setting. Mediterranean deep-water reefs are associated with temperatures ranging from 13.4 to 13.9 °C, salinities from 38.4 to 38.9 and dissolved oxygen from 3.75 to 4.54 ml L-1 (Freiwald et al. 2009). The temperatures in the deep-Mediterranean Sea are close to the upper limit for many cold-water corals living at bathyal depths (Freiwald et al. 2004). The occurrence of the two deep-water colonial scleractinian species living in the Mediterranean Sea, M. oculata and L. pertusa, appears to be a relict of a much more extensive distribution during the Pleistocene (Pérès 1985; Taviani et al. 2005). Most of the deep-water scleractinian species living in this basin are solitary (Zibrowius 1980). Only M. oculata and L. pertusa (that compose the so-called "white coral" community, Pérès and Picard 1958) and Desmophyllum dianthus can create deep-sea reefs.

More than 980 species have been reported from the Atlantic deep-water coral reefs (Buhl-Mortensen and Mortensen 2004), whilst overall, 222 species (19 still unidentified) were encountered in the Mediterranean corals of Santa Maria di Leuca (Mastrototaro et al. 2010). Although the Mediterranean deep-water coral systems are considered less diverse than the Atlantic ones (Pérès and Picard 1958; Zibrowius 2003), the data recently acquired demonstrate that this cannot be the case as some of the taxa investigated in the Atlantic have not yet been investigated in Mediterranean deep-water coral habitats. Although none of the benthopelagic species so far recorded occurs exclusively in the coral habitat, many of them can be collected in greater abundance within coral habitats than in surrounding areas of seabed. The most diverse taxa were Porifera (36 species), followed by Mollusca (35), Cnidaria (31), Annelida (24), Crustacea (23), and Bryozoa (19). A total of 40 benthopelagic fish species were also collected. Other taxa, such as brachiopods and echinoderms, included a lower number of species.

The most abundant taxa vary according to the sampling method used, but include cnidarians, bryozoans, mollusks, annelids, echinoderms, crustaceans, and fish. Megafauna (cephalopods, decapods, crustaceans, and fish) of the Mediterranean coral area showed a larger size, biomass, and abundance inside than outside the coral area (D'Onghia et al. 2009; Mastrototaro et al. 2010; Figure 4). Cephalopods, crustaceans, and fish can be attracted by the structural complexity of the deep-water coral reefs.

These habitats are used both as feeding and spawning grounds by several fish species including the deep-water sharks (Etmopterus spinax), hakes, rockfish, greater fork beard, gurnards and blackspot seabream and species of economic interested such as the shrimps Aristeus antennatus and Aristaeomorpha foliacea (D'Onghia et al. 2003; 2009). Off Malta thick fossil coral frameworks with overgrowing coral assemblages mainly consisting of M. oculata and L. pertusa associated with Corallium rubrum and gorgonians have been recently observed (Freiwald et al. 2009).

Figure 4. Specimens of Aristeus antennatus (left half) and Geryon longipes (right half) collected in the deep-Mediterranean Sea.

Cold Seeps

The Mediterranean seeps appear to represent a rich habitat, in terms of species richness (e.g., gastropods), including sponges and crabs or size larger than in sediment not influenced by seepage. Seep communities in the Mediterranean Sea include endemic chemosynthetic species and associated fauna differ from the other known seep communities in the world at the species level (Sibuet and Olu 1998; Sibuet and Olu-Le Roy 2002). The isolation of the Mediterranean seeps from the Atlantic Ocean (Messinian crisis) has likely contributed to create cold seep assemblages with composition and structure different from those of the Atlantic Ocean. The first biological evidence was from the top of the Napoli mud volcano, located at 1,900 m depth on the Mediterranean ridge in the subduction zone of the African plate (Corselli and Basso 1996). Here Lucinidae and Vesicomyidae shells were observed and a new Lucinidae bivalve species, Lucinoma kazani, associated with bacterial endosymbionts was described (Salas and Woodside 2002). In the South Eastern Mediterranean Sea, communities of polychaetes and bivalves were also found associated with cold seeps and carbonates near Egypt and the Gaza Strip at depths of 500-800 m (Coleman and Ballard 2001). Large fields of small bivalves (dead and alive), large size siboglinid tube worms, large sponges, and associated endemic fauna have been observed in various cold seep habitats associated with carbonate crusts at 1,700-2,000 m depth. Mud volcano fields were explored along the Mediterranean Ridge (Napoli, Milano, Urania, Maidstone mud volcanoes), and the Anaximander mounds south of Turkey (including the large Amsterdam mud and the smaller Kazan or Kula mud volcanoes; Olu-Le Roy et al. 2004; Zitter et al. 2005). These systems host 5 species of bivalves harboring bacterial symbionts and other symbiont-bearing taxa previously described from the Nile fan (Sturany 1896). Sulfur-oxidizing bacteria were identified in Vesicomyidae and Lucinidae bivalves and in Siboglinidae tubeworms (Duperron et al. 2007; 2009; Olu-Le Roy et al. 2004). A new species of Siboglinidae polychaete, the tubeworm colonizing cold seeps from the Mediterranean Ridge to the Nile deep-sea fan, has been described (Hughes and Craford 2008). Finally, cold seeps recently discovered in the Marmara Sea (Zitter et al. 2008) have also revealed chemosynthesis-based communities that showed a considerable similarity to the symbiont-bearing fauna of eastern Mediterranean cold seeps (Ritt et al. 2010).

Deep Hypersaline Anoxic Systems

Numerous Deep Hypersaline Anoxic Basins (DHABs) have been discovered in the Eastern Mediterranean Sea, the Red Sea, and the Gulf of Mexico. The six DHABs of the Eastern Mediterranean Sea (L'Atalante, Urania, Bannock, Discovery, Tyro, and La Medee) are located on the Mediterranean Ridge. The Mediterranean DHABs lie at depths ranging from 3,200 and 3,600 m and contain brine, the origin of which has been attributed to the dissolution of 5.9- to 5.3-million-year-old Messinian evaporates (Hsü et al. 1977). Brines enclosed in these basins are characterized by a high density, which hamper the mixing with overlying oxic seawater and result in a sharp chemocline and anoxic conditions. The combination of nearly saturated salt concentration and corresponding high density and high hydrostatic pressure, absence of light, anoxia, and a sharp chemocline make these basins some of the most extreme habitats on Earth. In all six hypersaline basins, bacterial diversity was higher than archaeal diversity, and the Urania basin displayed the lowest overall diversity. Gamma-, Delta- and Epsilon-Proteobacteria, Sphingobacteria, candidate division KB1, and Halobacteria are found consistently in these basins. Many of the dominant archaeal sequences belonged to the new subdivision MSBL1. Phylogenetic analyses based on 16S rRNA gene sequences revealed that microbial communities found in the brines are not found in normal seawater (van der Wielen et al. 2005). Such differences are probably related to the different geochemical conditions of the different basins together with their physical separation from each other, and isolation from the oxygenated deep-water layers for possibly millions of years. The extreme conditions of these anoxic basins have been considered too harsh for hosting metazoan life, but very recently, the first metazoa living in the permanently anoxic conditions were discovered in the L'Atalante basin (Danovaro et al. 2010b). The anoxic sediments were indeed inhabited by three species new to science of the Phylum Loricifera (Spinoloricus nov. sp., Rugiloricus nov. sp. and Pliciloricus nov. sp.). Using different tecniques, Danovaro et al. (2010b) provided evidence that these organisms were metabolically active and showed specific adaptations to the extreme conditions of the deep basin, such as the lack of mitochondria, and a large number of hydrogenosome-like organelles, associated with endosymbiotic prokaryotes.

Hydrothermal Vents

On the basis of the present knowledge, hydrothermal vents are virtually absent from the deep-Mediterranean Sea; the only hydrothermal vents with described biological assemblages occur in shallow depths of <100 m (Dando et al. 1999). Consequently, a profound difference between these and the described oceanic deep-sea vents is the occurrence of photosynthetic primary production. Also, the species that inhabit shallow-water Mediterranean hydrothermal vents are not endemic to these habitats, but represent a subgroup of the most tolerant species in the ambient fauna. The only published evidence for deep-sea hydrothermalism in the Mediterranean Sea consists of indicators of extinct activity observed on the peak of Marsili Seamount in the Tyrrhenian Basin at ~450-500 m depth (Uchupi and Ballard 1989).

Major Threats to Deep-Sea Biodiversity in the Mediterranean Sea

The deep-Mediterranean Sea is increasingly affected by episodic and chronic impacts related to the dumping of waste material (although it is forbidden by the Barcelona convention) and the wrecks of ships some of which retained to contain toxic or even radioactive barrels. Munitions and bombs have been also discharged at sea, especially during the activities in Kosovo, and their dumping in open waters (mostly in the southern sector of the Adriatic Sea) is contributing to seafloor contamination. Chemical contaminants such as persistent organic pollutants (POPs), toxic metals (e.g., Hg, Cd, Pb, Ni), radioactive compounds, pesticides, herbicides, or pharmaceuticals are also accumulating in deep-sea sediments (Richter et al. 2009).

The intensive shipping activity is another major cause of deep-sea habitat alteration. An estimate dated back to >3 decades years suggests that vessel-generated refuse discarded into the Mediterranean Sea were close to 325,000 tonnes. In the last decades since, the number of mercantile and other vessels sailing in the waters of the Mediterranean Sea have increased dramatically resulting in a increase of litter released on the deep-sea floorof the Mediterranean Sea (Galgani et al. 1995; 1996; Galil et al. 1995), which were dominated by plastic items (even>90%). In the Ionian Sea, from mesopelagic down to abyssal depths the most common litter items were paint chips (44%) and plastics (36%).

Another major threat to the benthic fauna is the presence of lost or discarded fishing gear, such as nets and long lines, which continue ghost fishing and damage some fragile deep-sea habitats. Recent studies have detected the presence of significant levels of dioxins in the commercial shrimp Aristeus antennatus (Rotllant et al. 2006), significant levels of POPs in mesopelagic and bathypelagic cephalopods (Unger et al. 2008) and of heavy metals in deep-sea shark species (e.g., the Portuguese dogfish Centroscymnus coelolepis, unpublished data, Figure 5).

The Mediterranean Sea has an important commercial fishing activity, which is now banned at depths >1000 m, but is very intense down to this depth. Little is known about the effects of deep-water trawling on the habitat and benthic fauna. Pioneer studies have shown that intense commercial trawling may trigger sediment gravity flows with an increase in near bottom turbidity (Martín et al. 2008; Palanques et al. 2006).

Figure 5. Two specimens of Centroscymnus coelolepis collected at ca 3,000 m depth from the deep Western Mediterranean Sea.

IMPACT OF CLIMATE CHANGE IN THE DEEP-MEDITERRANEAN SEA

The thermohaline circulation oxygenates the deep and bottom layers in the Mediterranean Sea. This vertical circulation is forced by the deep water formation processes occurring under favorable meteorological conditions in the Gulf of Lions and the northern Adriatic (Béthoux et al. 2002; Ivanov et al. 2004). However, events in the past 20 years demonstrated the instability of the process. An abrupt change in hydrography and large-scale circulation of the deep waters of the Eastern Mediterranean Sea resulted from a unique, high-volume influx of dense waters from the Aegean Sea during the 1990s (CIESM 2000). The event named 'Eastern Mediterranean Transient' (EMT) (Roether et al. 2007), caused significant changes in deep-sea biodiversity (Danovaro et al. 2004). Extreme scenarios of climate change predict changes in the site of the deep water formation and a weakening of the thermohaline circulation, which could result in changes in the oxygenation and biogeochemical cycles in the bottom layers of the deep-Mediterranean Sea (Danovaro et al. 2001). Recently, episodic/catastrophic events have been described as one of the main environmental factors for faunal disturbance and thus one of the main potential drivers of the deep-sea biodiversity (Levin and Dayton 2009; Levin et al. 2010; Smith et al. 2008). Limited information is available to date but the effects of these episodic events on the deep-Mediterranean Sea appear relevant (Cap de Creus Canyon, Western Mediterranean; Canals et al. 2006; Company et al. 2008; Pusceddu et al. 2010). Another potential effect of climate change is related to energy transport from surface waters to the seafloor (Smith et al. 2008; 2009) related with changes in primary production and water column stratification, which could alter also the deep-sea food chain altering the quantity, quality and temporal variability in supply of organic matter.

REFERENCES

Acosta, J., Canals, M., Lòpez-Martìnez, J., Munõz, A., Herranz, P., Urgeles, R., Palomo, C., Casamor, J. L. (2002). The Balearic Promontory geomorphology (Western Mediterranean): morphostructure and active processes. *Geomorphology* 49: 177-204.

Anderson, T. R., Rice, T. (2006). Deserts on the Sea floor: Edward Forbes and his azoic hypothesis for a lifeless deep ocean. *Endeavour* 30(4): 131-137.

Bamstedt, U. (1978). Studies on the deep-water pelagic community of Korsfjorden, western Norway: Seasonal variation in weight and biochemical composition of Chiridius armatus (Copepoda), Boreomysis arctica (Mysidacae), and Eukrohnia hamata (Chaetognatha) in relation to their biology. *Sarsia* 63: 145-154.

Béthoux, J. P., Durrieu de Madron, X., Nyffeler, F., Tailliez, D. (2002). Deep water in the western Mediterranean: peculiar 1999 and 2000 characteristics, shelf formation hypothesis, variability since 1970 and geochemical inferences. *Journal of Marine Systems* 33-34: 117-131.

Bianchelli, S., Gambi, C., Pusceddu, A., Danovaro, R. (2008). Trophic conditions and meiofaunal assemblages in the Bari Canyon and the adjacent open slope (Adriatic Sea). *Chemistry and Ecology* 24 (S1): 101-109.

Bianchelli, S., Gambi, C., Zeppilli, D., Danovaro, R. (2010). Metazoan meiofauna in deep-sea canyons and adjacent open slopes: a large-scale comparison with focus on the rare taxa. *Deep Sea Research*. Part I 57: 420-433.

Bianchi, N., Morri, C. (2000) Marine Biodiversity of the Mediterranean Sea: Situation, Problems and Prospects for Future Research. *Marine Pollution Bulletin* 40 (5): 367-376.

Buhl-Mortensen, L., Mortensen, P.B. (2004) Symbiosis in deep-water corals. *Symbiosis* 37: 33-61.

Canals, M., Puig, P., Durieu de Madron, X., Heussner, S., Palanques, A., Fabres, J. (2006) Flushing submarine canyons. *Nature* 444: 354-357.

Capezzuto, F., Carlucci, R., Maiorano, P., Sion, L., Battista, B., Giove A., Indennidate A., Tursi A, D'Onghia A. (2010). The bathyal benthopelagic fauna in the NW Ionian Sea: structure, patterns and interactions. *Chemistry and Ecology* 26 (S1): 199-217.

Charette, M. A., Smith W. H. F. (2010). The Volume of Earth's Ocean. *Oceanography* 23 (2): 112-114.

Christiansen, B. (1989). Acanthephyra sp. (Crustacea: Decapoda) in the Eastern Mediterranean Sea captured by baited traps. *Senckenbergiana maritima* 20: 187-193.

CIESM (2000). The Eastern Mediterranean climatic transient : its origin, evolution and impact on the ecosystem. CIESM Monograph N. 10, Trieste, March/April 2000. 86 pp.

CIESM (2003). Mare Incognitum? Exploring Mediterranean Deep-sea biology. CIESM Monograph N 23, Heraklion, 2 - 5 July 2003. 128 pp.

Clarke, A. (1983). Life in cold water: The physiological ecology of polar marine ectotherms. Oceanography and Marine Biology: *An Annual Review* 21: 341-453.

Coleman, D. F., Ballard, R. D. (2001). A highly concentrated region of cold hydrocarbon seeps in the southeastern Mediterranean Sea. *Geo-Marine Letters* 21:162-167.

Company, J. B., Puig, P., Sardà, F., Palanques, A., Latasa, M., Scharek, R. (2008). Climate control on deep-sea fisheries. *PLoS ONE* 3(1).e1431 doi:10.1371/journal.pone.0001431.

Corselli, C., Basso, D. (1996). First evidence of benthic communities based on chemosynthesis on the Napoli mud volcano (Eastern Mediterranean). *Marine Geology* 132: 227-239.

D'Onghia, G., Maiorano, P., Sion, L., Giove, A., Capezzuto, F., Carlucci, R., Tursi, A. (2009). Effects of deep-water coral banks on the abundance and size structure of the megafauna in the Mediterranean Sea. *Deep Sea Research* Part II, 57: 397-411.

D'Onghia, G., Mastrototaro, F., Matarrese, A., Politou, C.-Y., Mytilineou, Ch. (2003). Biodiversity of the upper slope demersal community in the Eastern Mediterranean: preliminary comparison between two areas with and without trawl fishing. *Journal of Northwest Atlantic Fishery Science*, 31: 263-273.

Dando, P. R., Stüben, D., Varnavas, S. P. (1999). Hydrothermalism in the Mediterranean Sea. *Progress in Oceanography* 44: 333-367.

Danovaro, R., Dell'Anno, A., Martorano, D., Parodi, P., Marrale, D. and Fabiano, M. (1999b). Seasonal changes in the biochemical composition of deep-sea nematodes: Bioenergetic and methodological considerations. *Marine Ecology Progress Series*, 179: 273–283.

Danovaro, R., Dinet, A., Duineveld, G., Tselepides, A. (1999a) Benthic response to particulate fluxes in different trophic environments: a comparison between the Gulf of Lions-Catalan Sea (Western-Mediterranean) and the Cretan Sea (Eastern Mediterranean). *Progress in Oceanography*, 44: 287-312.

Danovaro, R., Dinet, A., Duineveld, G. Tselepides, A. (1999c). Benthic response to particulate fluxes in different tropic environments: A comparison between the western and eastern Mediterranean. *Progress in Oceanography*, 44: 287-312.

Danovaro, R., Dell' Anno, A., Fabiano, M., Pusceddu, A., Tselepides, A. (2001). Deep-sea ecosystem response to climate changes: The Eastern Mediterranean case study. *Trends in Ecology and Evolution* 16: 505–510.

Danovaro, R. (2003). Organic inputs and ecosystem efficiency in the deep Mediterranean Sea. *Chemistry and Ecology*, 19: 391-398.

Danovaro, R., Dell'Anno, A., Pusceddu, A. (2004). Biodiversity response to climate change in a warm deep Sea. *Ecology Letters*, 7: 821-828.

Danovaro, R., Gambi, C., Lampadariou, N., Tselepides, A. (2008). Deep-sea nematode biodiversity in the Mediterranean basin: testing for longitudinal, bathymetric and energetic gradients. *Ecography*, 31: 231-244.

Danovaro, R., Canals, M., Gambi, C., Heussner, S., Lampadariou, N., Vanreusel, A. (2009a) Exploring patterns and hot spots of benthic biodiversity on the slopes of European margins. *Oceanography* 22(1): 16-25.

Danovaro, R., Corinaldesi, C., Luna, G. M., Magagnini, M., Manini, E., Pusceddu, A. (2009b) Prokaryote diversity and viral production in deep-sea sediments and Seamounts. *Deep Sea Research* Part II, 56: 738-747.

Danovaro, R., Company, J. B., Corinaldesi, C., D'Onghia, G., Galil, B., Gambi, C., Gooday, A. J., Lampadariou, N., Luna, G. M., Morigi, C., Olu, K., Polymenakou, P., Ramirez-Llodra, E., Sabbatini, A., Sardà, F., Sibuet, M., Tselepides, A. (2010a). Deep-Sea biodiversity in the Mediterranean Sea: the known, the unknown, and the unknowable. *Plos One* 5(8): e11832. doi:10.1371/journal.pone.0011832.

Danovaro, R., Dell'Anno A., Puscedd A., Gambi C., Heiner I. Kristensen M. R. (2010b). The first metazoa living in permanently anoxic conditions. *BMC Biology* 2010, 8:30 http://www.biomedcentral.com/1741-7007/8/30.

De Stigter, H. C. (1996). Recent and fossil benthic foraminifera in the Adriatic Sea: distribution patterns in relation to organic carbon flux and oxygen concentration at the Seabed. *Geologica Ultraiectina* 144: 254.

Duperron, S., Fiala-Médioni, A., Caprais, J.C., Olu, K., Sibuet, M., (2007). Evidence for chemoautotrophic symbiosis in a Mediterranean cold seep clam (Bivalvia: Lucinidae): comparative sequence analysis of bacterial 16S rRNA, APS reductase and RubisCO genes. *FEMS Microbiology Ecology,* 59: 64-70.

Duperron, S., de Beer, D., Zbinden, M., Boetius, A., Schipani, V., Kahil, N., Galil, F. (2009). Molecular characterization of bacteria associated with the trophosome and the tube of Lamellibrachia sp., a siboglinid annelid from cold seeps in the eastern Mediterranean. *FEMS Microbiology Ecology*, 69 (3): 395-409.

Falk-Petersen, S. (1981). Ecological investigations on the zooplankton community of Balsfjorden, northern Norway: Seasonal changes in body weight and the main biochemical composition of Thysanoessa inermis (Krøyer), T. raschi (M. Sars), and Meganyctiphanes norvegica (M. Sars) in relation to environmental factors. *Journal of Experimental Marine Biology and Ecology,* 49: 103-120.

Freiwald, A., Fossa, J. H., Grehan, A., Koslow, T., Roberts, J. M. (2004). Cold-water Coral Reefs. Cambridge: *UNEP-WCMC Biodiversity Series* No 22. 88 pp.

Freiwald, A., Beuck, L., Rüggerberg, A., Taviani, M., Hebblen, D. (2009). The white coral community in the Central Mediterranean Sea revealed by ROV surveys. *Oceanography* 22 (1): 36-52.

Gage, J. D., and Tyler, P. A. (1991). Deep Sea biology: a natural history of organisms at the deep-sea floor. Cambridge, UK: Cambridge University Press. 504 pp.

Galgani, F., Jaunet, S., Campillo, A., Guenegan, X., His, E. (1995). Distribution and abundance of debris on the continental shelf of the north-Western Mediterranean Sea. *Marine Pollution Bulletin* 30: 713-717.

Galgani, F., Souplet, A., Cadiou, Y. (1996). Accumulation of debris on the deep Sea floor off the French Mediterranean coasy. *Marine Ecology Progress Series* 142: 225-234.

Galil, B. S., Goren, M. (1994). The deep Sea Levantine fauna, new records and rare occurrences. *Senckenbergiana maritima* 25: 41-52.

Galil, B. S., Golik, A., Türkay, M. (1995). Litter at the bottom of the Sea: a Sea bed survey in the Eastern Mediterranean. *Marine Pollution Bulletin* 30: 22-24.

Galil, B. S., Zibrowius, H. (1998). First benthos samples from Eratosthenes Seamount, Eastern Mediterranean. *Senckenbergiana maritima* 28: 111-121.

Geistdoerfer, P., Rannou, M. (1972). Poissons benthiques récoltés en Méditerranée occidentale par le N.O. Jean Charcot (campagne Polymède). *Bulletin du Museum National d'Histoire Naturelle Series* 3 (25): 101–110.

Gili, J. M., Bouillon, J., Pagès, F., Palanques, A., Puig, P. (1999). Submarine canyons as habitats of prolific plankton populations: three new deep-sea Hydrodomedusae in the Western Mediterranean. *Zoological Journal of the Linnean Society* 125: 313-329.

Gili, J. M., Pagès, F., Bouillon, J., Palanques, A., Puig, P., Heussner, S., Calafat, A. Canals, M., Monaco, A. (2000). A multidisciplinary approach to the understanding of hydromedusan populations inhabiting Mediterranean submarine canyons. *Deep Sea Research* Part I 47: 1513-1533.

Greene, H. G., Wiebe, P. H., Burczynski, J., Youngbluth, M. J. (1988). Acoustical Detection of High-Density Krill Demersal Layers in the Submarine Canyons off Georges Bank. *Science* 241: 359-361.

Hughes, D. J., Crawford, M. (2008). A new record of the vestimentiferan Lamellibrachia sp. (Polychaeta: Siboglinidae) from a deep shipwreck in the eastern Mediterranean. *Marine Biodiversity Records,* 1, e21 doi:10.1017/S1755267206001989.

Harrold, C., Light, K., Lisin, S. (1998). Organic enrichment of submarine-canyon and continental-shelf benthic communities by macroalgal drift imported from nearshore kelp forests. *Limnology and Oceanography* 43: 669-678.

Hopkins, C. C. E., Sargent, J. R., Nilssen, E. M. (1993). Total lipid content, and lipid and fatty acid composition of the deep-water prawn Pandalus borealis from Balsjord, northern Norway: Growth and feeding relationships. *Marine Ecology Progress Series* 96: 217–228.

Hovland, M. (2008). Deep-water Coral Reefs: unique biodiversity hot-spots. Chichester, UK: Springer. 278 pp.

Hsü, K. J., Montadert, L., Bernoulli, D., Cita, M. B., Erickson, A., Garrison, R. E., Kidd, R. B., Meleres, F., Muller, C., Wright, R. (1977). History of the Mediterranean salinity crisis. *Nature* 267: 399-403.

Ivanov, V. V., Shapiro, G. I., Huthnance, J. M., Aleynik, D. L., Golovin, P. N. (2004). Cascades of dense water around the world ocean. *Progress in Oceanography* 60: 47-98.

Jones, E. G, Tselepides, A., Bagley, P. M., Collins, M. A., Priede, I. G. (2003). Bathymetric distribution of some benthic and benthopelagic species attracted to baited cameras and traps in the deep Eastern Mediterranean. *Marine Ecology Progress Series* 251: 75-86.

Jorrisen, F. J. (1988). The distribution of benthic foraminifera in the Adriatic Sea. *Utrecht Micropaleontological Bulletins* 37: 1-174.

Klausewitz, W. (1989). Deep-sea and deep water fish of the Eastern Mediterranean, collected during the METEOR-Expedition 1987. *Senckenbergiana maritima* 20: 251-263.

Krom, M. D., Kress, N., Brenner, S., Gordon, L. I. (1991). Phosphorus limitation of primary productivity in the Eastern Mediterranean. *Limnology and Oceanography* 36: 424-432.

Lehtonen, K. K. (1996). Ecophysiology of the benthic amphipod Monoporeia affinis in an open-sea area of the northern Baltic Sea: Seasonal variations in body composition, with bioenergetic considerations. *Marine Ecology Progress Series* 143: 87-98.

Levin, L. A., Etter, R. J., Rex, M. A., Gooday A. J., Smith C. R., Pineda J., Stuart C. T., Hessler R. R., Pawson D. (2001). Environmental influences on regional deep-sea species diversity. *Annual Review of Ecology and Systematics* 32: 51-93.

Levin, L. A., Dayton, P. K. (2009). Ecological theory and continental margins: where shallow meets deep. *Trends in Ecology and Evolution* 24: 606-617.

Levin, L. A., Sibuet, M., Gooday, A. J., Smith, C. R., Vanreusel, A. (2010). The roles of habitat heterogeneity in generating and maintaining biodiversity on continental margins. *Marine Ecology* 31: 1-5.

Luna, G. M., Dell'Anno, A., Giuliano, L., Danovaro, R. (2004). Bacterial diversity in deep Mediterranean sediments: relationship with the active bacterial fraction and substrate availability. *Environmental Microbiology* 6: 745-753.

Martín, J., Puig, P., Palanques, A., Masqué, P., García-Orellana, J. (2008). Effect of commercial trawling on the deep sedimentation in a Mediterranean submarine canyon. *Marine Geology* 252: 150-155.

Massiotta, R., Cita, M. B., Mancuso, M. (1976). Benthonic foraminifers from bathyal depths in the Eastern Mediterranean. *Maritime Sediments*, Special Publication 1: 251-262.

Mastrototaro, F., D'Onghia, G., Corriero, G., Matarrese, A., Maiorano P, Panette, P., Gherardi, M., Longo, C., Rosso, A., Sciuto, F., Sanfilippo, R., Gravili, C., Boero, F., Taviani, M., Tursi, A. (2010). Biodiversity of the white coral and sponge community off Cape Santa Maria di Leuca (Mediterranean Sea). *Deep Sea Research* Part II 57: 412-420.

Miliou, H., Moraitou-Apostopoulu, M., Argyridou, M. (1992). Biochemical composition of Tisbe holothuriae (Copepoda: Harpacticoida) and its differentiation in relation to developmental stages. *Ophelia* 36: 159-166.

Myers, N., Mittermeier, R. A., Mittermeier, C. G., da Fonseca Gustavo, A. B., Kent, J. (2000). Biodiversity hotspots for conservation priorities. *Nature* 403: 853-858.

Olu-Le Roy, K., Sibuet, M., Fiala-Médioni, A., Gofas, S., Salas, C., Mariotti, A., Foucher, J.-P., Woodside, J. (2004). Cold seep communities in the deep eastern Mediterranean Sea: composition, symbiosis and spatial distribution on mud volcanoes. *Deep Sea Research* Part I 51: 1915-1936.

Palanques, A., Martín, J., Puig, P., Guillén, J., Company, J. B., Sardà, F. (2006). Evidence of sediment gravity flows induced by trawling in the Palamós (Fonera) submarine canyon (northwestern Mediterranean). *Deep Sea Research* Part I, 53: 201-214.

Pérès, J.M. (1985). History of the Mediterranean biota and the colonization of the depths. In Margalef, R., (Ed.). Key Environments: Western Mediterranean. pp. 198-232. Pergamon Press, Oxford, UK.

Pérès, J. M., Picard, J. (1958). Recherches sur les peuplements benthiques de la Mediterranée Nord -Orientale. *Annales de l' Institute Océanographie Paris* 34: 213-281.

Polymenakou, P. N., Bertilsson, S., Tselepides, A., Stephanou, E. G. (2005) Links between geographic location, environmental factors and microbial community composition in sediments of the Eastern Mediterranean Sea. *Microbiology Ecology* 49: 367-378.

Psarra, S., Tselepides, A., Ignatiades, L. (2000). Primary productvity in the oligotrophic Cretan Sea (NE Mediterranean): Seasonal and interannual variability. *Progress in Oceanography* 46: 187-204.

Puig, P., Palanques, A., Guillen, J., García-Ladona, E. (2000). Deep slope currents and suspended particle fluxes in and around the Foix submarine canyon (NW Mediterranean). *Deep Sea Research* I 47: 343-366.

Puig, P., Ogsto, A.S., Mullenbach, B. L., Nittrouer, C. A., Sternberg, R. W. (2003). Shelf-to-canyon sediment transport processes on the Eel Continental Margin (Northern California). *Marine Geology* 193: 129-149.

Pusceddu, A., Gambi, C., Zeppilli, D., Bianchelli, S., Danovaro, R. (2009). Organic matter composition, meiofauna and nematode biodiversity in deep-sea sediments surrounding two seamounts. *Deep Sea Research* Part II 56: 755-762.

Pusceddu, A., Mea, M., Gambi, C., Bianchelli, S., Canals, M., Sanchez-Vidal, A., Calafat, A., Heussner, S., Durrieu De Madron, X., Avril, J., Thomsend, L., Garcìa, R., Danovaro, R. (2010). Ecosystem effects of dense water formation on deep Mediterranean Sea ecosystems: an overview. *Advances in Oceanography and Limnology in Oceanography and Limnology* 1(1): 67-83.

Ramirez-Llodra, E., Ballesteros, M., Company, J. B., Dantart, L., Sardà, S. (2008). Spatio-temporal variations of biomass and abundance in bathyal non-crustacean megafauna in the Catalan Sea (North-western Mediterranean). *Marine Biology* 153: 297-309.

Ramirez-Llodra, E., Company, J. B., Sardà, F., Rotllant, G. (2009). Megabenthic diversity patterns and community structure of the Blanes submarine canyon and adjacent slope in the Northwestern Mediterranean: a human overprint? *Marine Ecology* 1-16.

Ramirez-Llodra, E., Brandt A., Danovaro R., De Mol B., Escobar, E., German, C. R., Levin, L. A., Martinez Arbizu, P., Menot, L., Buhl-Mortensen, P., Narayanaswamy, B. E., Smith, C. R., Tittensor, D. P., Tyler P. A., Vanreusel, A., Vecchione, M. (2010). Deep, diverse and definitely different: unique attributes of the world's largest ecosystem. *Biogeosciences* 7: 2851–2899.

Rex, M. A. (1981). Community structure in the deep-sea benthos. *Annual Review of Ecology Evolution And Systematics* 12: 331-353.

Richter, T. O., de Stigter, H. C., Boer, W., Jesús, C. C., van Weering, T. C. E. (2009). Dispersal of natural and anthropogenic lead through submarine canyons in the Portuguese margin. *Deep Sea Research* Part I 56: 267-282.

Ritt, B., Sarrazin, J., Caprais, J. C., Noel, P., Gauthier, P. C., Henry P., Desbruyères D. (2010). First insights on the cold-seep benthic communities in the Marmara Sea. *Deep Sea Research* Part I 57: 1120-1136.

Roether, W., Klein, B., Manca, B. B., Theocharis, A., Kioroglou, S. (2007). Transient Eastern Mediterranean Deep waters in response to the massive dense-water output of the Aegean Sea in the 1990s. *Progress in Oceanography* 74: 540-571.

Rotllant, G., Abad, E., Sardà, F., Ábalos, M., Company, J. B., Rivera, J. (2006). Dioxin compounds in the deep-sea rose shrimp Aristeus antennatus (Risso, 1816) throughout the Mediterranean Sea. *Deep Sea Research* Part I 53: 1895-1906.

Salas, C., Woodside, J. (2002). Lucinoma kazani n. sp.(Mollusca, Bivalvia): evidence of a living community associated with a cold seep in the Eastern Mediterranean Sea. *Deep Sea Research* Part I, 49: 991-1005.

Sardà, F., Cartes, J.E. (1997). Morphological features and ecological aspects of early juvenile specimens of the aristeid shrimp Aristeus antennatus (Risso, 1816). *Marine Freshwater Research* 48: 73-77.

Sardà, F., Company, J. B., Bahamon, N., Rotllant, G., Flexas, M. M., Sánchez, J. D., Zúñiga D., Coenjaerts J., Orellana D., Jordà, G. Puigdefábregas, J. Sánchez-Vidal A., Calafat, A., Martín, D.,Espino, M. (2009) Relationship between environment and the occurrence of the deep-water rose shrimp Aristeus antennatus (Risso, 1816) in the Blanes submarine canyon (NW Mediterranean). *Progress in Oceanography* 82: 227-238.

Sardà, F., D'Onghia, G., Politou, C.-Y., Tselepides, A. (2004). Mediterranean Deep-sea Biology. Monographs Scientia Marina 63. 204 pp.

Sargent, J. R. and Whittle, K. (1981). Lipids and hydrocarbons in the marine food web. In Longhurst, A. R. (Ed.). Analysis of marine ecosystems. pp. 491-533. Academic Press: London, UK.

Sargent, J. R., and Henderson, R. J. (1986). Lipids. In Corner, E. D. S. and O'Hara, S. C. M. (Eds.). The biological chemistry of marine copepods. pp. 59-108. Clarendon Press: Oxford, UK.

Sibuet, M., Olu, K. (1998). Biogeography, biodiversity and fluid dependence of deep-sea cold-seep communities at active and passive margins. *Deep Sea Research* Part II 45: 517-567.

Sibuet, M., and Olu-Le Roy, K. (2002). Cold seep communities on continental margins: structure and quantitative distribution relative to geological and fluid venting patterns. In Wefer, G., Billett, D., Hebbeln, D., Jorgensen, B., Schlüter, M., van Weering, T. (Eds.). Ocean Margin Systems. pp. 235-251. Springer: Berlin, DE.

Smith, C. R., De Leo, F. C., Bernardino, A. F., Sweetman, A. K., Arbizu, P. M. (2008). Abyssal food limitation, ecosystem structure and climate change. *Trends in Ecology and Evolution* 23: 518-528.

Smith, K. L., Ruhl, H. A., Bett, B. J., Billet, D. S. M., Lampitt, R. S., Kaufmann, R. S. (2009). Climate, carbon cycling, and deep-ocean ecosystems. *Proceedings of the National Academy of Sciences* 106: 19211-19218.

Soltwedel, T. (2000). Metazoan meiobenthos along continental margins: a review. *Progress in Oceanography* 46: 59-84.

Stanley, D.J., and Wezel, F.C. (1985). Geological evolution of the Mediterranean basin. New York, WA: Springer. 589 pp.

Stefanescu, C., Morales-Nin, B., Massutí, E. (1994). Fish assemblages on the slope in the Catalan Sea (western Mediterranean): influence of a submarine canyon. *Journal of Marine Biological Association UK* 74: 499-512.

Sturany, R. (1896). Zoologische Ergebnisse VII. Mollusken I (Prosobranchier und Opisthobranchier; Scaphopoden; Lamellibranchier) gesammelt von SM Schiff "Pola" 1890-18. Denkschriften der Kaiserlichen Akademie der Wissenschaften, *Mathematische-Naturwissenschaftlischen* Classe 63, 1-36: 31-32.

Taviani, M., Remia, A., Corselli, C., Freiwald, A., Malinverno, E., Savini A., Tursi A., CORAL Shipboard Staff. (2005). First geo-marine survey of living cold-water Lophelia reefs in the Ionian Sea (Mediterranean basin). *Facies* 50: 409-417.

Trincardi, F., Foglini, F., Verdicchio, G., Asioli, A., Correggiari, A., Minisini, D., Piva, A., Ridente, D., Taviani, M. (2007). The impact of cascading currents on the Bari Canyon System, SW-Adriatic Margin (Central Mediterranean). *Marine Geology* 246: 208-230.

Tselepides, A., Papadopoulou, N., Podaras, D., Plaiti, W., Koutsoubas, D. (2000). Macrobenthic community structure over the continental margin of Crete (South Aegean Sea, NE Mediterranean). *Progress in Oceanography* 46(2–4): 401–428.

Tudela, S., Sardà, F., Maynou, F., Demestre, M. (2003). Influence of submarine canyons on the distribution of the deep-water shrimp (Aristeus antennatus, Risso 1816) in the northwestern Mediterranean. *Crustaceana* 76: 217-225.

Tursi, A., Mastrototaro, F., Matarrese, A., Maiorano, P., D'Onghia, G. (2004). Biodiversity of the white coral reefs in the Ionian Sea (Central Mediterranean). *Chemistry and Ecology* 20: 107–116.

Uchupi, C., Ballard, A. (1989). Evidence of hydrothermal activity on Marsili Seamount, Tyrrhenian Basin. *Deep Sea Research* Part I 36: 1443-1448.

Unger, M. A., Harvey, E., Vadas, G. G., Vecchione, M. (2008). Persistent pollutants in nine species of deep-sea cephalopods. *Marine Pollution Bulletin* 56: 1486-1512.

Vanney, J. R., Gennesseaux, M. (1985). Mediterranean seafloor features: overview and assessment. In Stanley, D. J. and Wezel, F.-C. (Eds.). Geological evolution of the Mediterranean Basin. pp. 3-32. New York, WA: Springer.

van der Wielen, P. W. J. J., Bolhuis, H., Borin, S., Daffonchio, D., Corselli, C., Giuliano, L., D'Auria, G., de Lange G. J., Huebner, A., Varnavas, S. P., Thomson, J., Tamburini, C., Marty, D., McGenity, T. J. Timmis, K. N., BioDeep Scientific Party (2005). The enigma of prokaryotic life in deep hypersaline anoxic basins. *Science* 307: 121–123.

Vetter, E.W. (1994). Hotspots of benthic production. *Nature* 372: 47.

Vetter, E. W., Dayton, P. K. (1998). Macrofaunal communities within and adjacent to a detritus-rich submarine canyon system. *Deep Sea Research* Part II 45: 25-54.

Wezel, F. C. (1985). Structural features and basin tectonics of the Tyrrhenian Sea. In Stanley, D. J. and Wezel, F.-C. (Eds.). Geological evolution of the Mediterranean Basin. pp. 153-194. Springer. New York.

Wright, W. C., Rupert, F. P. (1981). Late Neogene and Recent Bathyal Foraminifera of Mediterranean. *American Association of Petroleum Geologists Bulletin* 65: 1009.

Yacobi, Y. Z., Zohary, T., Kress, N., Hecht, A., Robarts, R. D., Waiser, M., Wood, A. M., Li, W. K. W. (1995). Chlorophyll distribution throughout the south Eastern Mediterranean in relation to the physical structure of the water mass. *Journal of Marine Systems* 6: 179-190.

Zibrowius, H. (1980). Les Scléractiniaires de la Méditerranée et de l'Atlantique nord-oriental. *Mémoires de l'Institut Océanographique Monaco* 11: 1-227.

Zibrowius, H. (2003). The "White Coral Community", Canyon and Seamount Faunas of the deep Mediterranean Sea. Project Report for the preparation of a Strategic Action Plan for the Conservation of Biological Diversity in the Mediterranean Region (SAP BIO). 39 pp.

Zitter, T. A. C., Huguen, C., Woodside, J. M. (2005). Geology of mud volcanoes in the eastern Mediterranean from combined sidescan sonar and submersible surveys. *Deep Sea Research* Part I 52: 457-475.

Zitter, T. A. C., Henry, P., Aloisi, G., Delaygue, G., Çagatay, M. N., Mercier de Lepinay, B., Al-Samire, M., Fornacciari, F., Tesmer, M., Pekdeger, A., Wallmann, K., Lericolais, G. (2008). Cold seeps along the main Marmara Fault in the Sea of Marmara (Turkey). *Deep Sea Research* Part I 55: 552-570.

Zúñiga, D., Flexas, M. M., Sánchez-Vida, A., Coenjaerts, J., Calafat, A., Jordà, G., García-Orellanae, J., Puigdefàbregas, J., Canals, M., Espinod, M., Sardà, F., Company, J. B. (2009). Particle fluxes dynamics in Blanes submarine canyon (Northwestern Mediterranean). *Progress in Oceanography* 82: 239-251.

INDEX

A

Abyssal, 174, 176, 177, 183, 191
Abyssal basin, 176
Acanthephyra eximia, 176
access, 146
accessibility, 145, 146
acid, 99, 134, 135, 137, 140, 143, 145, 165, 188
activation, 152
active site, 137, 146
adaptations, 164, 183
adenosine triphosphate, 163
adhesion, 157, 158, 162, 165
adhesion force, 157, 162
Adriatic, 179, 183, 184, 185, 187, 188, 191
Adriatic Sea, 183, 185, 187, 188
advancements, viii, 171
Aegean, 180, 184, 190, 191
Aegean Sea, 180, 184, 190, 191
aerobic, 150, 155, 160, 161, 167
aerobic anoxygenic phototrophic bacteria, 167
Africa, 44, 45, 118, 132
agar, 133, 136
age, vii, viii, 1, 3, 8, 13, 20, 50, 63, 64, 65, 69, 71, 78, 79, 80, 81, 83, 97, 98, 101
agent, 162
Al, 171, 172, 173, 174, 176, 177, 178, 179, 180, 182, 183, 184, 192
Alaska, 63, 65, 67, 68, 72, 79, 80, 83, 95, 96, 97, 98, 161
Albian-Maastrichtian planktic foraminifera, vii, 1
Alboran, 180
algae, 68, 75, 76, 81
alkane, 165
alternative energy, 154
amino acid, 135, 136, 137, 138, 140, 142, 143, 144, 145, 146, 147, 152, 164
amino acids, 136, 142, 143, 152, 164
ammonia, 152, 156, 158, 159, 161
ammonium, 153
amphipods, 172
anaerobic, 149, 150, 154, 155, 163, 164, 166, 168, 169
Ancona, 171
Annelida, 180
anoxia, 182
anoxic, viii, 171, 174, 176, 182, 187, 192
anthropogenic, 190
antibiotic, 164
antibody, 159
antigen, 159
antisense, 141
aquaria, 64, 65, 82, 83
aquarium, vii, 63, 65, 78, 82, 83, 95
aquarium observations, vii, 63, 78, 83
archaea, 149, 150, 151, 152, 154, 158, 159, 161, 162, 164, 165, 166, 168, 169
archaeal, 149, 150, 151, 153, 154, 157, 159, 160, 162, 163, 164, 169
Arctic water, 150, 175
Aristaeomorpha foliacea, 181
Aristeus antennatus, 178, 181, 183, 190, 191, 192
Asia, 70, 71, 83, 118
assessment, 79, 163, 192
assimilation, 159
Atlantic, 172, 176, 180, 181, 186
Atlantic Ocean, 150, 151, 156, 159, 172, 176, 181
atmosphere, 152
atomic force, 156, 157, 159, 161, 162, 163, 165, 166, 170
Atomic Force Microscopy, (AFM), 156, 157, 158, 161, 162, 166, 168
ATP, 160
autolysis, 136, 143
autotrophic, 152, 160, 161

B

Bacillus, 136, 156
bacteria, 132, 133, 134, 135, 137, 143, 144, 151, 154, 155, 156, 157, 160, 161, 162, 163, 164, 165, 166, 167, 169, 182, 187
bacterial, 156, 157, 159, 162, 164, 165, 167, 169
Bacterial biomass, 173
bacterial pathogens, 156
bacteriochlorophyll, 161
bacteriophage, 169
bacterium, viii, 131, 132, 133, 147, 156, 158, 160, 168
Baltic Sea, 189
banks, 64, 68, 180, 186
Bannock, 182
Barcelona, 183
base, 8, 9, 13, 17, 20, 21, 32, 38, 41, 43, 44, 46, 47, 48, 70, 71, 94, 106, 129, 150, 165
bathyal, 176, 177, 179, 180, 185, 189, 190
Beijing, 96, 161
benthic, 172, 173, 174, 175, 176, 177, 178, 179, 183, 186, 187, 188, 189, 190, 192
Benthic Organisms, 172, 173
benthic species, 172
Benthos, viii, 171, 172, 188, 190
bicarbonate, 153
bioavailability, 159, 172
biochemistry, 132, 152
biodiversity, vii, viii, 57, 171, 172, 173, 175, 176, 177, 178, 179, 184, 187, 188, 189, 190, 191
biofilms, 154, 155
Biogeochemical, 185
biomarker, 156
biomarkers, 155, 160, 164, 165, 167, 168
biomass, 151, 152, 158, 173, 177, 178, 180, 190
biosphere, vii, 171
biota, 174, 189
biotic, 178
bivalves, 179, 181, 182
Black Sea, 150, 151, 156, 166
body composition, 189
body weight, 79, 80, 187
bones, viii, 131, 132, 133, 147
boreholes, 51
branching, 151
bryozoans, 180

C

Ca^{2+}, 137, 139, 141, 143, 144
California, 190
Campanian-Maastrichtian age, viii, 101
carbohydrate, 134, 168, 173
carbon, vii, 99, 150, 152, 153, 154, 155, 156, 157, 158, 159, 160, 161, 162, 163, 164, 165, 166, 168, 172, 173, 178, 187, 191
carbon cycling, 164
carbon dioxide (CO2), 153, 154, 157
carbon fixation, 153, 154
carbon fluxes, 173
Caribbean, 8, 23, 27, 36, 44, 45, 60, 102, 116, 117, 118, 119, 128
Caryophylla calveri, 179
case study, 125, 164, 186
casein, viii, 131, 134, 136, 138, 139, 140, 145, 146, 147
Catalan Sea, 186, 190, 191
Catalonia, 178
Cd, 183
cell, 156, 159, 163, 167
cell surface, 163
cellular energy, 152
Centroscymnus coelolepis, 184
chemical, 64, 159, 175, 176
chemical characteristics, 175
Chemotherapy, 164
China, 66, 96, 132, 149, 160, 161
Chinese, 160, 161
chlorophyll, 172, 192
chlorophyll-a, 172
chromatography, 159
circulation, 184
classification, vii, viii, 1, 3, 22, 49, 50, 55, 56, 57, 59, 61, 101, 102, 103, 123, 125, 126, 127, 139, 141
cleavage, 132
climate, 57, 155, 172, 184, 186, 191
climate change, 155, 172, 184, 186, 191
clusters, 109, 144
cnidarians, 180
coastal ecosystems, 172
coding, 136
codon, 136, 142, 143
coenzyme, 159
cold seeps, 149, 150, 153, 154, 161, 164, 171, 176, 182, 187, 192
cold-water corals, 180
collagen, viii, 131, 132, 133, 135, 138, 140, 145, 146, 147
College Station, 53, 57, 60, 62, 127
colloids, 162
Colombia, 55, 116, 118, 122, 125
colonisation, 176
colonization, 176, 180, 189

combined effect, 173
commercial, 67, 96, 97, 183, 184, 189
communities, viii, 131, 150, 160, 161, 162, 163, 166, 167, 168, 169, 180, 181, 182, 186, 188, 189, 190, 191, 192
community, viii, 131, 150, 152, 153, 160, 162, 169, 173, 176, 180, 185, 186, 187, 189, 190, 191
comparative analysis, 179
complexity, 177, 181
components, 156
composition, 64, 75, 99, 164, 172, 173, 176, 177, 179, 181, 185, 186, 187, 188, 189, 190
compositions, 164
compounds, 164, 183, 190
connectivity, 179
conservation, 178, 189
consumption, 64, 153, 154
contaminants, 183
contamination, 183
continental, 160, 190
Continental margin,, 180
Continental shelf, 180, 187
Continental slopes, 177, 178
control, 165
convention, 183
conversion, 152, 156, 168
Copepods, 191
copper, 167
coral, 176, 180, 186, 187, 189, 192
coral reefs, 180, 181, 192
corals, viii, 171, 174, 178, 179, 180, 185
correlation, 94, 123
correlation coefficient, 94
Coryphaenoides armatus, 176
coverage, 152
covering, vii, 6, 146, 171
Cr, 191
crabs, 132, 181
crenarchaeota, 149, 150, 151, 152, 153, 154
Cretan Sea, 173, 177, 186, 189
Crete, 191
Crustaceans, 176, 180
cultivation, 133, 168
culture, 136, 137, 139, 149, 152, 162
currents, 178, 180, 190, 191
cyanobacteria, 157, 165
cycles, 150, 154, 159, 185
cycling, vii, 153, 164, 191
cytometric techniques, 163
cytometry, 156, 160, 162

D

data set, 79, 134
debris, 178, 187
Decapods, 176, 180
Deep basins, viii, 171, 176
Deep Hypersaline Anoxic Systems, 182
deep sea, 175, 178
deep water, 150, 184, 185, 188, 190
deep-sea biodiversity, viii, 171, 172, 173, 177, 178, 184
Deep-sea ecosystems, viii, 171, 173, 179
Deep-sea floor, 175, 176, 183, 187
deep-water, viii, 171, 172, 176, 180, 181, 182, 184, 185, 186, 188, 191, 192
Deep-water coral, viii, 171, 174, 176, 180, 181, 185, 186
degenerate, 141
degradation, vii, viii, 131, 135, 147, 163
Delta, 164, 182
demersal, 178, 186
Dense water formation, 190
density, 182
deposition, 151, 176
deposits, 59
depression, 122
depth, vii, 8, 63, 64, 65, 67, 70, 76, 83, 93, 98, 133, 151, 152, 153, 171, 173, 174, 175, 176, 177, 178, 179, 180, 181, 183, 184
Depth distribution, 180
derivatives, 140, 145, 146, 147
Desmophyllum cristagalli, 179
Desmophyllum dianthus, 180
detection, 150, 168
diarrhea, 64
diatoms, 13, 159
diet, 81
differentiation, 165
diffusion, 150
dioxins, 183
direct measure, 168
Discovery, 182
dissolved organic carbon, 155, 163, 166
dissolved oxygen, 180
dissolved oxygen (DO), 180
distribution, vii, viii, 4, 9, 10, 12, 14, 15, 16, 18, 19, 23, 24, 27, 30, 31, 32, 38, 42, 43, 44, 45, 46, 47, 48, 49, 63, 64, 67, 68, 83, 95, 96, 98, 101, 104, 106, 110, 113, 114, 116, 118, 122, 125, 151, 160, 161, 162, 163, 165, 167, 171, 176, 180, 187, 188, 189, 191, 192
disturbance, 185
divergence, 165

diversification, 169
diversity, 2, 4, 130, 150, 155, 156, 158, 159, 160, 162, 163, 165, 169, 174, 178, 179, 182, 187, 189, 190, 192
DNA, 132, 133, 134, 136, 137, 142, 151, 158
DNA polymerase, 142
domain structure, 145
dominance, 151, 161, 179
double keeled planktics, viii, 101
drainage, 178
drawing, 66, 67
dumping, 183
ester, 96
ethylene, 139
Etmopterus spinax, 181
euphotic zone, 151
Europe, 27, 44, 45, 46, 47, 60, 118, 128
euryarchaeota, 149, 150, 152
Eurythenes gryllus, 176
evidence, 47, 57, 96, 152, 153, 155, 170, 181, 183, 186, 190
evolution, 1, 2, 3, 22, 27, 30, 49, 50, 56, 57, 102, 122, 125, 126, 154, 155, 156, 170, 186, 191, 192
extracellular matrix, viii, 131

E

Eastern Mediterranean (EMED), 172, 173, 179, 182, 184, 186, 188, 189, 190, 192
Eastern Mediterranean Sea, 172, 173, 182, 184, 186, 189, 190
Eastern Mediterranean Transient (EMT), 184
Echinoderms, 176, 180
ecology, 186
ecosystem, vii, 147, 151, 171, 173, 176, 177, 178, 186, 190, 191
eel, 152
Egypt, 118, 123, 124, 128, 182
El Niño, 96
elastin, 138, 140
electrolyte, 156
electron, 2, 6, 102, 133, 166
electrophoresis, 164, 169
electrostatic force, 166
elongation, 26, 29, 30, 38, 44, 49
encoding, 136, 137, 145, 166
endemic, 172, 181, 183
endemic species, 172
endemism, 178
Endosymbiotic, 183
energy, 150, 152, 154, 161, 172, 173, 176, 185
Energy, 172, 173, 176, 185
energy supply, 176
England, 53, 124
environment, vii, 131, 135, 147, 150, 155, 163, 164, 166, 168, 169, 173, 176, 177, 191
environmental conditions, 151
environmental factors, 177, 185, 187, 189
environmental variables, 178
environments, 13, 61, 124, 131, 149, 150, 155, 157, 158, 161, 167, 178, 186
enzymatic activity, 138, 139
enzyme, viii, 131, 132, 137, 138, 139, 140, 141, 145, 146
Eratosthenes Seamount,, 188

F

facies, 58, 128
Fairbanks, 97
families, 54, 135, 179
fat, 64, 66, 96
fatty acids, 135, 165
fauna, 54, 95, 97, 124, 128, 172, 173, 176, 179, 181, 183, 184, 185, 188
fish, 64, 65, 66, 68, 69, 71, 72, 73, 74, 75, 78, 80, 81, 82, 83, 84, 95, 96, 97, 98, 132, 176, 179, 180, 181, 188
Fish, 176, 179, 180, 188, 191
fisheries, 64, 71, 97, 178, 186
Fishes, 174, 176
fishing, 68, 71, 72, 81, 82, 96, 183, 184, 186
Fishing, 183, 186
fixation, 153, 154
flagellum, 133
Fluctuations, 172
fluid, 159, 191
fluorescence, 151
food, 81, 82, 96, 168, 169, 172, 173, 176, 180, 185, 191
food chain, 185
food sources, 172
food web, 168, 169, 191
foraminifera, 176, 179, 187, 188
force, 155, 156, 157, 159, 161, 162, 163, 165, 166, 170
formation, 134, 154, 158, 184, 185, 190
formula, 78, 80, 81, 94
fossil, 181, 187
fractionation, 159, 167
France, 44, 45, 53, 54, 57, 59, 60, 61, 66, 127, 128
freshwater, 162, 172
fructose, 134

G

gas hydrates, 149, 150, 153, 154, 162, 165
Gastropods, 176, 181
Gaza Strip, 182
gel, 158, 169
gelatin, viii, 131, 134, 138, 140, 145, 146, 147, 157
gene, 151, 152, 158, 162, 169
gene expression, 169
gene transfer, 158, 162
genes, 136, 145, 150, 152, 156, 159, 162, 187
genetics, 132
genome, 134, 152, 157, 164, 168
genomics, 152, 156, 157, 159, 165
genre, 54, 61, 124, 129
genus, vii, viii, 1, 2, 3, 6, 22, 23, 25, 26, 28, 29, 34, 43, 45, 49, 50, 54, 55, 56, 61, 65, 101, 123, 124, 125, 126, 128, 131, 132, 133, 134, 135
geochemical, 152, 167
geological history, 176
geology, vii
geomorphology, 185
Germany, 118, 129, 136
Geryon longipes, 181
Gibraltar, 180
gill, 72, 92
glass, 176
glucose, 134
glue, 110
glycerol, 134, 156, 162, 164
glycine, 136, 138, 139, 141, 146
GOES, 167
gold, 156
gonads, 80, 81
Gorgonians, 181
Gradient, 172
grain size, 178
gravity, 184, 189
grouping, 102
groups, 149, 150, 152, 164
growth, viii, 56, 64, 78, 79, 83, 96, 125, 134, 151, 152, 167
Growth, 188
growth rate, viii, 64, 78, 79
GSA, 155, 157
Gulf Coast, 55, 59, 127
Gulf of Lion, 177, 178, 180, 184, 186
Gulf of Mexico, 60, 150, 154, 158, 160, 162, 163, 165, 166, 169, 182

H

habitat, viii, 68, 69, 171, 176, 177, 178, 180, 181, 182, 183, 184, 188, 189
Hake, 178
halos, 136
Harpacticoida, 189
harvesting, 152
heavy metals, 183
height, 71, 73, 176
heterogeneity, 178, 179, 189
heterotrophic, 153, 154, 160, 168
Hg, 183
high fat, 64
history, vii, 63, 64, 69, 79, 83, 97, 173, 176, 187, 188, 189
homogeneity, 136, 137, 146
host, 146, 162, 169, 182
hot spring, 151, 165
hotspots, 176, 178, 187, 189, 192
House, 98, 155, 159, 164
Human impact, viii, 171
Hungary, 118, 127
Hunter, 155
hyaline, 22, 23, 26, 27, 29, 31, 32, 38, 41, 43, 44, 46, 47, 48, 104, 106, 109, 113, 114, 116, 118, 122
hybridization, 132, 133, 151
Hydrate, 150
hydrocarbons, 154, 158, 191
Hydrodynamic, 175, 178, 179
hydrogen, 160
hydrolysis, 134, 140, 145
hydrophilic, 160
hydrophobic, 160
Hydrothermal, 167
hydrothermal activity, 192
hydrothermal vents, viii, 149, 171, 176, 183
hyperthermophilic, 150, 160
hypothesis, 152, 159, 179, 185

I

image, 6, 66, 156, 168
image analysis, 156
imagery, 167
imaging, 157, 166
imitation, 173, 188, 191
Impact, 184, 186, 191
in situ, 151, 158
in situ hybridization, 151
in vitro, 142
India, 118, 123, 128

indicators, 164
indigenous, 156
individuals, 67, 68, 69, 71, 76, 83, 179
induction, 156, 160
inferences, 176, 185
initiation, 30, 49, 142
inositol, 134
insertion, 145, 147
insight, 152
integration, 159
interactions, 154, 163, 168
interannual, 189
interpretation, 167
invertebrates, 172
Ionian, 172, 183, 185, 191, 192
Ionian Sea, 172, 183, 185, 191, 192
ionization, 159
Iowa, 164
islands, 179
isolation, 132, 162, 181, 182
isotope, 53, 158, 160, 162, 164, 165
Israel, 118, 122, 123
Italy, 23, 27, 44, 45, 46, 47, 58, 171

J

Japan, 64, 65, 66, 67, 68, 69, 73, 75, 78, 79, 81, 82, 83, 94, 95, 96, 97, 98, 127, 129, 131, 133, 135, 136
Jordan, 65, 66, 67, 71, 94, 96, 97, 118, 130
Jung, 166
juveniles, 68, 69, 81

K

Kadophellia bathyalis, 179
Kenya, 132
keratin, 138, 140
King, 157
Kosovo, 183
Kuril Islands, 65, 67, 68, 69, 71, 75, 78, 81, 97, 98, 99

L

La Medee, 182
lactose, 134
land, 178
languages, 96, 98
larva, 68, 69, 76
lead, 30, 190
lectin, 168
leucine, 153
Levantine, 188
life cycles, 178
light, 6, 12, 14, 15, 16, 18, 19, 69, 152, 161, 182
Light, 182, 188
Ligurian, 180
lipids, 156, 157, 159, 160, 164, 166, 167, 168, 169, 172
liquid chromatography, 159
litter, 183
localization, 159
location, 150
Lophelia pertusa, 180
Loricifera, 182
Luo, 148, 160
lysis, 159

M

Madrepora oculata, 180
magnitude, 11, 17, 21, 151, 154
majority, 2, 150, 151, 152
Malta, 181
maltose, 134
Management, 178
mannitol, 134
mantle, 81, 83
mapping, 155, 169
marine ecosystems, 191
marine environment, 132, 135, 150, 157, 161, 163
marine fish, 67, 96, 97, 99
Mars, 167
mass, 143, 146, 153, 159, 192
mass spectrometry, 159
materials, 63, 163, 165, 170
matter, 157, 162, 163, 166, 168, 169, 190
measurement, 163, 168, 170
measurements, 23, 29, 77, 79, 84, 158
meat, 64, 66, 82, 96
mechanical properties, 163
Mediterranean basin, viii, 171, 175, 176, 178, 187, 191
Mediterranean Sea, viii, 171, 172, 173, 174, 175, 176, 177, 178, 180, 181, 182, 183, 184, 185, 186, 187, 189, 190, 192
Meganyctiphanes norvegica, 187
Meiofauna, 173, 177, 185, 190
Meiofaunal assemblages, 177, 178, 185
membrane permeability, 164
membrane potential, 163, 164
membranes, 166
Merluccius merluccius, 178

Messinian, 181, 182
Metabolic, 152
metabolic pathways, 150
metabolism, 152, 153, 158, 168
metal ions, 139
metals, 183
metazoa, 182, 187
Meteor expedition, 174
methane, 149, 150, 151, 152, 154, 155, 159, 160, 162, 163, 164, 165, 168
methane oxidation, 154, 164, 168
methanogenesis, 152, 159
methylene, 152
Mexico, 54, 61, 63, 116, 125, 129, 150, 154, 158, 160, 162, 163, 165, 166, 169, 182
Mg^{2+}, 139
microaerophilic, 151
microbial, 155, 157, 158, 160, 161, 163, 165, 166, 167, 168, 169
Microbial, 156, 160, 161, 163, 167, 168, 182, 189
microbial communities, 160, 161, 167, 168, 182
microbial community, 189
microorganisms, 132, 154, 158, 165, 169
microscope, 2, 6, 50, 102, 155, 156, 157, 161, 166
microscopy, 156, 157, 159, 161, 162, 163, 165, 166, 167, 170
Middle East, 118, 122
migration, 55, 69, 96
minerals, 154, 166
miniature, 172
Missouri, 7
mitochondria, 183
mixing, 182
model, 65, 172
modeling, 163
modifications, 24
molecular mass, 135, 136, 137, 138, 140, 142, 143, 145, 146
molecular weight, 135, 140
molecules, 138, 153
Mollusca, 180, 190
Mollusks, 180
Monaco, 188, 192
Mongolia, 132
Moon, 96, 148
Morocco, 11, 49, 57
morphology, vii, 2, 49, 56, 63, 122, 125, 154, 156, 167
morphometric, 71
mortality, viii, 64, 79, 80, 83, 99, 165, 167, 169
mosaic, 177
Moscow, 63, 83, 96, 97, 98, 99
movement, 160, 169
museums, 96
mussels, 132
Mysids, 176

N

NaCl, 133, 134, 136, 146
National Academy of Sciences, 157, 191
natural environment, 166
natural selection, 155
Nematoda, 177
Nematode, 173, 177, 179, 187, 190
Nematodes, 172, 173, 179, 186
Neogene, 192
neon, 74
New Jersey, 158, 162
New South Wales, 95
New York, 158, 163, 167
Nicaragua, 102, 117, 118, 119
Nile, 124, 182
nitrification, 153, 154, 170
nitrite, 134
nitrogen, 132, 135, 147, 150, 153, 154, 163
NOAA, 97, 98, 99
nonthermophilic archaea, 149
North America, 58, 67, 68, 69, 70, 71, 72, 73, 128
North Atlantic, 151, 153, 162, 168
Northern Ireland, 127
Norway, 185, 187, 188
nucleic acid, 165
nucleotide sequence, 133, 136, 142, 143
nutrient inputs, 173
nutrients, 172, 173
nutrition, 161

O

observations, 150
Ocean, 172, 177, 181, 185, 186, 188, 191
oceans, 150, 151, 172, 176
octopus, 74, 160
offshore, 172, 180
Oklahoma, 61
Oligotrophic, 172, 189
oligotrophic conditions, 172
Open slopes, 177, 178, 185
operations, 51, 123
organelles, 183
Organic enrichment, 188
organic growth, 99
organic matter, 157, 162, 163, 166, 168, 169, 173, 178, 179, 185

Organic Matter, 173, 178, 179, 185
Organic pollutants, 183
organism, 132, 173
overlap, 71, 132
oxidation, 150, 154, 155, 159, 163, 164, 168
oxidize, 152
oxygen, 151, 172, 187

P

Pacific, vii, 9, 17, 36, 39, 53, 54, 58, 63, 64, 65, 66, 67, 68, 69, 71, 72, 73, 74, 75, 82, 83, 84, 92, 95, 96, 97, 98, 99, 102, 115, 116, 119, 120, 122, 132, 133, 150, 151, 153, 155, 156, 160, 161, 163, 166, 168, 169, 177
Paradigm of marine diversity, 174
parallel, 6
parasite, 82
particles, 166
Particles, 173, 177
pathogenic, 157
pathogens, 156
pathways, 150, 178
Pb, 183
PCBs, 167
PCR, 133, 137, 141, 145, 150, 167
Pelagic, 185
peptidase, 145, 146
peptides, 140, 142
perception, 157
permeability, 164
Peru, 150, 155, 164
pesticides, 183
petroleum, 58, 128, 154, 160, 169, 192
pH, 132, 133, 134, 136, 138, 139, 141, 146, 147, 158
pH values, 158
pharmaceuticals, 183
Philippines, 83
phosphate, 138, 163
Phosphorus, 188
photic, 173
photic layer, 173
photic zone, 150
photographs, 6, 66, 67, 83, 103
photosynthesis, 161, 170
photosynthetic, 155, 166, 168
Photosynthetically available radiation 400-700 nm (PAR), 188
phototrophic, 155, 158, 160, 167
phylogenetic tree, 134, 144
phylotypes, 150
physical structure, 192
physicochemical, 163
physicochemical methods, 163
Physiological, 134, 163
physiology, 152
phytoplankton, 156, 158, 165, 167
plankton, 150, 158, 178, 188
Plankton, 178, 188
planktonic, 172
plano-convex test, viii, 101, 113, 114, 118, 122, 123
plasmid, 136, 137, 146
plastics, 183
Plastics, 183
Pleistocene, 180
Pola, 174, 191
Poland, 46, 47, 56, 66, 125, 128, 129
polar, 133, 151, 186
pollutants, 156, 183, 192
pollution, 185, 187, 188, 192
polyacrylamide, 169
polychaete, 182
Polymède campaign, 174
polymerase, 137
polymerase chain reaction, 137
polypeptide, viii, 131, 132
polyps, 179
polysaccharides, 165, 166
poor, 161
POPs, 183
population, 151
populations, 178, 179, 188
Porifera, 180
porosity, 2, 3, 9, 11, 21, 24, 49, 102, 113, 122
potassium, 163
preservation, 7, 11, 13, 20, 21, 162
primary production (PP), 183, 185, 186, 187, 188, 189, 191, 192
primary productivity, 188
Princess Alice expedition, 174
principles, 50
probability, 80, 102
probability distribution, 80
probe, 165
Processes, 175, 184, 185, 190
Prochlorococcus, 157, 158, 165, 167, 168
production, 151, 152, 158, 159, 169, 187, 192
productivity, 167
profit, 178
prokaryotes, 159, 173, 183
prokaryotic, 151, 153, 156, 159, 164, 169
promote, 152
promoter, 142, 143
prosperity, 66
protein sequence, 143
proteinase, 145

proteins, 132, 138, 140, 152, 155, 164, 172
proteobacteria, 162, 182
proteomics, 154, 158, 165
proteorhodopsin, 162, 169
protocol, 158, 159
Pseudomonas, 154
Puerto Rico, 23, 27, 44, 45, 61, 116, 128
purification, 139, 140

Q

Queensland, 59

R

radiation, 131, 175
reactivity, 167
reading, 78, 81, 136
recognition, 2, 159
reconstruction, 103, 159, 168
recovery, 140
recrystallization, 11, 17, 20, 21
recycling, 132, 135
Red Sea, 182
redox, 164, 165, 167
reduction, 150, 154, 167
refractory, 163
regression, 65, 77, 78
regression analysis, 65
regulation, 163
relict, 179, 180
Remote operated vehicle (ROV), 187
repair, 169
researchers, 80
reservation, 162
residues, 137, 144
resolution, viii, 2, 3, 4, 9, 50, 101, 102, 122, 164, 166
resources, 98, 99
response, 172, 186, 190
restriction enzyme, 136
restructuring, 68
rhodopsin, 155
ribosome, 142, 143
rings, 78, 79
rods, 133
Romania, 8, 39, 44, 45, 46, 51, 128
Russia, 63, 66, 73, 83, 99

S

salinity, 167, 172, 176, 178, 188
salinity crisis, 188
salmon, 64, 68, 72, 73, 74, 81
salt concentration, 182
sampling, 151
samplings, 134, 174
SAP, 192
Sargasso Sea, 155, 158, 168
Scandinavia, 124
scavengers, 131
scyphozoan, 179
SDS-PAGE, 135, 136, 137, 140, 143
Sea of Marmara, 192
sea-level, 125
Seamounts, viii, 171, 174, 176, 179, 187, 190
seasonal flu, 172
seasonal variations, 189
seawater, 162, 169, 182
secretion, 140
sediment, vii, 13, 132, 133, 147, 150, 152, 154, 164, 172, 178, 181, 184, 189, 190
sediment transport, 178, 190
sedimentation, 11, 12, 178, 189
sedimentation rate, 178
sediments, vii, 1, 7, 8, 9, 11, 13, 17, 20, 21, 30, 49, 51, 57, 60, 109, 115, 119, 120, 127, 133, 149, 150, 151, 153, 154, 158, 159, 160, 161, 162, 163, 164, 165, 167, 168, 169, 171, 174, 179, 182, 183, 187, 189, 190
sequence, 187
sequencing, 168
series, 161, 164
serine, 135, 139, 141, 143, 144
sexual dimorphism, 81
shape, vii, 1, 2, 4, 5, 6, 26, 29, 41, 45, 48, 50, 104, 106, 109, 113, 114, 116, 118, 122, 176
shipping, 183
Shipping activity, 183
shores, 75
showing, 103, 105, 108, 119, 120, 133, 134
shrimp, 178, 183, 190, 191, 192
Siboglinid, 182, 187
Sicilian, 180
Sicilian Channel, 180
signal peptide, 136
Sills, 176
Sinai, 123
skeleton, 132
skilfish, vii, 63, 64, 65, 66, 67, 68, 69, 70, 71, 72, 73, 75, 76, 77, 78, 79, 80, 81, 82, 83, 84, 92, 95, 96, 97, 98
skin, 132
software, 137
soil, 166

soil particles, 166
solution, 26, 156
South America, 116, 118, 122
South Dakota, 58
South Pacific, 177
spatial, 175, 176, 177, 189
species composition, 176, 179
species richness, viii, 171, 174, 177, 179, 181
specificity, 162
spectroscopy, 165
spine, 78
sponge, 132, 150, 152, 165, 189
Sponges, 176, 178, 181
spore, 156
stability, 139, 141
standard error, 77, 78
standard length, 94
Staphylococcus, 164
Staphylococcus aureus, 164
starch, 134
stars, 132
state, 7, 20, 82, 113
states, 82, 160, 163
stereomicroscope, 2
stock, 79, 96
stomach, 81
storage, 172, 173, 178
strain, 170
Strait of Gibraltar, 176, 180
Strait of Sicily, 176
strategies, 158, 166, 172, 173
stratification, 172, 185
stress, 146, 167
strontium, 53
structural protein, 132
structure, 96, 108, 122, 154, 168, 169, 173, 177, 181, 185, 186, 190, 191
Structure, 173, 177, 181, 185, 186, 190, 191, 192
structuring, 178
Submarine canyons, viii, 171, 176, 178, 185, 188, 190, 192
substitutions, 134
substrate, 136, 138, 140, 144, 145, 146, 147, 178, 180, 189
substrates, 132, 138, 140, 145, 146, 179
succession, 8, 9, 11, 13, 17, 20, 21, 124, 131
sucrose, 134
sulfate, 150, 154, 164
sulfur, 153
summer, 156
supercritical, 159
surface layer, 76, 163
surface ocean, 165
surface properties, 154
Switzerland, 23, 27, 44, 45, 46
SYBR Gold, 167
SYBR Green I, 163, 167
symbiont, 150
symbiosis, 185, 187, 189
synthesis, 168
systems, 158

T

Taiwan, 98
tanks, 81
taxa, 3, 49, 102, 177, 178, 179, 180, 182, 185
taxonomic revision, vii, 1, 56, 125, 126
taxonomy, 1, 2, 4, 6, 56, 59, 101, 125, 126, 128, 129
techniques, 6, 163
technologies, 149
temperature, 84, 91, 131, 134, 135, 138, 139, 141, 147, 149, 150, 166, 172, 176, 178
Temporal Variability, 173, 185
temporal variation, 190
tendon, 133
termination codon, 142
thermal sediments, 149
thermal springs, 149
thermal stability, 141
Thermohaline, 184
thermohaline circulation, 184
thermophiles, 150
threats, viii, 171, 183
TIREM, 161
tissue, 131
toxic metals, 183
traits, 97, 169
transition, 150
transmission electron microscopy, 166
transport, 178, 185, 190
transportation, 82, 178
Trawling, 184, 189
Trieste, 186
triglycerides, 99
Trinidad, 7, 23, 27, 54, 124
Turbidity, 184
Turkey, 118, 182, 192
turnover, 57, 172, 178, 179
Tyro, 182
Tyrrhenian Basin, 176, 183, 192
Tyrrhenian Sea, 174, 192

U

U.S. Department of Commerce, 97, 98, 99
UK, 187, 188, 189, 191
ultrastructure, viii, 2, 9, 11, 21, 27, 49, 50, 101, 102, 113, 122
UNEP United Nations Environment Programme, 187
United Nations Environment Programme UNEP, 187
United States (USA), 36, 37, 39, 40, 46, 47, 49, 53, 54, 55, 57, 58, 59, 60, 61, 63, 64, 65, 66, 83, 95, 96, 97, 99, 102, 116, 118, 122, 124, 126, 127, 128, 148, 155, 156, 157, 158, 159, 160, 163, 164, 165, 166, 170
Urania, 182
USSR, 61, 95

V

valuation, 169
values, 152, 158
variability, 156
variables, 178
variations, 160, 166, 189
Venezuela, 53
vessels, 183
Vietnam, 66
viral, 159
viruses, 155, 156, 158, 161, 162, 163, 167, 168, 169
visualization, 165
vulnerability, 172

W

Washington, 6, 51, 53, 54, 57, 58, 59, 60, 61, 83, 99, 103, 126, 128, 174
Washington expedition, 174
waste, 183
water, viii, 76, 82, 83, 149, 150, 151, 153, 154, 157, 158, 166, 168, 171, 172, 174, 175, 176, 179, 180, 181, 182, 183, 184, 185, 186, 187, 188, 190, 191, 192
water mass, 192
Water Temperature, 176
wealth, 66
wells, 102, 123
Western Europe, 180
Western Mediterranean (WMED), 172, 173, 176, 179, 184, 185, 187, 188, 189
Western Mediterranean Sea, 172, 173, 179, 184, 187
whales, viii, 131, 132, 135
worms, 132, 182

X

Xenophyophora, 176

Y

yeast, 133, 134, 136, 168
yield, 143

Z

zooplankton, 81, 187